Praise for *The Republican War on Science*

"A frankly polemical survey of scientific finding and procedures in collision with political operations."

—*New York Times Book Review*, Editor's Choice 12/25/05

"The American conservative movement, as Chris Mooney points out in this fiercely anti-Republican book, has brought together two powerful constituencies—big industry and the religious right—both of which have an interest in skewing scientific advice so that it says what they want to hear." —*The New Statesman*

"*The Republican War on Science* by Chris Mooney: There's a man who doesn't hide his views. But Mr. Mooney needs to be listened to. Among other things, he wrote a piece for *The American Prospect* a few months ago titled 'Thinking big about hurricanes: It's time to get serious about saving New Orleans.' Alas, nobody listened."

—Paul Krugman recommends RWOS on his interactive page at NYTimes.com

"Mooney makes a strong case that science policy is often shaped by partisan expedience and ideology, that there often is a "war" on science— or at least an unhealthy disregard for it." —*The Weekly Standard*

"Mooney performs a useful service by researching all the details and interviewing as many of the protagonists as possible. He also enriches the narrative with much historical context, tracing over decades a gradual politicization of science that has culminated in the present farce."

—*The Guardian*

"[Mooney] is a talented and energetic young Washington correspondent for *Seed*, an excellent and relatively new popular-science magazine. In writing a book about science-policy-making in America today, Mooney has bravely tackled a gigantic and complex topic."

—*Washington Post* 9/18/05

"*The Republican War on Science* . . . does score some major hits when it takes on ideological campaigns against embryonic stem cell research and for intelligent design."   —*New York Sun*

"The connections Mooney discusses are crucial, because they provide proof that these [antiscience] actions are politically and economically motivated, rather than based on principled scientific worries."
—*Science Magazine* 10/7/05

"Chris Mooney's book detailing the Bush administration's attitude toward science will either horrify or annoy you. Either way, though, it's an essential piece of detective work on the nature of science policy-making."
—*Newark Star Ledger*

"Chris Mooney has written a stinging indictment of the Republican Party's attitudes toward science, focusing particularly on the manipulative and dismissive thinking and policies of the current administration."
—*The Christian Century* 11/15/05

THE
REPUBLICAN

WAR
ON SCIENCE

CHRIS MOONEY

BASIC
BOOKS

A Member of the Perseus Books Group
New York

Hardcover edition first published in 2005 by Basic Books,
A Member of the Perseus Books Group
Paperback edition first published in 2006 by Basic Books

Designed by Brent Wilcox
Text set in Adobe Garamond

The Library of Congress catalogued the hardcover edition as follows:
Mooney, Chris.
    The Republican war on science / Chris Mooney.
        p.   cm.
    ISBN-13 978-0-465-04675-1
    ISBN 0-465-04675-4 (hardcover : alk. paper)
    1. Science—Political aspects—United States.   2. Conservatism.   3. Republican
Party (U.S. : 1854–)   I. Title.
Q175.52.U5M66   2005
509'.73—dc22

                                                                    2005004889

Paperback: ISBN-13: 978-0-465-04676-8; ISBN 0-465-04676-2

EBA 06 07 08 10 9 8 7 6 5 4 3 2 1

CONTENTS

PREFACE TO THE PAPERBACK EDITION

I wish I could say that since the original publication of the hardcover edition of *The Republican War on Science* in the fall of 2005, the situation I denounced at that time had visibly improved. If anything, it has gotten worse. New attacks on science have emerged continually from the Bush administration, from Republicans in Congress, and at the state level (the latter most prominently involving the teaching of evolution). In light of these developments, I believe the argument presented in this book, which centrally attributes today's glut of politically motivated outrages against science to modern conservative ideology and attempts to appease key Republican interest groups, holds up very well. But because so much has happened, I would be remiss to release this paperback without exploring more recent events and their significance to the increasingly high-profile saga of science and politics in the United States.

This preface, along with updates at the end of Chapters 7 through 13, seeks to accomplish that task. But first, it seems fitting to reflect here on why *The Republican War on Science* has attracted such a high level of attention. As a first-time author, I was a bit staggered. I am confident that I wrote a strong and needed book, but I also believe that something else was afoot. Call it zeitgeist: Science policy rarely counts as a high-profile issue, but in recent years, the notion that the Bush administration might in some sense be "antiscience" has increasingly resonated. It has become a kind of Ur-narrative regarding this presidency, I suspect, because it reaffirms central suspicions about Bush nourished even by many who once supported him: that he's in a bubble walled off from reality; that he takes matters on faith; that he allows ideology to trample expert opinion; that he staffs the government with cronies who run it incompetently.

In other words, widespread concerns about the mistreatment of science by Bush's administration cannot be fully understood except in the context of related worries about the overselling of the Iraq war based on dubious intelligence, or about our government's pathetically inept response to Hurricane Katrina. After all, one key case study of science abuse on the part of top Bush administration officials lies in their repeated promotion of the dubious notion—as a rationale for preemptive war, no less—that Iraq's confiscated aluminum tubes were intended for centrifuges and uranium enrichment rather than for rocketry (a claim that nuclear experts almost uniformly rejected). Does that sound familiar? If so, it's because the administration has shown a similar reliance on scientific outlier perspectives on any number of other issues.

Or consider New Orleans and Hurricane Katrina. President Bush himself arguably misrepresented the state of knowledge when he so confidently declared, just after the storm, "I don't think anybody anticipated the breach of the levees." In fact, countless engineering and hurricane experts knew that New Orleans's underfunded levee systems might not withstand a direct hit from a major storm; in the event, even a sideswipe from Katrina did the city in. No wonder that when the hardcover edition of this book appeared at around the time of the destruction of much of New Orleans, my audiences repeatedly asked me to interpret recent events in light of my "war on science" thesis. (Perhaps my story also resonated because my mother's home, in the city's Lakeview neighborhood, had just been swamped by ten feet of floodwater.)

The politics of science manages to snuggle up alongside these higher-profile subjects of war and disaster because they're all part of a package, a collection of anecdotes that together speak the same message: The president doesn't seem to care about what's going on in the "reality-based community." If mistreatment of science reverberates as an issue, it is because it is emblematic of why so many Americans oppose George W. Bush to begin with. They think that he is unfit to lead and that those he appoints cannot competently administer a government with such a wide range of duties, virtually all of which require some form of expertise (often scientific) if they are to be carried out properly. Bush's meeting with the controversial novelist Michael Crichton to discuss global warming—according to journalist Fred Barnes, the two "talked for an hour and were in near-total agreement"—typifies the president's disregard for the critical role of legitimate expertise in decision-making.

So prominent have complaints about the administration's "war on science" become that even president Bush has responded, albeit obliquely, to them. To be sure, Bush did not explicitly acknowledge the criticisms that have become so widespread (and that this book epitomizes). Nevertheless, I believe that Bush sought at least in part to neutralize his detractors by announcing, in his 2006 State of the Union address, a new plan to promote science education and shore up America's scientific competitiveness. Then Bush got himself photographed peering into a microscope at a high school in Dallas (in Texas, at least, Bush and science are still chums).

But while Bush professes admirable goals for American science, they are hard to take seriously from a president who has been science's worst enemy when it really counts. Consider Bush's personal record—not his administration's record, but his own statements and actions—on the three most politically fraught scientific topics of the day: evolution, embryonic stem cell research, and global climate change. On evolution, Bush has endorsed teaching pseudoscientific "intelligent design" in public high school science classes (thereby shattering any credibility he might otherwise have had to talk about science education). On stem cells, as shown in Chapter 1, Bush misled the nation in 2001 about the scientific basis for his policy, and after it became clear that his promised "more than 60" embryonic stem cell lines did not exist, never bothered to amend his statement or to revise that policy. And then there's global warming, where Bush has falsely claimed that a "debate" exists over whether the globe is warming due to natural or manmade causes.

On matters of science, then, the president's credibility can be expressed as an empty set. As for the rest of his administration, the problems begin at the top but then filter down into the agencies of the federal government. Almost no major agency with science as a significant part of its portfolio or mandate has gone untouched by charges of politically motivated abuse of science during Bush's presidency. What began in 2002 as a list of allegations about the Department of the Interior (see Chapter 14) has expanded into charges concerning the National Cancer Institute, the CDC, FDA, EPA, NOAA,. . . the litany of abbreviated federal agencies goes on and on.

In this context, it can hardly be deemed irrelevant that these agencies, collectively, overflow with hundreds more political appointees than roamed the government at the end of the Clinton years. According to Princeton University political scientist David E. Lewis, the spike upwards under Bush has been on the order of about 350 appointees. Some of these appointees have sought, in very heavy-handed fashion, to keep government scientists (or "science moles," as one conservative put it) in line. So if we seek to understand why so many science fights have erupted at so many diverse agencies during the Bush administration—and why they continue to break out with considerable regularity—the role and the zeal of Bush administration political appointees becomes essential.

Perhaps somewhat unfairly even for this bunch, their archetype has become George Deutsch, the 24-year-old NASA public relations official who reportedly described his job as being to "make the president look good" and who sought to rein in famed climate scientist James Hansen (who had merely attempted to warn that we may lose the polar ice sheets if we don't move quickly to address climate change). Hansen got the better of Deutsch in that episode, going public in early 2006 and drawing a new wave of attention to the fraught relationship between the Bush administration and its own scientists (dramatically undercutting Bush's newly announced competitiveness initiative in the process). Meanwhile, Deutsch resigned, even as the *New York Times* revealed that, contrary to his résumé, he had not graduated from Texas A&M University. With stories like this, can anyone seriously wonder why Americans have become so worried about the way this administration treats not just science in general, but its own scientist employees?

Seriously addressing such concerns for perhaps the first time, the Bush administration—or at any rate, NASA—responded seriously to the Deutsch–Hansen affair with an avowal that it would relax restrictions on interactions between government scientists and the media. But the spirit of Deutsch lives on in episode after episode of conflict between federal scientists and administration political appointees, as well as in the broader political fights over science that continue to resound in American public life. So many of these conflicts have emerged since the final text was set for the hardcover edition of this book that I felt it important to provide updates for the paperback edition. The broad story of abuse of science told here has not appreciably changed; the newly accumulated details, however, make the argument considerably stronger, especially since most of them merge seamlessly into the book's preexisting framework.

Therefore, at the end of each of the main chapters covering present-day events (Chapters 7–13), I have set aside a section to fill the reader in on subsequent developments (this preface itself is, in a sense, an extension of Chapter 14). These updates aren't too lengthy, but they give you a flavor for how the stories told in the chapters have progressed. Meanwhile, the text has also been revised to fix any typographical or other errors discovered in the previous edition, as well as to include minor editorial changes. (For example, I have frequently replaced the rather imprecise and ill-defined phrase "politicization of science" with something more specific.) For those wishing to compare editions—an activity guaranteed to overcome insomnia—I have made a complete list of changes to the text available at the book website, www.waronscience.com.

In closing this preface, I would like to address what I view as one problem with of the first edition of *The Republican War on Science*: its rather clipped description of possible solutions to the crisis I had spent the entire book describing. I can't tell you how many readers responded that the book made them exceedingly angry but provided little in the way of a release valve for the built-up steam. In part, that springs from the very nature of the problem of science abuse: It is inherently difficult to fight against. The incentives to attack politically influential scientific information are vast, and those who opt for this strategy have a built-in advantage over science's defenders. It's much easier to sow confusion and misinformation than it is to generate new and reliable knowledge. It's much easier to spin the media than to correct errors once they've been broadly disseminated. It's much easier simply to make something up than to do the hard work required to prove or disprove a hypothesis.

Nevertheless, that's no reason for the allies and defenders of science to give up. When science falls under political attack, responsible people—scientists especially—cannot simply stand on the sidelines. They must speak out and defend the knowledge they have brought into the world.

Since the book's first publication, then, I have thought a great deal about how scientists might better defend their knowledge and their work. My ideas can be read in more detail in a column on the subject written for *Seed* magazine entitled "Learning to Speak Science," available online at http://www.seedmagazine.com/news/2006/01/learning_to_speak_science.php. But to summarize, I believe that in order to fight back, scientists must learn to emphasize a very different set of skills from those that they generally use in strictly scientific pursuits. Too many scientists have grown accustomed to the security of their labs and university communities, occasionally lamenting the American public's poor understanding of science but doing little in a concerted way to improve it. And small wonder: American science rewards the publication of peer-reviewed research but offers little incentive for scientists to communicate and translate what they know to the public. So scientists in the United States have little practice when it comes to crafting a message or winning a political debate.

Scientists need to become better advocates, not for a particular political party, but for the integrity of science itself. In the process, they must learn to speak to the public in a language it understands and to value individuals in their ranks who excel at communication. At the same time, they must avoid messages that backfire politically. For example, many scientists are atheists, and when they come to the defense of the teaching of evolution, they bring their godlessness along with them and attack or disparage the religious beliefs held by a large proportion of Americans. I am second to none in defense of the right to a lack of religious belief, but if science advocates are smart, they will recognize that the evolution battle is a cultural one, not a scientific one. Conservative Christians believe that the theory of evolution is destroying religious belief and leading to moral chaos. In this situation, there's nothing strategically dumber than reinforcing a damaging and divisive stereotype.

I want to stress that in advancing this argument about public communication by scientists, I am not blithely assuming that a better-informed public will magically solve all of our problems for us. It won't. Many of the distortions of science described in these pages have attained such a level of sophistication that only experts can properly set the record straight about what science actually says and what it doesn't. And no matter how well educated about science the American public becomes, we will never have an electorate consisting wholly of Ph.D.s. For a similar reason, in most cases I do not expect that many politicians will have to pay a price for their attacks on science at the ballot box. Most voters simply have too many other things on their minds—war, their pocketbooks—to get deeply involved in matters of science policy.

But that doesn't mean scientists can't make a difference. They can bring their message straight to policymakers, the people who most need reliable information. They can correct and, if necessary, embarrass journalists who fall prey to scientific-sounding distortions. Finally, at a time when the attacks on evolution or climate science take the form of strategic PR campaigns, scientists must launch counter-campaigns to blunt the impact of widespread misinformation on the media and public. Better strategic organization on the part of scientists certainly won't supplant the need for better science education and public understanding; these efforts must be complementary. But at least the battle would finally have been joined by scientists, in a concerted and politically informed way, in the arena where it truly matters: the media and public square.

If you're worried about the state of science in American today, then I have little doubt that the following pages will make you outraged and angry. But remember that that's just the first step. Restoring the integrity of science to our government and public life will depend on commitments and contributions from all who hold to the Enlightenment-inspired belief that, if we can just get the science right, we're at least *somewhat* more likely to get the policy right as well—that, in other words, we not only can, but must, use our knowledge to improve our future. That is my own conviction, and I'm proud to stand up and defend it, again, through the issuance of this paperback. I hope you will join me.

Chris Mooney
April 2006

ACKNOWLEDGMENTS

I am not a scientist, and don't pretend to be. Instead, I am a journalist with a background in political writing whose published work has often covered controversies at the intersection of science and politics. In those articles, I have relied heavily on the willingness of scientists, doctors, bioethicists, and science policy analysts to share their views with me and explain basic issues. It is often claimed that scientists don't like journalists much. They worry, supposedly, that we will oversimplify complex questions, or fail to describe their views with the proper qualifications, or go drama-seeking and pretend to uncover a raging scientific debate where none actually exists. Yet if a widespread distrust of the media does exist in the scientific community, I am glad to say that it hasn't interfered with my research in any way.

Let me also explain my principles for reporting on science. In my opinion, far too many journalists deem themselves qualified to make *scientific* pronouncements in controversial areas, and frequently in support of outlier or even fringe positions. In contrast, I believe that journalists, when approaching scientific controversies, should use their judgment to evaluate the *credibility* of different sides and to discern where scientists think the weight of evidence lies, without presuming to critically evaluate the science on their own. Such has been my approach, and I have tried to stick to it as faithfully as possible. Nevertheless, I take responsibility for any and all errors that may have crept into this work despite my best efforts.

This book would not have been possible without the help and cooperation of a wide range of scientists, science policymakers, and other scholars and experts who have helped me understand where consensus lies, where uncertainty still lingers, and the merits of different pieces of scientific work. At the end of the book, I have credited the scientists and other experts on whose input I relied. In some cases, you won't find these experts actually quoted in the chapters themselves. But their perspectives nevertheless helped to inform and enrich the text.

Beyond scientists and other interviewees, many kindred spirits have helped me along as I strove to complete this book. First, I must thank Arthur Caplan, perhaps our nation's "most-quoted bioethicist," who originally suggested the idea that grew into this text. Since then, Caplan has played a key role in our country's raging science wars, on the side of the angels. In March 2004, he organized more than 170 bioethicists to protest the politicization of the President's Council on Bioethics.

This book would never have materialized without the steadfast support and guidance of my agent, Sydelle Kramer, who stuck with me through a tentative first book proposal and showed me how the publishing world works. Likewise, I would like to thank my editor, Bill Frucht, of Basic Books, for seeing the potential of this project and helping me bring it to completion, and my publicists at Basic—Christian Purdy, Holly Berniss, Jamie Brickhart, Paul Colbert—for their diligent work getting the book (and me) out and around. In addition, many thanks to William Hawkins for invaluable advice on public speaking and the media, and to Adam Bly and the crew at *Seed* Magazine for bringing me into the fold, believing in the book, and always being willing to celebrate it.

Many individuals read and offered comments on this book in its various stages of development. For helpful comments and critical readings, I'd like to thank Nicholas Confessore, Ben Fritz, Jeet Heer, John Judis, Bryan Keefer, Eli Kintisch, Christopher Kirchhoff, Michael Mooney, Sally Mooney, Matthew Nisbet, Brendan Nyhan, Elena Saxonhouse, Nicholas Thompson, and a very old friend who asked not to be named. For expert comments and readings, my thanks go to Glenn Branch, Allan Brandt, Devra Lee Davis, Josh Dunsby, Barbara Forrest, David Guston, Stanley Henshaw, Patricia Labosky, Peter Moyle, Roger Pielke,

Jr., Patricia Princehouse, Kevin Trenberth, David Vladeck, David Wright, and a number of other John Does who know who they are. Although these individuals provided technical readings, they are in no way responsible for my arguments or conclusions, and may well disagree with some or all of them. And again, I retain full responsibility for any errors that may have crept into the text despite their guidance.

For research help, meanwhile, I must thank David Meyer (who also aided in fact-checking), Nicholas Bobys, and Susie Bachtel, who let me rifle through old Office of Technology Assessment papers at her home for an entire morning. I must also thank all of the Tryst servers whose steady stream of coffee facilitated the writing of this book on many a late night in Adams Morgan.

I would also like to thank, collectively, all the readers of my blog, "The Intersection," at ScienceBlogs.com/Intersection. This crowd stuck with me for over a year as I tested ideas for this book and kept track of political interferences with science. In particular, my readers often helped me refine my thinking on subjects where I was a bit muddled. I soon realized that in many cases they knew more about the subject matter than I did, and I occasionally left posts asking for comments on still undeveloped ideas. Enlightenment usually came quickly.

Finally, I would like to dedicate this book to the memory of my grandfather, the biologist Gerald A. Cole, who inspired a love of science in me that ultimately grew into a desire to undertake a "two cultures"-merging project like this. "Paw" never knew I was writing this book, but I like to think he would have appreciated the end product. And so, hopefully, would the man he liked to call "our friend Chuck."

Chuck Darwin, that is.

PART ONE

WHERE IT BEGINS

The success of science depends on an apparatus of democratic adjudication—anonymous peer review, open debate, the fact that a graduate student can criticize a tenured professor. These mechanisms are more or less explicitly designed to counter human self-deception. People always think they're right, and powerful people will tend to use their authority to bolster their prestige and suppress inconvenient opposition. You try to set up the game of science so that the truth will out despite this ugly side of human nature.

STEVEN PINKER

CHAPTER 1

THE THREAT

IN THE SUMMER OF 2001, long before his reelection and even before he became a "wartime president," George W. Bush found himself in a political tight spot. He responded with a morsel of scientific misinformation so stunning, so certain to be exposed by enterprising journalists (as indeed it was), that one can only wonder what Bush and his handlers were thinking, or whether they were thinking at all. The issue was embryonic stem cell research, and Bush's nationally televised claim—that "more than sixty genetically diverse" embryonic stem cell lines existed at the time of his statement—counts as one of the most flagrant purely scientific deceptions ever perpetrated by a U.S. president on an unsuspecting public.

Bush's assertion, made on August 9, 2001, came as the president sought to escape a political trap of his own making. Campaigning in 2000, Bush told the U.S. Conference of Catholic Bishops that taxpayer money "should not underwrite research that involves the destruction of live human embryos." The statement threw a bone to Bush's pro-life followers, who view the ball of about one hundred fifty cells constituting a five-day-old embryo as deserving of the same moral and legal protections as fully developed human beings. Accordingly, these religious conservatives consider embryonic stem cell research—the study of excess embryos donated for research from in vitro fertilization clinics—ethically abhorrent.

But some prominent Republicans, such as Utah senator Orrin Hatch, favored the research because of its scientific promise. As the issue came to a head in the summer of 2001, Bush publicly agonized over what to do. Finally, he opted for a supposed compromise: he would allow federal funding, but only for research on preexisting cell lines. That way—at least arguably—the government would not be complicit in the destruction of any more embryos, and Bush would have kept his campaign promise.

From August 9, 2001, forward, Bush declared, the federal government would fund research on embryonic stem cell lines only "where the life and death decision has already been made." For this rather arbitrary decision to sound plausible, though, Bush needed to turn scientists loose on a largish-sounding number of lines—something on the order of, say, sixty. Having such an impressive figure to cite "made this decision possible," a senior administration official told *Time* shortly after Bush's speech.

Scientists, though, expressed instant skepticism. The count didn't come from the published and peer-reviewed literature; instead, it arose from a global telephone survey conducted by the National Institutes of Health (NIH). It gradually became clear that the NIH figure referred to stem cell *derivations*: every known case in which scientists had removed the inner cell mass of an early embryo, or blastocyst, before the Bush deadline. But while derivations represent attempts to produce a cell line, these attempts do not always succeed; *lines* must reliably grow and divide in the laboratory so that scientists can study them and ship them to colleagues. In an interview, Stanford professor emeritus of medicine and Nobel laureate Paul Berg vividly explained the problem with many of the Bush cell "lines": "At some point, somebody took a blastocyst from an IVF clinic and cracked it open and poured everything into a vial and stuck it into a liquid nitrogen tank. In which case, we don't know if it's a line. And most of them died, and that's why there are so few now."

The Bush White House either didn't know or didn't care about the distinction between derivations and lines. Lacking a permanent science adviser at the time—and not bothering to consult with acting science adviser Rosina Bierbaum, a Clinton administration holdover—Bush went on national television and announced to Americans, roughly a third of whom had tuned in to his August 9 speech, a policy based on science fiction. As of this writing, more than four years later, only twenty-two available lines qualify for federal funding, and scientists consider many of those almost useless.

You might call it a textbook example of how bad scientific information leads, inexorably, to bad policy.

Bush's approach to the issue of embryonic stem cell research—his very first political test—showed a deep disregard for the role of scientific information in political decision-making. Rather than seriously analyzing the number of stem cell lines in existence and assessing their viability for research, the White House cherry-picked a questionable number to justify a desired result. It used science as window dressing, as public relations. That is sadly typical of how our current president, and the political movement of which he serves as figurehead, approach scientific information generally.

Rallied by the Union of Concerned Scientists, a liberal-leaning group based in Massachusetts, many of the nation's leading scientists, and by no means only liberals, have sweepingly denounced the Bush administration's misuses of science. Beyond the issue of embryonic stem cell research, they have accused the administration of skewing science on global warming, mercury pollution, condom effectiveness, the alleged health risks of abortion, and much else. These charges, many of them substantiated in this book, have received considerable attention. But the broader story behind them has not. Few have recognized the deep connection between the Bush administration's treatment of science on issues like embryonic stem cell research on the one hand, and recent American political history on the other.

Unlike his father, George W. Bush is no Republican moderate. Rather, he owes his allegiance to the modern American conservative movement, which over the past fifty years has gone from the political fringe to a position of dominion over the Republican Party, not to mention the entire U.S. government. In the process, the modern Right has adopted a style of politics that puts its adherents in increasingly stark conflict with both scientific information and dispassionate, expert analysis in general. Small wonder, then, that Bush's presidency has been characterized by unprecedented distortions of scientific information.

To be clear, this is no indictment of every member of the Republican Party. Moderate Republicans like Senator John McCain have often fought back against conservative distortions of science, particularly on the issue of global warming, and it was a Republican president, Dwight Eisenhower, who first installed a scientific advisory apparatus in the White House. Alas, today moderate Republicans simply aren't running the show.

At its most basic level, the modern Right's tension with science springs from conservatism, a political philosophy that places a strong value upon preserving traditional social structures and institiutions. The dynamism of science—its constant onslaught on old orthodoxies, its rapid generation of new technological possibilities—presents an obvious challenge to more static worldviews. From Galileo to Darwin and beyond, this conflict has played out repeatedly over the course of history.

Yet conservative philosophy alone cannot explain the sweeping controversy over science that has emerged during George W. Bush's presidency. Another ingredient must go into the mix: raw politics. During its rise to political triumph and domination of the Republican Party, the modern conservative movement has relied heavily on two key constituencies with an overriding interest in the outcomes of scientific research in certain areas: industry and the religious Right. Companies subject to government regulation regularly invoke "science" to thwart federal controls and protect the bottom line. Religious conservatives, meanwhile, seek to use science to bolster their moralistic agenda. The Bush White House, in true modern conservative fashion, has bent over backward for both groups.

Other factors, too, contribute to a standoff between the conservative community and the scientific one. Modern conservatism's broad distrust of "big government" worsens the tensions with science, much of which either depends on federal funding or takes place at government agencies. Early in Bush's first term, one conservative even warned that "science moles" lurking in the federal bureaucracy might sabotage the president's agenda.

The Right's oft-expressed disdain for "liberal" higher education—epitomized by Bush strategist Karl Rove's smirking definition of a Democrat as "somebody with a doctorate"—fans the conflict as well. Today's "red state" conservatives nourish a deep suspicion of the nation's urban and coastal liberal enclaves, home to many leading universities and research centers. To counter mainstream science, economics, and political analysis coming out of these universities (as well as "liberal" think tanks such as the Brookings Institution), the Right has created its own favorable sources of expertise and analysis—think tanks like the Heritage Foundation, the American Enterprise Institute, and many others.

Finally, conservatives' views on science have been shaped as much by their political enemies as by their friends. In particular, the environmental movement—a core Democratic constituency—draws regularly on science to demonstrate the harms of various forms of environmental degradation and to demand stronger government regulation. In political spats with environmentalists, conservatives have thus learned to attack not just policies favored by green groups, but also the scientific information used to support those policies—which they repeatedly denounce as "junk science." To hear the modern Right tell it, you would think that environmental science, as conducted at America's leading universities, suffers from endemic corruption on a scale reminiscent of Tammany Hall.

Given all of these tendencies, a grand clash between modern American science and modern American conservatism may well have been inevitable.

Conservative politicians would of course not agree that they are guilty of systematic abuse of science. In fact, they have long accused their *opponents* of this crime. Conservative Republican leaders drone on about "sound science" (a counterweight, in their eyes, to environmentalist "junk science"). GOP lawmakers in the House of Representatives, led by Representative Chris Cannon of Utah, have even sought to form a "Sound

Science Caucus" to ramp up the role of "empirical" and "peer-reviewed" data and information in laws such as the Endangered Species Act (discussed at greater length in Chapter 10).

These politicians have found a support base for their actions in conservative think tanks. In 2003, the Maryland-based Annapolis Center for Science-Based Public Policy, a conservative-leaning group funded in part by industry, feted Cannon for his "sound science" initiative at its annual dinner. And Annapolis is just one of many science policy think tanks on the Right, a niche sector of the ideas industry in which conservatives have dominated liberals. Other such outlets include the George C. Marshall Institute, a hotbed of global warming doubters and contrarians, and the Center for Science and Public Policy at Frontiers of Freedom, a group founded by former Wyoming Republican senator Malcolm Wallop, an early proponent of the quest to develop space-based laser technologies (an agenda ultimately embraced in Ronald Reagan's controversial "Star Wars" program). Borrowing a line from the conservative *Fox News,* the Center for Science and Public Policy promises—shades of Orwell—a "fair and balanced" approach to scientific information.

Conservative intellectuals have even preemptively introduced a critique of left-wing science politicization into the public discourse. In books with titles such as *Politicizing Science: The Alchemy of Policymaking* (from the Marshall Institute) and *Silencing Science* (from the libertarian Cato Institute), they bemoan the abuse of science by the Left, particularly environmental groups and environmental scientists. Long before the science community voiced its collective outrage over the abuses of the Bush administration, conservatives had cleverly established a diametrically opposed argument.

Let's be fair: those on the political left have undoubtedly abused science in the past. While the best environmental groups marshal good science to make their case, more radical groups have occasionally allowed ideology to usurp fact. In objecting to genetically modified foods, for instance, Greenpeace has suggested that these "Frankenfoods" pose human health risks due to the "inherently risky process" by which they are made. Yet in a 2004 report, the Institute of Medicine of the National Academy of Sciences, the government's leading independent adviser on science and technology, refused to treat food created through genetic engineering as inherently more dangerous than food created through other forms of genetic modification such as conventional breeding, adding that "to date, no adverse health effects attributed to genetic engineering have been documented in the human population."

Clearly, many who preach the human "dangers" of genetically modified foods have clothed their moral or policy objections in scientific attire. The New York University food expert Marion Nestle has written that environmentalists and consumer advocates use the food safety issue as a "surrogate" for a host of broader social concerns. They argue about health risks because they lack a venue in which to air their real objections, about corporate control of the food supply, for instance.

We have also seen distortions of science from another camp traditionally associated with the political left: the animal rights movement. Notorious for its attacks on medical researchers, the movement has also made questionable scientific claims to advance its agenda. In arguing against research on animals, some animal rights activists have asserted that these studies aren't medically necessary because alternatives like computer modeling could teach scientists everything they need to know, an argument that Donald Kennedy, executive editor-in-chief of the journal *Science*, calls "a remarkable piece of science fiction."

It must also be acknowledged that much of science emerges from the liberal-leaning academic world. In an interview, Harvard's celebrated cognitive psychologist Steven Pinker, author of *The Blank Slate: The Modern Denial of Human Nature*, explained to me how this political reality tends to wall off certain areas of inquiry that might be seen as supporting conservative viewpoints: "When it's academics who wield the power, the political bias will be on the Left."

Though they often have little control over government science policy-making, academics wield plenty of power in universities. As a result, Pinker notes, topics such as the genetic underpinnings of human behavior have often gone unstudied out of a "general left-of-center sensibility that anything having to do with genes is bad." The notion that a tendency toward violence might have genetic roots, for example, raises among many on the Left the fear of eugenics-style solutions.

So give conservatives a few points, here and there, on the question of left wing science abuse. It certainly exists, and later chapters will highlight further examples on issues like mercury pollution and embryonic stem cell research (where advocates have sometimes hyped the possibility of quick cures for diseases). In fact, in politicized fights involving science, it is rare to find liberals entirely innocent of abuses. But they are almost never as guilty as the Right.

The counterargument against conservative "sound science" proponents begins with a fundamental scientific litmus test: evolution. Religious conservatives have aligned themselves with "creation science," a form of religiously inspired science mimicry that commands the allegiance of nearly half the American populace, according to polls. While campaigning for the presidency in 1999, George W. Bush supported the teaching of creationism alongside evolution in public schools, which alone obliterates his claim to be a defender of science.

And now, backed in part by conservative Republicans like Pennsylvania senator Rick Santorum, creationism has been brought up to date by a far slicker form of antievolutionism, one that doesn't depend on absurdities like dinosaurs being herded onto Noah's ark and goes under the name of "intelligent design." (We will have more to say on this in Chapter 11.) *National Review,* arguably the leading conservative journal, routinely publishes articles by defenders of intelligent design on its website.

Other religiously impelled abuses of scientific information, supported by conservatives, include claims that social science supports the effectiveness of abstinence education, that condoms aren't very good at preventing HIV and other sexually transmitted diseases, that abortion increases the risk of breast cancer or mental illness in women, that adult stem cells have more research promise than embryonic ones or can even replace them for scientific purposes—the list goes on and on. But it is only half the story. If powerful religious commitments lead to politically driven abuses of science, so does an unswerving commitment to the bottom line.

From tobacco to energy companies, business interests have long had a fundamental stake in the outcome of scientific research in areas that could potentially imperil their profits by inspiring government regulation. In many cases, these companies have sponsored research through the support of independent or university-based scientists. There is nothing fundamentally wrong with such funding, provided it gets distributed with no strings attached and is disclosed in publication. Nevertheless, studies suggest that ties to industry can significantly influence the outcomes of research. This means at the very least that we should weigh such ties when assessing a piece of scientific work.

Tobacco provides an illustrative example. A 1998 analysis of over one hundred review articles on the health risks of secondhand smoke, published in the *Journal of the American Medical Association,* found that the odds of an article reaching a "not harmful" conclusion were "88.4 times higher" if its authors had tobacco industry affiliations. In many cases, the study noted, the authors had not disclosed their funding sources.

Free-market conservatives have shown a marked tendency to align themselves with industry "science," even in cases where it represents a clear outlier viewpoint that warrants considerable skepticism, and to dismiss or ignore concerns about conflicts of interest. Examples include not only research on climate change and secondhand smoke, but also studies downplaying the risks of mercury pollution, the herbicide atrazine, and other environmental and health dangers. The conservative faith in industry and unrestrained capitalism seems to fuel a parallel assumption that industry-sponsored science—like the free market itself—stands above reproach.

Conservative politicians have even tried to pass laws altering the way government agencies use scientific information, reforms that would favor industry's tobacco-style agenda of using favorable "science" to thwart regulation. Examples include the Gingrich-era push for "regulatory reform," the so-called Shelby amendment, and the Data Quality Act, all discussed later in this book. Proposed legislative changes to introduce "sound science" and "data quality" into the process of Endangered Species Act enforcement, discussed in Chapter 10, only add to the stratagem.

In short, the Right's critique of left-wing science politicization rings rather hollow when you consider that conservatives have shown little concern about the systematic, and often far cruder, war on science conducted by their own ideological allies, especially industry and pro-life religious conservatives.

Why should we care about such manipulation? First and most obviously, when conservatives distort health information—claiming that condoms don't work very well in protecting against sexually transmitted diseases, for example—such abuses can quite literally cost lives. Similarly, when they deny the science of global warming, their refusal to consider mainstream scientific opinion fuels an atmosphere of policy gridlock that could cost our children dearly (to say nothing of the entire planet). That includes children not just in low-lying New Orleans, where I myself grew up, but in low-lying Bangladesh and other nations across the globe.

And science abuse threatens not just our public health and the environment, but the very integrity of American democracy, which relies heavily on scientific and technical expertise to function. At a time when more political choices than ever before hinge upon the scientific and technical competence of our elected leaders, the disregard for scientific consensus and expertise—and the substitution of ideological allegiance for careful assessment—can have disastrous consequences.

Most politicians aren't scientists. Their job is to debate the meaning and implications of policies, to reach political consensus despite moral and ideological differences. In doing so, they need to rely on the best scientific knowledge available and proceed on the basis of that knowledge to find solutions. But the whole process derails when our leaders get caught up in a spin game over who can better massage the underlying data.

When scientific information becomes merely something to be manipulated to achieve a political end, the quality and integrity of the political process inevitably suffer.

Furthermore, such politicization has a corrosive effect on the American public's trust in science. If Americans come to believe you can find a scientist willing to say anything, they will grow increasingly disillusioned with science itself. Ultimately, trapped in a tragic struggle between "liberal" science and "conservative" science, the scientific endeavor itself could lose legitimacy.

The Right's war on science also spreads a massive amount of misinformation, and sometimes even fosters outright ignorance. Consider the creationist quest to thwart the teaching of the theory of evolution to public school children. Science has managed to answer one of the most profound questions around—where does the human species come from?—but religious conservatives don't want anyone to know about it. And with the spread of ignorance and pseudoscience comes a decline in critical thinking—a lapse in our collective capacity to cut through all the lies and distortions and determine which ideas we should trust.

Finally, science political abuse cheapens democratic debate among our citizens. When political argument descends into scientific disputation, each side races to find supporting "facts." This creates a breeding ground for disingenuousness, as fringe and pseudoscientific arguments make their way into the public sphere, generating sterile confusion rather than illumination. Meanwhile, with the underlying facts at issue, disputants have far more difficulty finding common ground. Instead, competing claims multiply while political positions harden and become entrenched.

The history of another country, one Americans don't much like comparing themselves with, illustrates the grave dangers of yoking political ideology to dubious science. In the 1930s under Joseph Stalin, the quack "scientist" Trofim Lysenko, who promoted himself through party newspapers rather than rigorous experiments, rose to prominence and took control of Soviet biological, medical, and agricultural research for several decades. Lysenko used his power to prosecute an ideologically driven crusade against the theory of genetics, which he denounced as a bourgeois affront to socialism. In short, his political presuppositions led him to embrace bogus scientific claims.

In the purges that followed, many of Lysenko's scientist critics lost their jobs and suffered imprisonment and even execution. By 1948 Lysenko had convinced Stalin to ban the study of genetics. Soviet science suffered immeasurable damage from the machinations of Lysenko and his henchmen, and the term "Lysenkoism" has since come to signify the suppression of, or refusal to acknowledge, science for ideological reasons.

In a democracy like our own, Lysenkoism is unlikely to take such a menacing, totalitarian form. Nevertheless, the threat we face from conservative abuse of science—to informed policymaking, to democratic discourse, and to knowledge itself—is palpably real. And as the modern Right and the Bush administration flex their muscles and continue to battle against reliable, mainstream conclusions and sources of information, this threat is growing.

We cannot fully grasp the dilemma we face, however, without first understanding how we arrived at this point. Thus far, the broad story of how today's political Right developed its animus with modern science has been incompletely told at best. By examining a range of scientific issues in depth, while also explaining the emergence of modern conservatism and the reasons for its challenged relationship with modern science, this book explains how our nation gave rise to a political movement whose leaders, to put it bluntly, often seem not to care what we in the "reality-based community" know about either nature or ourselves.

CHAPTER 2

POLITICAL SCIENCE 101

B EFORE EXPLORING THE HISTORY of the modern Right, its present-day conflicts with science, and the threat to science and policy alike if current trends continue, we need a strong foundation for the discussion. Any extended look at the relationship between politics and science demands some intellectual calisthenics at the outset.

Most crucial is defining "science" itself. Though science may provide us with rock-solid facts, these facts, in and of themselves, do not constitute *science*. Instead, science amounts to a *process*—institutionalized at leading universities, research facilities, and scientific journals worldwide—for systematically pursuing knowledge about nature and, in the social sciences, ourselves. As its core, this process features the testing and retesting of hypotheses to ensure that they withstand the most withering scrutiny.

That's the theory, anyway. In real life, though, things get messy, and the "scientific method" rarely matches this idealized description. Scientists are human. They have plenty of foibles, and in some cases outright myths they tell about themselves. They also have values and agendas that factor heavily into their research decisions. Moreover, the inquiries and investigations of scientists take place in a social and cultural context that shapes both their underlying assumptions and even (at least to some extent) how they measure and interpret nature itself.

These caveats, which have emerged from the field of science and technology studies, have dramatically changed the way we think about the guys in the white coats. But this isn't the place to weigh the extent to which science is "socially constructed," rather than reflecting an ultimate reality that is "out there"—a question scholars have debated endlessly. For our purposes, it is enough to note that despite its contingent nature, science has nevertheless uncovered what few would dispute are reliable conclusions: The earth is about 4.5 billion years old. It orbits the sun. It is spherical. And so on.

Science also *works*. Its practitioners have managed to split the atom, an unavoidable and terrifying reality of modern life. It seems safe to assume that scientists have learned a thing or two about the fundamental structure of the world if they are capable of destroying much of it.

For these reasons, the insights of science studies have hardly undermined the credibility of science as a fount of useful information crucial to modern political decision-making. Instead, this body of research teaches us the value of humility. Science isn't infallible; it is only as good as we are, and often we aren't all that good. Much of what we think we "know" today may turn out wrong. But if it does, scientists stand the best chance of telling us so, thanks to the process of peer review, hypothesis testing, and internal criticism that they use to check themselves. In other words, science still probably provides the best tool we have for understanding nature—a "candle in the dark," as Carl Sagan once put it. And it is something we should cherish.

Ironically, those guilty of misusing science frequently claim to share this belief. Because science commands such respect in our society, everybody wants a piece of it. The number of "-ologies" has increased at an absurd rate. Among others, we now encounter "parapsychologists" (those who investigate psychic powers), "sindonologists" (those who study the Shroud of Turin), and "cryptozoologists" (those who chase fantastic creatures such as Bigfoot and Nessie). But many of these investigators don the trappings of science yet fail to respect the standards of evidence, rigor, peer review, and the rejection of ideas that have been found wanting. At heart, they're true believers.

When pressed, ideological appropriators of science will rarely relinquish a cherished idea, no matter how many times it has been convincingly debunked. They seek to adopt the veneer of science, but not the critical rigor that should accompany it. All scientists have personal opinions, but practitioners of "science as advocacy" differ dramatically in degree, if not in kind. They flatly refuse to amend their views or follow the evidence where it leads (assuming they are even interested in evidence in the first place).

Simultaneously, science intersects with the political process at countless nodes. First, the United States government generously funds scientific research through the National Institutes of Health, the National Science Foundation, and other agencies, as well as through direct congressional appropriations (so-called "academic pork"). Political considerations may shape the funding of particular projects; perhaps a legislator wants a new research institute in his district. Politics can also create pressure to defund studies because the research occurs in a controversial area (such as examinations of the lives of sex workers). Complaints arise regularly over funding choices, and scientists continually grasp for more money. Yet with some exceptions, both political parties have found common ground when it comes to ensuring that the scientific enterprise in the United States remains strong.

This book isn't about battles over the funding of science, because that's not where the Right's war on science truly makes itself felt. Any politician can stand up and call for "more research." That's the easy part. But what does that politician do when science, federally funded or otherwise, points toward inconvenient conclusions? Does he listen, or does he try to deny and distort the information? All too often, conservative Republicans have taken the second course. I therefore focus my attention on the most controversial area where science interacts with politics: providing input into decision-making.

As befits our technological age, American political leaders constantly seek to justify their choices with reference to hard data. To gather information, they consult with scientists and technical experts. But politicians don't consume scientific information dispassionately; rather, they constantly seek to use science for advocacy purposes. What politicians often want to hear from scientists is, "Well, the science says that you must do what you wanted to do anyway," observes physicist Robert Frosch, administrator of the National Aeronautics and Space Administration

(NASA) under President Carter, and later vice president in charge of research laboratories at General Motors.

Because of this tension, preexisting ideological commitments exert considerable pressure on pure science. Politicized science isn't a new phenomenon, any more than paranormalist fringe science is. But it has recently reached crisis levels in the United States as the modern conservative movement—and the administration of George W. Bush—has shown a systematic willingness to misrepresent or even concoct its own "science" to skew debates of fundamental consequence to the nation.

But what does it mean to politicize science? What constitutes political science "abuse" in the first place? Here is my definition: any attempt to inappropriately undermine, alter, or otherwise interfere with the scientific process, or scientific conclusions, for political or ideological reasons. To count as inappropriate, such incursions must undermine the integrity of science by turning it into just another tool of political advocacy. We can never hope to divorce science from politics entirely, but a buffer zone must exist to ensure that science doesn't devolve into politics by other means.

What does all of this mean in practice? As the controversy over science and politics mounted during George W. Bush's first term (it would only heighten during his second), some tried to catalogue the various types of science misuses and abuses. In any such exercise, groupings may overlap. Still, a number of different problems seem to recur.

Undermining science itself. First, we should worry about those who would cheapen the scientific method. Creationists, for example, have often derided evolution as "just a theory," confusing the colloquial sense of the word "theory" with its very different scientific meaning. Far from mere hunches, scientific theories require repeated confirmation by independent investigators and broad acceptance by the scientific community. This creationist argument thus undercuts the very nature of scientific knowledge in an ideological quest to unseat Darwin's theory.

Beyond all-out assaults on science, remaining political abuses break down into two different categories: interference with individual scientists or the scientific process and attempts to alter or slant the results of science based on political considerations, rather than on the weight of scientific evidence. In the first category can be found suppression, targeting individual scientists, and rigging the process.

Suppression. Quashing scientific reports for political reasons constitutes a fundamental assault on the integrity of science, a process that thrives on openness and seeks to uncover truth regardless of the consequences. Numerous examples of this phenomenon emerged during the presidency of George W. Bush, and during earlier presidencies as well. As discussed in Chapter 4, the Reagan administration stalled the release of a report on acid rain by its very own White House Office of Science and Technology Policy because the report challenged the administration's inaction on the issue.

When stalling and suppression fail, political actors may also seek to edit unpublished reports to make their content more palatable. Such ideologically driven scrub jobs also violate the integrity of science.

Targeting individual scientists. Obviously, we don't want our scientists teaching terrorists how to make bombs or chemical weapons. But such extreme cases notwithstanding, the blocking of ordinary scientific exchange has no justification. Such interference intrudes on the integrity of science as a process in which independent investigators share ideas in an open quest for knowledge.

Political actors should never place unreasonable gag orders on what scientists in government can say or with whom they can communicate. Neither should they force government scientists to state conclusions that they don't actually accept or believe, or otherwise prevent them from presenting their honest scientific opinions. During the administration of president George Herbert Walker Bush, for example, the White House budget office broke this rule, deliberately altering the scientific testimony of NASA climate expert James Hansen to weaken his conclusions.

Finally, one of the most disturbing phenomena discussed in this book involves attacks on individual scientists aimed at discrediting their work. Legitimate scandals may arise in science, and criticism certainly has its place. Yet the political targeting of scientists often occurs in the absence of any evidence of scientific wrongdoing, and represents intimidation rather than oversight.

Rigging the process. The suppression of research and the targeting of individual scientists occur when particular scientific results offend political sensibilities. But science abuses can also arise earlier in a policy debate, as political actors seek to control the *input* into a scientific deliberation in order to shape the ultimate output.

Packing a scientific advisory committee with ideologues is one example of such rigging. So is the attempt by legislators to change the scientific rules that government agencies must follow simply in order to achieve politically desirable outcomes. Both types of meddling undermine scientific integrity by seeking, for political purposes, to rig the outcome of a scientific assessment.

When it comes to interference with scientific results, rather than the scientific process, we see a different array of problems:

Errors and misrepresentations. The simplest and most clear-cut science offense is a deliberate misstatement of fact, or the issuance of a demonstrably false claim. George W. Bush's assertion about the existence of "more than 60" embryonic stem cell lines provides a perfect example. Within this category are also included errors of omission—for example, citing one favorable study rather than the full range of relevant studies on a given question.

Most common, however, are *misrepresentations* or *distortions* of scientific work—essentially, the "spinning" of science. The flagrant twisting of research findings to humor a particular political view violates the integrity of science by treating its conclusions as mere political fodder, rather than useful information to be considered in its full context.

Consider one high-profile example, discussed in detail in Chapter 7. Conservatives have repeatedly misrepresented the findings of a 2001 report from the highly respected National Academy of Sciences on climate change. The report responsibly discusses, at various points, lingering uncertainties in climate science. Nevertheless, it also confirms the robust conclusion of the United Nations' Intergovernmental Panel on Climate Change that humans are likely (i.e., the conclusion has a fairly high degree of scientific certainty attached to it) contributing to rising global average temperatures. In a classic case of misrepresentation, conservatives have cited the uncertainties while ignoring or dismissing the affirmative conclusions.

Magnifying uncertainty. In political science debates, one specific form of misrepresentation occurs so frequently that it needs its own category. And that is the hyping and exaggerating of scientific uncertainty, frequently with the goal of preventing political action.

Science never provides absolute certainty about the world. We can always imagine a future study reaching a different conclusion than all the others, and oftentimes new discoveries raise more questions than they actually answer, thus increasing controversy rather than reducing it.

Since scientific uncertainty can never be fully dispelled, it hardly provides a good excuse for ducking political action. If it did, nothing would ever get done. Yet in policy fights with a strong scientific component, conservatives have touted uncertainty to precisely this end. Moreover, they have strategically magnified uncertainty itself, effectively misrepresenting what scientists actually know. Some industry groups have even gone so far as to "manufacture" uncertainty by strategically attempting to sow doubt about mainstream conclusions.

A fairly sophisticated discussion of how policymakers should deal with scientific uncertainty came from the late Rep. George Brown, ranking Democratic member of the House Committee on Science, in a 1996 report criticizing his Republican colleagues for demanding unreasonable levels of scientific certainty in decision-making:

> Science may be able to guide policymakers, but it cannot relieve policymakers of the obligation to make tough policy choices, choices that require a difficult balancing of competing interests. . . . [The] demand for absolute and incontrovertible truth prior to action is a choice to ignore science rather than be counseled by it and an abdication of the responsibility to use the best knowledge available at any given time to serve the common good.

As this passage suggests, policymakers who call for "more research" as a way of punting on a given issue may be guilty of misusing science. Whether the magnifying of uncertainty reaches the level of an abuse, of course, depends on the relative completeness of scientific knowledge on the issue in question. In some cases, Brown seems to suggest, policymakers may have to make a decision even in the face of considerable uncertainty. After all, science provides only one component of decision-making, and political action and inaction alike can have long-term consequences.

A variant on this type of abuse involves demanding different degrees of scientific certainty on different issues—essentially, recalibrating the certainty meter for political convenience. For example, conservatives clearly favor intense skepticism of the conclusions of mainstream climate science. In a 2001 letter to four Republican senators, President George W. Bush highlighted lingering scientific uncertainties as a reason to forestall action to cut carbon dioxide emissions, remarking on the "incomplete state of scientific knowledge of the causes of, and solutions to, global climate change." Yet Bush hardly abided by such stringent rules of evidence on other issues. In a caustic 2001 editorial, *Scientific American* compared Bush's approach to climate change with his support of ballistic missile defense, which most informed physicists doubt can perform its job adequately in current form due to technical shortcomings. "In one case, the president invokes uncertainty; in the other, he ignores it," noted the editorial. "In both, he has come down against the scientific consensus."

Relying on the outliers. One of the most common attempts to skew science occurs when politicians handpick experts whose views coincide with what they want to hear, even when the vast majority of scientists believe something else. Rather than cherry-picking specific scientific findings, these science abusers cherry-pick expertise itself.

Dissent has an important place in science. But the notion that policymakers should rely on scientific outliers to justify their decisions is preposterous. The best available consensus science should guide policy, not the most convenient science that politicians can find in a pinch.

At the extremes, fringe science overlaps with outright pseudoscience and quackery. Philosophers of science have struggled to find a firm line of demarcation between science and pseudoscience, and some have dismissed the entire endeavor as a "pseudo-problem." Still, we can safely use the term "pseudoscience" as long as we simply define it as bad science taken to an extreme.

Pseudoscientists make poorly substantiated or demonstrably false claims and refuse to relinquish them when shown the counterevidence. "Scientific" creationism and its 2.0 version, "intelligent design," provide the canonical examples of the conservative embrace of pseudoscience. Creationists and intelligent design proponents claim to act scientifically, but in fact they do little more than spread scientific-sounding arguments in defense of a biblical or religious agenda. It is doubtful whether any amount of evidence would change their minds.

Ginning up contrary "science." Some industries and interest groups have even plotted strategies for upsetting a consensus scientific view or shaping the development of scientific understanding in a way favorable to their interests. In other words, they have sought to generate "science" solely as a political device for advancing their objectives, regardless of actual truth. Often, the goal has been to "manufacture" scientific uncertainty to create a semblance of controversy where it doesn't actually exist.

The decades-long fight over the regulation of tobacco provided an especially fertile opportunity for the creation of contrary science. Take just one case study from the tobacco industry's annals of science manipulation. Documents released as a result of litigation in the state of Minnesota show that Big Tobacco paid a group of scientists thousands of dollars apiece to write letters to scientific journals, such as the *Journal of the American Medical Association,* the *Journal of the National Cancer Institute,* and the *Lancet,* disputing the findings of a 1992 Environmental Protection Agency report linking secondhand smoke to lung cancer. This led to charges that the industry had engaged in a "systematic effort to pollute the scientific literature," as Stanton Glantz, of the University of California, San Francisco, an expert on the health effects of secondhand smoke, put it at the time.

Dressing up values in scientific clothing. Politicians and policymakers should not pretend that their raw political choices flow from scientific realities. Claiming scientific justification for a purely political move undermines science by treating it as a source of post hoc justification, rather than a valuable input into the decision-making process.

Consider one particularly striking example, discussed at greater length in Chapter 13. In May 2004, the Food and Drug Administration ignored a 23–4 recommendation from its scientific advisers and refused to approve over-the-counter sales of the "morning after" pill Plan B. Commentators denounced the decision as a transparent attempt to appease pro-life religious conservatives who had lobbied against Plan B's approval.

Yet ironically, the FDA cited a scientific justification for ignoring its scientists. The agency supposedly needed more data on how the drug would affect young adolescents, even though such data had rarely, if ever, been demanded for other drug approvals and no evidence suggested that young teens would use the drug any differently than would older teens and adults. This misuse of science has had serious consequences. If made more readily available, Plan B could help prevent unintended pregnancies and thus reduce the number of abortions.

This catalogue of politicized interferences with science provides a skeleton key to the rest of this book. Except to take stances against inappropriate legislative interference with science and to advocate a strengthening of our government's science policy apparatus, the text takes no position on questions of pure policy. If a politician presents a fair picture of climate science, but nevertheless opposes the Kyoto Protocol on economic grounds, I leave it to economists to criticize him or her. If a president takes advice from a well-balanced panel of experts and then makes a contrary decision, that too is his or her prerogative, as long as the decision doesn't get bedecked in scientific garb.

Commentators across the political spectrum generally agree that science should inform, but not dictate, political choices, in much the same way that input from the intelligence community helps to inform military strategy and foreign policy. "I don't think there are very many scientists who are naive enough to think that science should always determine outcomes, but you shouldn't defend outcomes by distorting the science," says physicist John Holdren, director of the Science, Technology, and Public Policy Program at the Harvard Kennedy School. The tragedy of the present moment is that this relationship has all too frequently been upended. Political allegiances repeatedly shape the conservative movement's approach to scientific information itself, leading to direct interference with the scientific process. And that, in turn, both cheapens science and leaves us with inadequate, or even indefensible, policies on the books.

With this groundwork in place, we can now turn to the history of modern conservatism, and see how the Republican Party hemorrhaged moderates and alienated scientists as it became increasingly dominated by the political Right.

CHAPTER 3

FROM FDR TO NIXON

O N A MAY MORNING, Russell Train sits in a corner office of the massive Washington, D.C., World Wildlife Fund (WWF) building, surrounded by stacked boxes of his papers that will be heading out soon to the Library of Congress. Now in his mid-eighties, Train occupies a place of honor in the institution he helped create, having led one of the most distinguished environmental careers that America has yet seen. A lifelong Republican as well as a staunch environmentalist, Train helped found the U.S. branch of WWF in 1961. He then went on to serve as under secretary of the interior, chairman of the newly formed White House Council on Environmental Quality, and second administrator of the Environmental Protection Agency (EPA) under Richard Nixon (and later, Gerald Ford), before finally returning to head the wildlife organization.

In the early 1970s—the political height of the environmental movement—a Nixon administration anxious to burnish its green credentials welcomed Train into the Republican mainstream. Nixon may not have cared about clean air and water personally, but he knew which way the wind was blowing, creating the EPA in 1970 and signing the Endangered Species Act and a raft of other environmental laws. Today, however, the iconic WWF giant panda button that Train wears suggests a painful irony. Train has become a politically endangered species: a moderate, pro-environment Republican in a party run by its right wing.

Speaking slowly, deliberately, and often turning to face his window and think silently for long moments, Train reflects on his virtual exile from today's GOP. "The Republican Party as I used to know it was a more middle of the road kind of party," he recalls, in a deep and raspy voice whose accent seems to hint at his Rhode Island origins. The change over time has been "absolutely incredible," Train continues, especially on environmental matters.

But Train doesn't simply balk at the Republican policies of today. He rejects the way they're often implemented, with a willingness to place political considerations above scientific and regulatory expertise. Speaking of the administration of George W. Bush, Train criticizes the "constant interjection of the White House in regulatory matters," a theme he hammers repeatedly in his memoir, *Politics, Pollution, and Pandas.* In contrast, Train recalls a 1975 White House meeting with President Ford that he attended to discuss a pending decision on auto emissions standards. Ford said, "'We're interested in this, would like to know what you're going to do, but I want you to be totally comfortable in the fact that no effort whatsoever will be made to try to change your position in any way,'" Train remembers. "And they didn't. And that's just the way it was."

No wonder, then, that Train reacted with alarm after the *New York Times* reported in June 2003 that the Bush White House had force-edited the climate change section of a draft EPA report on the state of the environment, attempting to water down the text and even taking out references to conclusions by the respected National Academy of Sciences. In a much-quoted letter published in the *Times,* Train declared, "I can state categorically that there never was such White House intrusion into the business of the E.P.A. during my tenure." Later, Train would make common cause with the Union of Concerned Scientists (UCS) when the group denounced the Bush administration's science manipulations and abuses. During a February 2004 UCS conference call, Train stated of the politicization of science, "I don't see it as a partisan issue at all. If it becomes that way, I think it's because the White House chooses to make it a partisan issue."

Given his statements and actions, a scientist might call Russell Train an anomaly in need of explanation. His sharp criticisms of his own party's approach to science and the environment raise an inexorable question: How did the GOP change so much, and leave moderates like Train so far behind?

In broad outline, the story of Train's exile and the rise of the modern conservative movement are one and the same. The tale begins nearly half a century ago, a time when science enjoyed immense prestige in American life, but also when a little-noticed band of conservatives launched a political rebellion that would achieve its full realization only decades later. The ideological merger between business interests and religious conservatives that occurred in the 1970s and 1980s, under the mantle of the Republican Party, left middle-of-the-road types like Train scratching their heads. Train himself may have put it best in his interview with me, when we discussed the rise of the so-called religious Right as a force within the Republican Party.

"I don't understand the religious conservatives," he said. "They're so far out of my ken."

For anyone who doubts the crisis over science politicization today, let me take you back to a very different era: the 1940s. Regardless of what we may think of the results now, a strong partnership between science and our political leaders smoothed our way to victory in World War II and ushered in the nuclear age. Innovations in atomic weaponry as well as radar played a pivotal role in the defeat of the Axis powers, leaving a glow on the American scientific enterprise.

President Roosevelt, though, understood that science could do much more than win wars. He and the director of his wartime Office of Scientific Research and Development, Vannevar Bush, sought to mobilize scientists to provide continuing benefits during peacetime. In his famed 1945 report *Science: The Endless Frontier,* Bush called upon the government to invest heavily in the funding of scientific research at American universities. The impact of Bush's report has since been greatly exaggerated and even mythologized. Nevertheless, it captured the ethos of a time when a postwar nation applauded science for its role in helping to secure military victory, and government investment in university-based scientific research boomed.

The prestige of science only increased following the October 4, 1957, Soviet launch of *Sputnik*, a "technological Pearl Harbor" that jolted the nation and pushed President Eisenhower to convene the star-studded President's Science Advisory Committee (PSAC) and name the first White House science adviser, MIT president James Killian. The period represented the "apogee of presidential science advising," as Yale science historian Daniel Kevles put it to me.

President Kennedy carried on the romance with scientists. Surrounding himself with intellectuals and grasping the vital importance of scientific research and innovation to a great power, Kennedy established the White House Office of Science and Technology and made his science adviser, Jerome Wiesner, an electrical engineer who would later also serve as MIT president, part of his inner circle. On issues of science and technology, it became known that "Wiesner spoke for the president."

The cooperative relationship between science and politics during the Kennedy era shone through in the space program. At the very start of the administration, Wiesner and other PSAC members informed Kennedy that in considering manned spaceflight, he should be aware that unmanned probes could do a cheaper and safer job of exploring space and that a "crash program aimed at placing a man into orbit . . . cannot be justified solely on scientific or technical grounds." Yet the president wanted to put a man in orbit anyway and to proceed with the Apollo moon-landing program. An agreement resulted that satisfied both parties: Wiesner supported Kennedy's decision, and Kennedy, in turn, avoided claiming that Apollo would deliver vast amounts of scientific knowledge.

Kennedy's statements in favor of manned lunar exploration "were never justified as being based on science," recalls Stanford physicist W. K. H. ("Pief") Panofsky, who sat on PSAC at the time. In an interview in his office at the Stanford Linear Accelerator Center, Panofsky singled out this episode as an example of politicians *not* doing what we have seen so much of since then: dressing up political goals in the habit of science.

During this more innocent era of government-science cooperation, modern American conservatism first reared its head. From the start, the movement had its tensions with the scientific community. Conservative pooh-bah William F. Buckley, Jr., who founded the modern Right's leading magazine, *National Review,* in the mid-1950s, had risen to prominence by railing against Yale's secular intellectual caste in his classic *God and Man at Yale*. Buckley's critique explicitly targeted Yale's social scientists among other professors deemed too critical of religion.

Meanwhile, on the political front, the conservative movement's darling, Arizona Republican senator Barry Goldwater, argued against the New Deal and for a tough, militant anticommunism. When Goldwater ran for president in 1964, the movement that grew up to support him pulsed with right-wing anti-intellectualism—a deep distrust of the "Eastern establishment," the elite media, and the nation's leading universities that runs strong in the conservative movement to this day. Goldwater's candidacy deeply alienated the American intelligentsia—including scientists—and an early skirmish between the fledgling Right and the scientific community gave a hint of the conflict that would fully emerge decades later (albeit over a very different set of concerns). In a dress rehearsal for the science community's mobilization against George W. Bush, a group of leading scientists vehemently opposed Goldwater's election.

Consisting of prominent figures such as Eisenhower's second science adviser, George Kistiakowsky, and Kennedy adviser Jerome Wiesner, the anti-Goldwater group called itself Scientists and Engineers for Johnson-Humphrey. Their beef didn't concern misuses of science; instead, the scientists feared Goldwater's nuclear brinksmanship and opposition to arms control (a major political cause embraced by U.S. scientists during the Cold War years and afterward). The movement had an irresistible selling point: The creators of the bomb had aligned against someone they suspected might actually use it. The scientists' attack on Goldwater for being a "blustery, threatening man," as Manhattan Project contributor and Nobel laureate Harold Urey put it, helped doom his candidacy. It didn't hurt that Goldwater couldn't mobilize a convincing group of scientists on his own side.

But though it went down in defeat, Goldwater's campaign brought together, for the first time, the conservative activists who would ultimately achieve political victory. We often remember the 1960s as a liberationist era, but they also witnessed unprecedented conservative ferment. Disgust at a perceived corruption of American values by radical counterculturists stoked the conservative fire. During the 1970s, conservatives established a slew of activist organizations in Washington, D.C., such as the Conservative Caucus, founded in 1974. These groups complemented the older American Conservative Union, officially launched just days after Goldwater's defeat by his supporters.

Though conservatives remained divided—with the hardliners of the so-called New Right suspicious of more genteel conservatives like Buckley—they began to form a great ideological merger. Ultimately, modern conservatism would unite, under the same broad umbrella, worldly pro-business conservatives and cultural traditionalists fed up with hippies, feminism, and gay rights, and incensed by *Roe v. Wade* and the Supreme Court's banning of school prayer. Like the Goldwater movement from which it descended, the New Right also nourished a strong anti–East Coast, anti-intellectual animus that easily translated into distrust of the American scientific community.

The triumphs of the environmental and consumer movements of the late 1960s and early 1970s, and the attendant expansion of the federal regulatory state, spurred on the business community's political counterreaction. With actions like the banning of various chemical pesticides, unabashed regulators like Russell Train stirred up a hornet's nest. Rules by new agencies such as the Environmental Protection Agency and Occupational Safety and Health Administration necessarily required a firm scientific basis, and these bodies accordingly overflowed with technical experts specializing in science and risk assessment. Yet this, in turn, created a strong incentive for companies subject to potentially costly regulation to sponsor their own contrary science, a powerful technique for blocking or refuting proposed agency actions.

As an early but memorable example of this process, consider the battle over *Silent Spring*. In her 1962 bestseller, widely credited with awakening large-scale environmental sentiment, Rachel Carson denounced a number of chemical pesticides, including, most famously, DDT. In 1972, the EPA would ban the use of DDT, but not before the chemical industry had mounted a campaign to undermine Carson's science.

The onslaught began even before the book's publication, when the Velsicol Chemical Corporation wrote to Carson's publisher, Houghton Mifflin, challenging her statements about two chemicals manufactured

exclusively by Velsicol. A scientific review requested by the press vindicated Carson. After *Silent Spring* came out, the attacks intensified. One
critic dubbed Carson a "fanatical defender of the cult of the balance of
nature," while another called her book a "hoax." The Manufacturing
Chemists Association and National Agricultural Chemicals Association
went on a PR binge, mailing out pro-chemical articles and harsh reviews
of *Silent Spring.*

Critics often charged that Carson had used bad science in her arguments. And she had, after all, written a popular book, a polemic rather
than a dissertation. But many leading scientists emerged to defend Carson's claims, even as a 1963 PSAC study requested by President Kennedy
lent credence to her concerns and supported limitations on pesticide use.
Science magazine editorialized that the report "adds up to a fairly thoroughgoing vindication of Rachel Carson's *Silent Spring* thesis." Still, industry had successfully planted the notion that *Silent Spring*'s author
relied on environmental mysticism, rather than legitimate science. This
pattern—of attacking scientists whose work poses a challenge to industry
profits while drumming up contrary science—would recur on a wide
range of subsequent environmental and regulatory issues.

The preventive or "precautionary" nature of the new environmental
rules, such as the Nixon-era DDT ban, spurred on science conflicts. The
1970s mantra that government regulations would protect against future
health risks and environmental degradation represented "an enormous
psychological and social breakpoint in the use of science, and in the reliance on science now to actually try to forestall future harms," explains
Sheila Jasanoff, a leading science studies scholar at Harvard's Kennedy
School of Government.

The emphasis on prevention triggered battles over risk and risk assessment in which environmentalists increasingly found themselves depicted
as antiscientific worrywarts (just as Rachel Carson had been). In works
like Edith Efron's 1984 book *The Apocalyptics,* defenders of industry
spread a powerful and resilient meme: the notion that environmental
alarmists often overreact to minuscule or nonexistent health scares.
(The claim certainly rang true in some cases, but the antienvironmental
backlash more than matched any environmentalist excesses.) By thrusting

it into an adversarial courtroom setting, the potential for judicial review of agency science-based decisions only politicized science further.

The corporate reaction against regulation extended far beyond scientific disputes, of course. Believe it or not, industry had previously paid relatively little attention to goings-on in Washington, D.C. But in the 1970s, alarmed by a new wave of environmental, health, and safety rules, businesses erected a powerful lobbying apparatus, formed political action committees, and—most importantly for the later politicization of science—began to rethink the way they would spend their money in sponsoring research and intellectual inquiry. Major companies and trade associations hearkened to the fateful advice of Lewis D. Powell, later a Nixon Supreme Court nominee, and New York University political thinker and *Public Interest* editor Irving Kristol, a reformed Trotskyite turned "neoconservative," both of whom advised business leaders to mobilize to ensure their own survival. That included subsidizing the spread of pro-business ideas and arguments. "Corporate philanthropy should not be, and cannot be, disinterested," wrote Kristol.

Soon businesses turned to funding intellectuals and research organizations that would promote their own goals and interests, rather than mere objective study. Sponsored by corporate interests and conservative foundations, think tanks such as the American Enterprise Institute (which began life as a business association) and the Heritage Foundation enlisted right-leaning thinkers to provide handy "expertise" to conservative politicians and activists, frequently in contradiction to the findings of leading social scientists and university-based scholars. As John Judis details in his book *The Paradox of American Democracy: Elites, Special Interests, and the Betrayal of Public Trust,* an older generation of foundations had genuinely aspired to conduct disinterested social science research for the improvement of policy and the betterment of humankind. The new outfits, by contrast, shilled for a conservative political agenda.

The proliferation of think tanks created extremely propitious conditions for the politicization of science. After all, the cherry-picking of favorable expertise on a given policy question cannot occur unless that expertise actually exists in the first place. And the number of Washington,

D.C.–based think tanks—many of them on the political Right—has increased dramatically over the past fifty years. "At the end of World War II, only a handful of private policy think tanks were at work in Washington; at the end of the Cold War there were over one hundred, the largest ones spending tens of millions of dollars annually on the analysis of policy problems," science policy analyst Bruce Bimber has noted.

Another key development came with the birth of advocacy groups who would cite the Right's new experts and even bring their arguments into the courtroom. As a case in point, consider the Pacific Legal Foundation (PLF), a right-wing group founded in 1973. Organizations like PLF, notes Alan Crawford in his 1980 classic *Thunder on the Right,* "were started to counteract the growing influence of such organizations as the Sierra Club and the Environmental Defense Fund." As we will see in Chapter 10, arguing back against environmentalists would frequently bring PLF's lawyers into scientific territory.

The Right's new class of advocates and intellectuals weighed in on a range of issues, science policy included. In fact, a key inspiration behind the 1973 birth of the Heritage Foundation was the Senate's vote, two years earlier, to cut funding for the supersonic transport program (SST). A Nixon-sponsored project to develop a high-speed passenger jet, SST lost its support after critics successfully argued that the plane would create intolerable "sonic boom" disturbances and atmospheric pollution, and wouldn't be very cost-efficient as a form of transportation anyway. Working as a senatorial aide at the time, Heritage founder Paul Weyrich received, to his dismay, an American Enterprise Institute analysis of the program only *after* Congress had already killed it. The incident inspired Weyrich and another congressional aide, Edwin Feulner, to form Heritage with money from conservative sugar daddies Joseph Coors, the beer tycoon, and later, Richard Mellon Scaife. The new outfit would serve as a rapid-response organization promoting conservative policies.

In this incident, one can catch a glimpse of how the Right's quest to establish its own sources of favorable expertise would ultimately put the conservative movement fully at odds with the scientific mainstream. Simultaneously, the supersonic transport debate ignited a battle between Nixon and his own scientists that would culminate in a devastating blow to the role of independent, high-quality scientific advice—always an antidote to political science abuse—in government.

As head of Nixon's Council on Environmental Quality at the time, Russell Train publicly raised questions about the SST program's environmental impact. So did many of Nixon's own science advisers, most notably Richard Garwin, a PSAC physicist. After the Nixon administration refused to release a negative report on the SST by a panel that Garwin had chaired, the physicist testified against it before Congress, devastatingly charging that the information provided to legislators on the program had been "less than adequate, and in many cases distorted." (Appropriately enough, three decades later, Garwin, also a self-described Republican, would join Train in endorsing the Union of Concerned Scientists' statement on the Bush administration's systematic misuses of science.)

Train angered the Nixon White House, but got away with criticizing the SST. Garwin wasn't so lucky. The physicist's freelancing contributed to Nixon's decision, after his 1972 reelection, to dissolve PSAC and abolish the office of the presidential science adviser, a landmark moment in the relationship between scientists and government, and one that laid the groundwork for much of the politicization that came later.

By the close of Nixon's first term, his staff had come to see the scientists in their midst as "vipers." The scientists found themselves associated not only with opposition to the SST, but more broadly with the academic community's stance against the antiballistic missile system and the Vietnam War. In a White House that had trouble brooking dissent, they simply had to go. "Nixon did not want to hear the facts," Jerome Wiesner later wrote. "In a sense, he chose to kill the messenger." Though science's influence on policy would rebound somewhat under Gerald Ford and Jimmy Carter, the damage had been done. The American scientific establishment had been knocked from its pedestal, even as conservatives who found science at odds with their economic, religious, or political interests had caught on to a nifty trick: They could generate their very own scientific expertise.

CHAPTER 4

"CREATION SCIENCE" AND
REAGAN'S "DREAM"

RUSSELL TRAIN'S TERM AT EPA ended with Jimmy Carter's 1976 defeat of Gerald Ford. By then, the Republican Party had begun its lurch to the right, as indicated by Ford's decision to choose a conservative running mate, Bob Dole, rather than the moderate former EPA administrator William Ruckelshaus. We don't know how a second Ford administration would have governed. But we do know that the GOP's next champion, Ronald Reagan, pulled together and unified disparate and often fractious strains of the Right. In the process, Reagan brewed a political concoction—equal parts big business and religious conservatism—that proved highly toxic to the role of science in government.

Today, conservatives regard Reagan as little short of a deity. Yet amid the worshipful reflections that accompanied a national outpouring of grief at his death in the summer of 2004, few bothered to discuss the Reagan administration's disturbing treatment of science.

Not only did Reagan fail to acknowledge and speak out about the AIDS epidemic until 1987, at least in part out of deference to religious conservatives in his administration such as domestic policy adviser (and later religious Right leader) Gary Bauer, who recoiled at the notion of educating children about safe sex and condoms. The administration even forbade Surgeon General C. Everett Koop from saying anything about AIDS during the entirety of Reagan's first term, according to Koop's memoir. "I have never understood why these peculiar restraints were placed on me," Koop wrote.

Reagan's administration showed considerable solicitousness toward both regulated industries and newly influential religious conservatives, and a willingness to slant science to appease each. Initially, the Reagan team even considered doing without a presidential science adviser. Before the appointment of the little-known physicist George Keyworth, Reagan's budget director, David Stockman, reportedly remarked of White House scientists to a visitor, "We know what we want to do, and they'll only give us contrary advice."

From a scientific perspective, Reagan's single greatest offense may have been his antievolutionism. During his 1967–1975 California governorship, Reagan's self-appointed state board of education had pushed to weaken the teaching of evolution and endorsed creationism. Campaigning nationally in 1980, Reagan followed the California precedent. He pronounced that "great flaws" existed in evolutionary theory, and that public schools should therefore teach the "biblical story of creation" as well.

After Reagan's election, members of his administration also toed the line. Reagan science adviser George Keyworth refused to repudiate the teaching of creationism in public schools during his 1981 confirmation hearing. And in a 1987 interview with the *New Republic*'s John Judis, Reagan's secretary of education William Bennett took a similar line, waffling on the issue and ultimately pleading ignorance about the nature of creationism. In 1986, however, Bennett had declared that in his view, the selection of public school textbooks should involve the "judgment of the community," a tacit nod to creationist forces at the local level.

The Reagan administration's sympathies with creationism signaled a new development for the Republican Party and conservatism more generally. From this moment forward, many of the party's leaders willingly distorted or even denied the bedrock scientific theory of evolution, and encouraged pseudoscientific thinking, to satisfy a traditionalist religious constituency.

There is no mystery about why this happened: politics. Reagan said as much himself during his campaign: "Religious America is awakening," he declared. With respect to religious conservatives and the GOP, "the marriage started around 1980," explains John C. Green, a professor of political science at the University of Akron, in Ohio, who specializes in religion and politics. The relationship has grown stronger with every passing election, Green continues, but Reagan set a presidential precedent in catering to the religious Right, whose deacons, such as Jerry Falwell of the Moral Majority, had made antievolutionism a core concern. During the 1980s, even previously secular conservative intellectuals like Irving Kristol embraced various critiques of evolution, as if they deemed it a necessary political step toward sustaining a governing coalition. And perhaps they were right.

Ironically, Reagan turned toward creationism even as creationism itself turned increasingly toward science, or at least attempted to. America's indigenous creationist movement dates back to the Protestant fundamentalist awakening of the early twentieth century. It has always been religious in nature and inspiration. But over the last century, creationists showed a marked trend towards the appropriation of scientific trappings and the masking of outwardly religious forms of argumentation.

Even as they thumped their Bibles and denounced evolution, early American creationists sometimes made "scientific" arguments as well. Scopes trial advocate William Jennings Bryan even joined the American Association for the Advancement of Science in 1924. But not until the 1960s and 1970s did creationists consciously style themselves as practitioners of "creation science," purging their writings and arguments of scriptural references and consciously recruiting Ph.D.s who were also fundamentalist Christians to their side. Not content with merely denying science, they increasingly began to mimic and abuse it.

"Creation science" emerged for specific cultural and legal reasons. First, not even creationists could ignore the growing prestige of American science, especially following the scientific buildup that came as a reaction to Sputnik. But more importantly, they were losing in court. In the 1968 case *Epperson v. Arkansas,* the U.S. Supreme Court declared outright bans on the teaching of evolution unconstitutional. California creationists then tried a different tack, arguing on secular and scientific grounds for the incorporation of creationism *alongside* evolution in school science curricula. "What they had to do was pretend that it was a science and that it should be given equal time," explains Stephen Brush, a historian of science at the University of Maryland.

With the publication of John C. Whitcomb, Jr., and Henry Morris's *The Genesis Flood* in 1961, the formation of the Creation Research Society in 1963, and the explicit adoption of the term "creation science" in the early 1970s, creationism dramatically ramped up its scientific pretensions. Its proponents began to insist that you didn't have to believe in the Bible to see the "evidences" of creation, and that students could learn "creation science" without being preached to. In the 1974 publication *Scientific Creationism*, edited by movement leader Henry Morris, the Institute for Creation Research argued that creationism could be taught "with no references to the Bible or to religious doctrine." Hilariously, *Scientific Creationism* came in two versions: a secular version for public schools and a religious version that cited scripture in order to lead the student "into a comprehensive, coherent, and satisfying world-view centered in his personal Creator and Saviour, the Lord Jesus Christ."

Similarly, biochemist Duane Gish, a leading creationist, argued that the refutation of evolution could proceed "purely scientifically." Both the Creation Research Society and later the Institution for Creation Research strove to bring evangelical Christian Ph.D.s into the fold, providing a key infrastructure for the appropriation of "science" by antiscience fundamentalists.

In embracing this movement, Ronald Reagan committed multiple abuses of science. First, he misrepresented the nature of scientific knowledge, stating that evolution "is a scientific theory only, and [has] in recent years been challenged in the world of science." Furthermore, at least implicitly, Reagan endorsed the pseudoscientific content of "creation science."

"Creation scientists" rely heavily on the theory of "Flood geology," claiming that Noah's Flood laid down the fossil record, carved the Grand Canyon, and caused a wide array of other geological phenomena. They also sometimes argue that the second law of thermodynamics—which states that in closed systems, entropy will increase—precludes evolution, because it rules out the natural development of the type of organized complexity seen in living organisms. But both arguments amount to religious apologetics carried out through quasi-scientific argumentation. Though they sound scientific, you won't find serious physicists or geologists making these claims.

As far as entropy goes, the second law of thermodynamics applies only to closed systems in which the amount of energy is fixed; in such systems, entropy will increase and order decrease. But the earth constantly receives infusions of energy from the sun. It is an open system. Thus thermodynamics provides no challenge to evolution whatsoever.

Flood geology, meanwhile, emerged from the mind of Seventh-Day Adventist George McCready Price, who had little scientific training but felt that God had told him to enter the "unworked field" of evolutionary geology. Price's theory basically assumes, based on the Bible, the existence of a worldwide deluge, and then fits the facts of the fossil record to that assumption—precisely the opposite of how science should work.

Moreover, the specific claims of Flood geologists have been soundly refuted. For instance, these creationists claim that the occasional appearance of "older" sediment layers *above* "younger" ones in the geologic record refutes the mainstream geological idea that the earth's evolutionary history should be written in the order of rock layers and the fossils they contain. Yet the well-known geological phenomenon of "overthrusting," or "thrust faulting"—in which one block of rock climbs atop another—explains perfectly well how older rocks may sometimes wind up above younger ones.

If Reagan's endorsement of creationism threw a bone to religious conservatives, his administration did at least as much to appease industry on scientific questions. Though hardly so troubling as the administration of George W. Bush, in this respect, too, Reagan laid the groundwork for the abuses of our current president.

The pro-industry mood at the start of the Reagan administration was intoxicating. Antienvironmentalists James Watt and Anne Gorsuch (later Burford) took the helm at the Department of the Interior and the

Environmental Protection Agency. Vice President George H. W. Bush convened a task force on "regulatory reform" to address industry's complaints about oppressive regulations, even as the administration created, through executive order, a process at the White House Office of Management and Budget to second-guess the decisions made by federal regulatory agencies (many of them science-based).

Helping industry dodge regulations under the mantle of "regulatory reform" frequently overlapped with manipulating science. In one particularly egregious case in the summer of 1981, the Environmental Protection Agency held private "science forums" so that industry representatives could debunk concerns about the cancer-causing properties of the chemicals formaldehyde and DEHP. One critic called the process a "star chamber pitting EPA scientists against industry scientists." Despite the one-sided nature of the proceedings and the lack of independent review of industry's self-interested claims, the meetings seem to have derailed tougher regulations on both chemicals, a thorough corruption of the regulatory science process.

Similarly, Reagan's administration showed an inclination to politically manipulate the membership of obscure scientific advisory committees located deep within the federal bureaucracy. Early in the Reagan years, Congress uncovered an industry-friendly Environmental Protection Agency "hit list" of scientists who had been rated on the basis of their views. The list labeled objectionable scientists with epithets such as "a Nader on toxics" and "bleeding-heart liberal." Strikingly similar charges of advisory committee manipulation have since arisen against the Bush administration (see Chapter 14), and at many more agencies than just the EPA. The parallelism calls to mind Yogi Berra's famous line, "It's déjà vu all over again," Sheila Jasanoff has noted.

However, in some respects the Reagan administration seems to have cooled its early science-abusing ardor. The "hit list" scandal fed into other controversies that ultimately led to the resignation of Reagan's first EPA administrator, Anne Gorsuch (Burford). Russell Train himself attacked the direction taken by the EPA under Gorsuch in a February 1982 *Washington Post* op-ed piece; he lightened up on learning that his own predecessor at the EPA and another Republican moderate, William Ruckelshaus, would replace her. Upon his return to the EPA, Ruckelshaus pledged, "There will be no hit lists. . . . I intend that EPA will operate forthrightly and honestly." The mid-course correction suggests that federal scientific agencies largely retained their independence in the Reagan era, even if political types occasionally intruded upon it. By the administration of George W. Bush, that independence would have all but vanished.

Reagan's administration also flagrantly exploited scientific uncertainty to defend the goals of industry on the question of acid rain,* perhaps the most prominent environmental controversy of the 1980s and the subject of some eighty congressional hearings during the decade. During the 1970s, concerns about acidic precipitation and dry acid deposition, caused by the emission of sulfur dioxide and nitrogen oxides from human sources such as power plants and auto exhaust, began to mount. Reagan hardly distinguished himself on the issue while campaigning for the presidency, incorrectly asserting that Mount St. Helens produces more sulfur dioxide pollution than human sources.

Newspaper editorial pages slammed Reagan for the scientific gaffe at the time—"Ronald Reagan came out against Mount St. Helens on Wednesday," cracked the *Washington Post*—and they were right to worry. Once Reagan took office, it became clear that his administration had no qualms about ignoring and misusing science on acid rain.

Reagan's White House repeatedly invoked the concept of scientific uncertainty to undermine a developing consensus that human industrial emissions cause acid rain, which in turn causes various types of ecological damage, such as the killing of fish in acidified lakes and the inhibition of tree growth. The administration took the position that the science of acid rain remained inadequate to justify pollution controls on sulfur dioxide emissions, and that "more research" would first be required on the problem. In effect, the White House threw in its lot with coal and electric power interests, whose own scientists magnified uncertainty in a variety of

---

*The term "acid rain" was used differently by different parties to the debate. Depending on the user, it could mean, in order of least to most comprehensive, (a) acidic rain; (b) acidic precipitation (including not just rain, but snow, fog, and so forth); or (c) acid deposition (including both acidic precipitation and dry acid deposition). Uses of the term "acid rain" here refer to the broadest category, or what we might technically call *acid deposition*.

ways, questioning the connection between acid rain and environmental damage as well as industry's role in causing the phenomenon. "Industry challenged every possible link in the chain between emission and ecosystem effects," recalls Princeton physicist Michael Oppenheimer, who testified before Congress on acid rain multiple times during the 1980s as a senior scientist with the Environmental Defense Fund. "They argued about anything."

But in 1983, a number of scientific developments destroyed the credibility of this already dubious position. First, a panel of scientists convened by Reagan's own White House Office of Science and Technology Policy (OSTP) recommended that immediate steps should be taken to curb acid rain. Second, a report by the National Academy of Sciences clearly linked acid rain in the northeastern United States to industrial sources and car exhaust. "With two scientific reports published last week, the uncertainty about the causes and probable cures for acid rain was sharply reduced," noted the *New York Times* in July of that year.

The Reagan administration responded by denying the science. A backlash against the OSTP panel ensued within the administration; the group's final report—which in one draft acknowledged that "actions to reduce acid deposition will have to be taken despite incomplete knowledge"—was delayed and watered down. One unnamed administration official called the OSTP acid rain study a "catastrophe, which careened from embarrassment to embarrassment," and the incident helped inspire efforts by some White House staff to abolish OSTP entirely. Only in 1986 did Reagan himself actually concede that human sources cause acid rain. But later that same year, his energy secretary, John Herrington, called the acid rain problem "small" and continued to call for more research.

It may have been Reagan's right as a politician to ignore the acid rain issue (action on this score would fall to his successor, George H. W. Bush, who signed amendments to the Clean Air Act in 1990 to control sulfur dioxide and nitrogen oxide emissions). But Reagan's administration went a step farther. Even as scientists clearly and firmly defined the problem, Reagan's team continued to exaggerate scientific uncertainty about acid rain's causes and effects, and to disingenuously call for "more research," an excuse for regulatory inaction. This approach perfectly matched the industry line. The "sources of acid rain cannot be pinpointed," argued National Coal Association president Carl Bagge in 1983.

The high profile "Star Wars" controversy also pitted Ronald Reagan against the scientific community, leading to a dramatic rift that would reverberate in later years. The George C. Marshall Institute, founded in 1984 and now a leader in the presentation of a conservative perspective on science issues, actually delivered its very first pamphlet in defense of "Star Wars" and participated heavily in public debate on the issue (before later going on to challenge mainstream climate science).

"Star Wars," which Reagan referred to as his "dream," began with a little anticipated speech by the president on March 28, 1983. In it, Reagan called for a research and development program to determine ways of protecting the United States from nuclear missiles, thereby rendering such weapons "impotent and obsolete." The central idea, it soon became clear, involved futuristic space-based laser technologies. Although the Pentagon would not officially launch the Strategic Defense Initiative (SDI) until after Reagan's 1984 reelection, a great debate began over this massive and costly technological fishing expedition, and the scientific shortcomings of "Star Wars" soon became apparent.

The first problem concerned sheer technological feasibility. The idea of a space-based umbrella defense had bubbled up, in the late 1970s and early 1980s, from a number of different promoters, many of whom relied on borderline science fiction to justify their enthusiasm. These included Lawrence Livermore Laboratory nuclear physicist Edward Teller (deeply controversial, but a California pal who had Reagan's ear) and the aforementioned Wyoming senator Malcolm Wallop. Teller promoted Excalibur, the so-called bomb-pumped X-ray laser, which would be produced through a nuclear explosion in space and which was promised to the White House "on a five-year time scale." (The idea would later turn out to have been drastically oversold.) Wallop, meanwhile, favored chemically powered lasers, which he thought could be deployed in space by the mid-1980s. (Experts thought otherwise.)

Reagan science adviser George Keyworth, skeptical of these enthusiasts and the programs they were pitching, set up a White House Science

Council panel to look into the issue. The resultant report, completed not long before Reagan's famous speech in 1983, poured cold water all over the idea that directed-energy technologies could provide a source of new weaponry. Yet the administration ignored this expert advice, and Keyworth decided not to make a stand. After Reagan's speech, the loyal science adviser became a cheerleader for "Star Wars," cementing his deeply controversial reputation. "Keyworth's outspoken defense of SDI was increasingly at odds with the dominant opinion in the nation's scientific community," writes Gregg Herken in his history of presidential science advising, *Cardinal Choices.*

Besides the overselling of questionable technologies, "Star Wars" had an even more fundamental problem. Even if realistic space-based missile defense technologies came online someday, they would not serve their promised purpose. With the costs of failure so high if even one missile slipped through, and with the advantage in a nuclear war clearly resting with the aggressor (who could take relatively cheap countermeasures to frustrate any "Star Wars" system), many scientists specializing in arms control considered the notion of an umbrella defense against nuclear missiles sheer wishful thinking. "We all would like to meet the fundamental instinct of a human being, to defend your family, your home, your country," says Stanford theoretical physicist Sidney Drell, a "Star Wars" critic whose scientific advisory credits include serving on the President's Science Advisory Committee during the Johnson and early Nixon years. "But the realities of a nuclear era with lots of weapons were such that it was impossible."

Despite all of this, Reagan and members of his administration persisted in irresponsible rhetoric, repeatedly promising a perfect defense against nuclear weapons. Reagan even spoke of a "shield that could protect us from nuclear missiles just as a roof protects a family from the rain." Americans were, unfortunately, duped by the lie. Opinion polls showed that they supported the "Star Wars" program, but support remained contingent on the promise of a perfect defense. After all, who could object to one of those?

Meanwhile, further reports by the Union of Concerned Scientists, the American Physical Society, and Congress's Office of Technology Assessment (OTA) challenged SDI, and many members of the academic science community rose to protest the program. But this merely outraged "Star Wars" enthusiasts, rather than giving them pause. Soon full-fledged "science wars" broke out, as the George C. Marshall Institute, the Pentagon's Strategic Defense Initiative Organization, and others counterattacked against SDI's technical critics. OTA's challenges in particular struck the Reagan administration in the gut, coming as they did with the official imprimatur of Congress. In response to these reports, the administration and its conservative supporters engaged in a pitched battle against the scientific consensus, even attempting actual scientific suppression.

OTA's first and most controversial foray took the form of a 1984 study authored by physicist Ashton Carter, now a professor at Harvard's Kennedy School of Government, warning that "a perfect or near-perfect defense" represented an illusory goal that "should not serve as the basis of public expectation or national policy about ballistic missile defense." Carter's report enraged the Pentagon, which asked to have it disowned by the agency. Instead, an OTA expert review confirmed the study's conclusions. Meanwhile, the Heritage Foundation, highly committed to "Star Wars," used the incident as grist for a report charging OTA with political motives and the unauthorized release of classified information. The Heritage study complained at length about Carter's analysis, arguing that the time had come to "reassess" the Office of Technology Assessment, which is ironic given that OTA existed, in part, for the purpose of achieving a level of objectivity hardly to be expected from partisan think tanks like Heritage.

Two subsequent OTA studies, more comprehensive and thorough than the Carter report, proved no less disapproving of Reagan's "dream." In a 1985 assessment, OTA concluded that the goal of protecting the entire U.S. population would be "impossible to achieve if the Soviets are determined to deny it to us," but that SDI could ignite a new arms buildup on the Soviet side (thus creating greater instability and risk instead of a foolproof defense). And in 1988, OTA hit hard again with a study noting that "Star Wars" would stand a significant chance of "catastrophic failure" due to software glitches the very first—and, presumably, only—time it was used. The Pentagon held up the release of the report for months in an extensive classification review, and withheld three chapters entirely.

But despite this disturbing suppression of science, the Pentagon could not hold back the steady stream of critiques. Ultimately, Reagan's decision to adopt "Star Wars" came to be seen as a dramatic failure on his part to receive or listen to good scientific advice. Instead, Edward Teller often had a direct line to the president despite the fringe nature of his views. "Reagan had these personal relations with some individuals in the scientific community, and he sort of took their words as gospel, rather than seeking a broader scientific consensus," says Stanford's W. K. H. Panofsky, another "Star Wars" critic. In short, he cherry-picked his expertise.

Yet despite these troubling examples, Reagan-era science abuses do not appear to have been systemic or to have thoroughly infected the federal government. For one thing, a Democratic House of Representatives helped keep the executive branch in check. Moreover, sometimes individual actors within the administration flatly refused to use science as a political football despite the White House's best efforts.

The most stellar example, by far, is C. Everett Koop. Koop greatly peeved the Right, including many in the Reagan White House, by forthrightly advocating sex education and the use of condoms to prevent the spread of AIDS (once he was finally unleashed to speak on the topic). He also resisted pressure from the White House to produce a report on the physical and emotional aftereffects of abortion for women, an early strategic attempt to gather data on the topic for political purposes.

In a July 1987 memo to Gary Bauer entitled "New Prolife Strategy," then–White House policy analyst Dinesh D'Souza hit on a clever idea. Remarking on the effectiveness of previous surgeons general in the battle against smoking, D'Souza suggested having Koop produce a report on the health consequences of abortion. The hope was to change the focus of the abortion debate, shifting away from legal questions toward a health-oriented approach that would "rejuvenate the social conservatives." Soon afterward, in a speech to pro-lifers, Reagan called upon Koop to produce such a report.

Pro-life groups were infatuated with the idea of such a study, and it is not hard to see why. Proof of abortion's negative side effects—if such existed, or at any rate appeared to—would both discourage women from undergoing the procedure and also fuel lawsuits against abortion providers. And in the long run, such evidence could serve as a key device for overturning *Roe v. Wade*.

But despite his own opposition to abortion, Koop felt that pro-lifers had gone on a fishing expedition. In a letter to Reagan declining to produce the desired report, Koop wrote, "the scientific studies do not provide conclusive data about the health effects of abortion on women." In meetings with stakeholder groups, Koop added that psychological risks from abortion are "minuscule" when viewed from a public health perspective. Conservative Caucus chair Howard Phillips called Koop's refusal to produce a report "contemptible," but the surgeon general countered that "the very worst thing I could have done for the people who have been angry at me is to do what they wanted me to do. If I had put out the kind of report that was not scientific, that did not recognize the lack of physical evidence of what they wanted to know, it would have been attacked and destroyed by scientists and statisticians."

For all the unfortunate perversions of science from the Reagan years, then, we should acknowledge more hopeful episodes too. As we will see in the next chapter, Gingrich-era Republicans held a congressional hearing to dispute the extremely firm scientific theory that chlorofluorocarbons, or CFCs, deplete the stratospheric ozone layer that protects the planet from much of the sun's ultraviolet light. Yet while some Reagan administration officials sided with industry-friendly ozone depletion skeptics, Reagan's second-term EPA administrator Lee Thomas refused to cite scientific uncertainty to justify inaction on the issue. In 1987, the administration signed the U.S. on to the Montreal Protocol, an international agreement to limit CFCs.

It would fall to other conservative leaders, then, to make the fight against the scientific mainstream even more sweeping and comprehensive. Despite scattered incidents of the sort that might occur under any president, Reagan's successor, George H. W. Bush, a moderate and ally of Russell Train, was not one of them. True, Bush extended a Reagan-era ban on federal funding for fetal tissue transplantation research, which certainly rankled the scientific community and which President Clinton overturned. Yet in a nod to scientists, Bush elevated his science adviser, the late Yale physicist D. Allan Bromley, to the high-ranking position of "assistant to the president." Today, scientists largely remember the first President Bush as a friend.

When conservative firebrand Newt Gingrich swept into power during the "Republican Revolution" of 1994, however, the conservative movement crossed a new threshold, especially when it came to catering to industry on scientific questions. Most troubling of all—and in a move that may have come as partial payback for its negative reports on "Star Wars"—the Gingrich Republicans dismantled the congressional Office of Technology Assessment, which had been created in the wake of the supersonic transport controversy to provide members of Congress with their own source of independent scientific analysis.

Getting rid of an impartial scientific source like OTA greatly facilitated the politically driven distortions of science. Suddenly, Congress lacked an expert, bipartisan arbiter on scientific questions, an independent body to get to the bottom of the issues and weigh the evidence in a professional manner by convening leading experts and adhering to rigorous scientific peer review protocols. In the vacuum created by OTA's demise, "expertise" became even more of a political football, and science itself even more subject to manipulation and misrepresentation.

CHAPTER 5

DEFENSELESS AGAINST
THE DUMB

FOR ANALYSTS OF THE relationship between science and politics, House speaker Newt Gingrich presents a conundrum. On the one hand, this political revolutionary—who in 1994 led his party to control of the House of Representatives for the first time in four decades—presided over an era of stunning congressional science abuse. And yet Gingrich was no anti-intellectual rube. Far from it: He holds a Ph.D. (in history), has taught environmental studies, and has a reputation as a science fiction fan and ardent technophile.

In fact, after leaving Congress, Gingrich only increased his science boosterism. He became an evangelist for nanotechnology, and in a May 2002 testimony before the Senate called for not just doubling but *tripling* funding for the National Science Foundation. "The knowledge breakthroughs of the next twenty years will equal the entire twentieth century," Gingrich predicted.

The key to reconciling the two Gingriches—the science lover on the one hand, the science abuser on the other—lies in the speaker's odd notion of how politicians ought to get their science advice. The idea contains unmistakable echoes of the Gingrich Republicans' broader argument: that government should shrink, and that the private sector should take up the slack.

In defending his party's dismantling of Congress's Office of Technology Assessment (OTA), Gingrich has advocated what we might call a "free market" approach to scientific and technical expertise. In the speaker's conception, members of Congress should take the initiative to contact individual scientists and inform themselves, much as Gingrich himself did. Never mind that few members of Congress were such science buffs—or that some, like majority whip Tom DeLay, had had previous careers in fields such as pest extermination. "Gingrich's view was always, 'I'll set up one-on-one interactions between members of Congress and key members of the scientific community,'" recalls Bob Palmer, former Democratic staff director of the House Committee on Science. "Which I thought was completely bizarre. I mean, who comes up with these people, and who decides they're experts, and what member of Congress really wants to do that?"

This dubious approach helps explain the science politicizing bonanza of the Gingrich Congress. The dismantling of the Office of Technology Assessment contributed to a "free market" for scientific expertise all right—with alarming consequences. With OTA gone, Gingrich's troops didn't hesitate to invoke their own favored experts to undermine the scientific mainstream on ozone depletion and global warming. The attacks came as the new Republican majority sought to free up the market in another way as well: by ramming through a major "regulatory reform" bill that would have prescribed rigid and inflexible rules governing the use of science to protect public health and the environment.

As a rallying cry for this whole agenda, the Republicans loudly demanded that policy must rely on "sound science." But by this term, the new majority clearly didn't mean the distinguished work of OTA. Rather, the "sound science" crusade betrayed the incoming Republicans' close rapport with the tobacco industry, which had battled for decades to obscure the truth about dangers posed by its products, both to smokers themselves and to innocent bystanders. Big Tobacco and its allies had helped popularize the term "sound science" to describe an agenda that had little to do with scientific rigor, and everything to do with blocking government controls on industry by raising the burden of scientific proof required to justify action.

In short, the Gingrich Republicans picked up the torch of conservative science abuse that Reagan had carried, and used it to light the entire U.S. Congress on fire. Whipped by ideology and the sense that they had been in the minority for far too long, the newly triumphant Republicans wouldn't let their industry allegiances and enthusiasm for government downsizing take a back seat to anything. Their chance had finally arrived to remake Congress in their image. If science stood in the way, so much the worse for science.

The symbolic centerpiece of this agenda was the congressional Republicans' decision, in a stunning act of self-lobotomy, to dismantle their authoritative scientific advisory office. Obsessed with shrinking government, Gingrich's acolytes denounced OTA for being too slow in its assessments and (some added) suspect in its political orientation. The late Rep. George Brown, leading the Democratic minority on the House Science Committee at the time, memorably protested that the agency had served as Congress's "defense against the dumb," and continued, "it is shameful that OTA was defenseless against a very dumb decision by Congress."

Ironically, OTA's death did not come in its early days, as the office struggled and failed to prove itself. Rather, by the time of its demise, OTA had become a globally renowned agency, celebrated by *Washington Monthly* in a 1989 article entitled "How to Revolutionize Washington with 140 People."

Congress had created OTA in 1972, at a time of general distrust between the Nixon administration and the Democratic-controlled legislative branch over the supersonic transport and other issues. The era also saw mounting public concern over the dangers of pollution, nuclear energy, pesticides, and other technology-induced hazards. OTA, the thinking went, would both forecast coming technological quandaries and help Congress fact-check technical claims made by the various executive branch agencies. The office had a unique organizational structure: a twelve member board, comprising six members of Congress from each party, approved each OTA project. This arrangement theoretically ensured the agency's objectivity.

Nevertheless, partisan tensions hobbled the office from the outset. Because OTA's leading sponsor, Senator Edward Kennedy, of Massachusetts, headed the OTA board in 1977 when many considered him a presidential contender, conservatives suspected the office of being a "happy hunting ground of Kennedy apparatchiks" and "liberal technocrats," as William Safire put it in the *New York Times*. It didn't help that OTA blew through its first two directors in only a few years. When the folksy Tennessee physicist John (Jack) Gibbons, formerly of Oak Ridge National Laboratory, received a nod for the director's job, Michigan Democratic congressman John Dingell warned that he was the agency's last chance.

Gibbons promptly shook up the office's staff and rearranged OTA to focus on more immediate issues, rather than airy attempts at long-range technological forecasting. Between 1979 and 1993, under his guidance, the office also pursued a strategy of careful political neutrality, notes political scientist Bruce Bimber in his 1996 study of OTA, *The Politics of Expertise in Congress*. This approach gradually won the support of key Republican allies.

But when Ronald Reagan took office, the new administration endorsed *Fat City*, a 1980 book by conservative journalist Donald Lambro that identified OTA as one of Washington's many wasteful programs. Lambro called OTA an "unnecessary agency" that duplicated work performed by other parts of government. He also quoted an unfriendly member of Congress, who charged that most OTA reports ended up "in the warehouse gathering dust." In fact, OTA deliberately provided Congress with a second opinion so that it wouldn't simply have to trust the executive branch. As for its reports "gathering dust," an OTA rebuttal to *Fat City* noted that most of the office's studies had gone out of print at the Government Printing Office "despite frequent second and occasional third printings."

Fat City would not be the last time that conservatives attacked the agency. As noted in the previous chapter, OTA greatly angered the Reagan administration with its "Star Wars" reports, which made the office conservative enemies who would remember the affront later. Some even interpreted the 1995 killing of OTA as "Reagan's revenge," notes Gibbons.

Yet few OTA reports engendered controversy like its "Star Wars" work. Gibbons, who directed the office until becoming Bill Clinton's first science adviser in 1993, insisted that each study provide Congress with a range of well-informed policy options to choose from. "OTA produced a body of scientific information from which, then, the politics could be argued," says Rosina Bierbaum, who headed OTA's climate-change project in the 1980s and now serves as dean of the University of Michigan's School of Natural Resources and Environment. "And now, it doesn't seem to me like there's any consensus body of information that the Congress accepts."

OTA's twenty-four-year "consensus body of information" encompasses some 750 studies on topics ranging from acid rain to global climate change to the accuracy of polygraphs. Perhaps because the office vetted these documents so stringently, they have aged quite well. Following the anthrax attacks of late 2001, for example, a report prepared on behalf of Senator Ernest Hollings (D-SC) noted that OTA had studied the number of spores required to produce inhalation (or pulmonary) anthrax almost a decade earlier. "If this information had been readily available" during the crisis, Hollings's report noted, "it's conceivable that it even could have saved a life."

OTA developed a stellar international reputation as well. Scores of policymakers from overseas visited the office to learn how it worked, a series of interchanges that led to the creation of OTA analogues in European legislatures. When the U.S. Congress then did away with OTA in 1995, other nations were stunned. "That the leading technological state in the world, a democracy like us, should have abolished its own main means of democratic assessment left us aghast," wrote Lord Kennet, who created an umbrella group of mini-OTAs called the European Parliamentary Technology Assessment Network.

But the Gingrich Republicans, who had ridden the cleverly packaged "Contract with America" to victory, viewed matters very differently. OTA became a "sacrificial victim," observes Federation of American Scientists president and former OTA staffer Henry Kelly, because the new Congress wanted to prove its willingness to make budget cuts in its own house. Gingrich's move to consolidate power in the House of Representatives following the 1994 Republican sweep, combined with his own sense of himself as a science guru, may have also worked against OTA. As the office's last director, Roger Herdman, an M.D. and a Republican who now directs the National Cancer Policy Board at the National Academy of Sciences'

Institute of Medicine, told me, "There are those who said the Speaker didn't want an internal congressional voice that had views on science and technology that might differ from his."

Some of Gingrich's followers also considered the renowned agency a tool of the Left. According to Gingrich spokesman Rick Tyler, the Speaker felt that there was a tilt to OTA's reports: "In some cases, it was politicized work." Gingrich's lieutenant Robert Walker, who chaired the House Committee on Science at the time of OTA's demise, further argues that the agency's analyses simply took too long to prepare to be of use to Congress. But like many other science and environmental issues, OTA wound up dividing conservative and moderate Republicans. Amo Houghton, of New York, a classic GOP moderate, helped lead an almost-successful fight to save the agency under the slogan, "You don't cut the future." In an interview in late 2003 before his retirement from Congress, Houghton called cutting the office just plain "dumb." "It was not that much money," he added—$21.9 million the year of the office's demise—"and they were just looking for sort of symbolic targets."

Gibbons, for his part, has accused the Gingrich Republicans of political motives and decried the "callous treatment" OTA received. But he also interprets the agency's death in the context of the always fraught relationship between science advisers and those they advise. This "shoot the messenger" tradition goes back much farther than Nixon's defenestration of his science advisers. In a 2003 speech at Rice University, Gibbons drew a joking analogy between the fate of OTA and that of Socrates: "He gave advice to other people. He was poisoned."

Thus began the freewheeling politicization of expertise in Congress. OTA had scrupulously avoided making explicit policy recommendations, but its reports did sift through expert disagreements, rule out fringe scientific views, and challenge implausible technological assertions (including those associated with "Star Wars"). In OTA's absence, however, the new Republican majority in Congress freely called upon its own scientific "experts" and relied on analyses prepared by lobbyists and ideologically committed think tanks like the Heritage Foundation. A 2001 comment from "science geek" Gingrich, explaining OTA's death, said it all: "We constantly found scientists who thought what they [OTA reports] were saying was not correct."

As a case study in the new war on science, consider a series of major hearings held by the Republican-controlled Energy and Environment Subcommittee of the House Committee on Science, entitled "Scientific Integrity and Public Trust." The hearings—which began even as OTA prepared to close its doors in September 1995—fit nicely into the Gingrich Congress's broader attempt to expose the shoddy scientific basis for a wide range of environmental regulations, thereby demonstrating the need for "regulatory reform." Channeling allegations made by conservative think tanks, the House Republicans even charged that scientists had grown cozy with government regulators, addicted to federal funding, and highly prone to suppress or ignore dissenting views. Though it found little substantive support, the accusation betrayed the conservative movement's curious preference for private-sector scientific research over "public science," despite the fact that privately controlled research has myriad problems of its own (particularly when it comes to potential conflicts of interest between the pursuit of truth and protecting industry profits).

Presided over by Rep. Dana Rohrabacher, of California—who derided global warming concerns as "liberal claptrap"—the "scientific integrity" hearings covered three environmental issues of keen interest to industry: ozone depletion, climate change, and dioxin risks. Especially in the first two hearings, the Republicans evinced a highly dubious notion of how elected representatives should determine what science has to say. Rather than relying on major peer-reviewed scientific consensus documents—a policymaker's most reliable means of accessing scientific knowledge on a given question—they hosted adversarial "science courts" that pitted scientific outliers against the mainstream. After the fireworks died down, members of Congress, rather than scientists, were supposed to judge whose view was right.

Such an approach, noted Rep. George Brown in a highly critical report, showed that the incoming Republicans had "little or no experience of what science does and how it progresses." Instead, Rohrabacher and allies appeared to subscribe to the misguided notion that "scientific truth is more likely to be found at the fringes of science than at the center."

We can hardly overstate the absurdity of this view. Working scientists shouldn't follow their sense of where scientific "consensus" lies in preparing their research designs; if they try to upset long-held views, so much the better. But when it comes to the use of scientific information by nonexpert members of Congress, determining consensus is all-important. "Scientific knowledge *is* the intellectual and social consensus of affiliated experts based on the weight of available empirical evidence, and evaluated according to accepted methodologies," historian of science Naomi Oreskes, of the University of California, San Diego, has written. "If we feel that a policy question deserves to be informed by scientific knowledge, then we have no choice but to ask, what is the consensus of experts on this matter?"

In an interview, Robert Walker, Newt Gingrich's right-hand man and chair of the House Science Committee at the time, defended the very different approach taken on his watch. "Hearings are about trying to find out what the various points of view are so that you can adequately represent the totality of scientific evidence in whatever legislation you're doing," he told me. No one objects to free speech or a diversity of views, but Walker's argument fails to recognize that science isn't a democracy. Rather, it uses quality control—peer review—to rule out questionable interpretations and ensure that knowledge advances.

In attempting to upset and derail this process, congressional Republicans had undermined science itself, and set an alarming precedent for the use of questionable science to determine policy.

That became apparent in the first of the "scientific integrity" hearings, on the subject of ozone depletion. The Republicans could hardly have chosen a more unassailable field of environmental science to challenge. The hearing actually occurred in the same year that the original proponents of the hypothesis that chlorofluorocarbons (CFCs) deplete stratospheric ozone—Sherwood Rowland, of the University of California, Irvine, and Mario Molina, of MIT—won the Nobel Prize in chemistry.

Against this overwhelmingly accepted position, the Republicans pitted Dr. S. Fred Singer, of the Science and Environmental Policy Project (already on the record as disputing the dangers of acid rain and other environmental problems), and Dr. Sallie Baliunas, a senior scientist with the George C. Marshall Institute. The two challenged the science that had been used to justify an accelerated phaseout of CFCs in 1992. "There is no scientific consensus on ozone depletion or its consequences," Singer declared.

In this encounter, Singer and Baliunas depicted themselves as scientific "skeptics" heroically battling against received wisdom, and facing censorship and even suppression of their views. Doubters of the scientific mainstream on issues such as climate change and ozone depletion love to strike this pose, and for good rhetorical reasons. "Every good scientist is a skeptic through and through," notes Harvard biological oceanographer James McCarthy, an expert on the impacts of climate change.

But not every skeptic is necessarily a Galileo. Moreover, while prizing skepticism, science also has a place for the accumulation of knowledge and the acceptance of consensus conclusions that have themselves emerged from a process of exacting interrogation and challenge, which is precisely what today's "skeptics" on climate change and ozone depletion refuse to do. Their blanket skepticism renders them unwilling to accept the current state of scientific understanding, no matter how solid.* No wonder George Brown's report on the bizarrely named "scientific integrity" hearings was originally going to be titled "A Dip into the Skeptic Tank."

The CFC–ozone hypothesis had already been subjected to withering skepticism. When they originally published their idea in the scientific journal *Nature* in 1974—arguing that the long-lived chemicals contained in spray cans, refrigerators, and air conditioners would rise up into the stratosphere and release chlorine molecules that would, in a chain of chemical reactions, destroy the earth's protective ozone layer—Molina and Rowland were the scientific outliers. In a familiar pattern, CFC manufacturers like DuPont quickly challenged their assertions, and would maintain a stance of robust skepticism for more than a decade.

---

*For this reason, in this book I do not call those who dispute robust scientific consensus conclusions on ozone depletion and climate change "skeptics," or at least, not without preserving the scare quotes. Instead, following the example of a number of climate scientists including Stanford's Stephen Schneider, I call them "contrarians."

But the CFC–ozone theory withstood massive scientific scrutiny, including several studies by the National Academy of Sciences. Numerous ozone assessments organized by NASA as well as by the United Nations Environment Program and the World Meteorological Association also supported the consensus view. By the time Rowland and Molina won the Nobel Prize in 1995 (along with Paul Crutzen, of the Max Planck Institute for Chemistry, in Germany), they had convinced their colleagues that CFCs pose a severe threat to the ozone layer and, by implication, human health. The ozone depletion hypothesis had gone from fringe to mainstream, withstanding doubts at every turn. It hardly hurt that in 1985, scientists discovered alarming levels of ozone depletion (the so-called ozone hole) above Antarctica, and soon linked the phenomenon to CFCs.

It is in this context that we must weigh the wave of "skepticism" that greeted the CFC–ozone hypothesis following the 1987 Montreal Protocol, and George H. W. Bush's decision to institute an accelerated CFC phaseout in 1992. Surely, if the "skeptic" arguments truly carried such force, they would have managed to sway mainstream scientists by then. In fact, the work of the "skeptics" was dubious at best. For example, in her popular 1990 book *Trashing the Planet,* Dixy Lee Ray, a former governor of Washington as well as a zoologist, advanced a number of flawed contrarian arguments, perhaps most notably the suggestion that natural sources like volcanoes, rather than CFCs, might explain ozone depletion. Ray found a champion in radio host Rush Limbaugh, the popular voice of the Gingrich revolution, who labeled the CFC–ozone theory "balderdash" and "poppycock." Limbaugh called *Trashing the Planet* "the most footnoted, documented book" he had ever read.

But in his president's lecture to the American Association for the Advancement of Science (AAAS) in 1993, Sherwood Rowland exposed the sheer ineptness of these critics. In response to those asking, "Don't volcanoes cause the Antarctic ozone hole?" Rowland noted that volcanic eruptions give off large amounts of steam, which condenses into clouds higher up in the atmosphere. Any chlorine emitted in a volcanic eruption would dissolve in this water and fall back down to earth in the form of rain. Rowland further observed that aircraft measurements had shown little increase in stratospheric chlorine levels following the eruption of El Chicon, in

Mexico, in April 1982 or of Mount Pinatubo, in the Philippines, in June 1991. "Despite the misinformation problems," Rowland concluded, "an international scientific consensus has been achieved" on ozone depletion.

At the time of his AAAS speech, though, Rowland couldn't have known of the impending "Gingrich revolution," which would bring ozone depletion skepticism the endorsement of the party running Congress. On the eve of the House of Representatives' ozone hearing, Jack Gibbons, by then serving as White House science adviser, declared the decision to give a "few vocal skeptics" of ozone depletion equal standing with the scientific mainstream "incredible." "Healthy skepticism is an essential and treasured feature of scientific analysis," Gibbons added. "But willful distortion of evidence has no place at the table of scientific inquiry."

That didn't stop the House Republicans. Their ozone hearing put the new Congress's poor understanding of science—and simultaneous willingness to politicize it—on full display. Perhaps the most memorable moment came when one of Rohrabacher's allies, California Republican John Doolittle, responded to a colleague who asked what studies he could cite favoring his position by saying, "I'm not going to get involved in a mumbo-jumbo of peer-reviewed documents." Doolittle and Tom DeLay, the House majority whip, introduced separate bills to ease existing CFC regulations. In explaining the scientific basis for his position, DeLay confessed that "my assessment is from reading people like Fred Singer."

The notion that DeLay—who would later blame school shootings on the fact that we teach children that they are "nothing but glorified apes who are evolutionized out of some primordial soup of mud"—has any capacity to "assess" ozone science is something Jon Stewart's joke writers should look into. But it is no laughing matter for those who expect their elected representatives to rely on the best evidence that science has to offer. With the Republican Congress, and especially Rush Limbaugh, endorsing dubious science, millions found themselves exposed to such nonsense. The House Republicans' actions not only eroded respect for good science, but also seriously misinformed a wide swath of Americans.

The year 1995 also proved pivotal for our scientific understanding of global warming, an advance that brought with it powerful efforts to deny and obscure the new knowledge. Today, so much dubious science and outright nonsense circulates on this subject—bestselling author Michael Crichton's recent novel *State of Fear* being a case in point—that finding reputable information poses a herculean challenge to those unfamiliar with the terrain. For this disturbing state of affairs we must thank, at least in part, the Gingrich Congress.

The scientific concept of the "greenhouse effect," in which atmospheric gases such as carbon dioxide, methane, and water vapor cause planetary warming by trapping radiation that would otherwise escape into space, can be traced back more than a century. Near the close of the nineteenth century, Swedish scientist Svante Arrhenius examined the role of human industrial emissions in contributing carbon dioxide to the atmosphere, and calculated that doubling levels of that greenhouse gas could trigger a dramatic warming. But most other scientists at the time rejected this view for a variety of reasons, including the belief that the atmosphere was inherently self-regulating and that the oceans would absorb enough CO_2 to stave off any problems.

There the greenhouse question more or less rested until the 1950s, when oceanographers helped put the notion of global warming back on the map. The new research showed that oceans couldn't absorb nearly as much CO_2 as previously thought. Later, as the study of climate became a major international field of inquiry, researchers examining the carbon content of ancient air samples embedded in long, cylindrical ice cores— painstakingly extracted from the icecaps of Greenland and Antarctica— hit upon another key discovery. Over the course of the earth's history, CO_2 levels and temperatures correlated closely, with far higher atmospheric concentrations of the gas during warm periods than during glacial periods. Meanwhile, thanks to advances in computer technology, mathematical models of the climate system greatly increased in sophistication. A number of research groups began to simulate the possible effects of human greenhouse gas emissions on global average temperatures.

In the 1980s, the concept of "global warming" first truly entered the public consciousness. The decade set a record for temperatures and saw a rise in media coverage of the climate issue as well as hearings on the subject held by then-congressman Al Gore. Other pressing environmental problems also primed the public to think about the consequences of atmospheric pollution. Acid rain focused attention on the unintended consequences of industrial emissions, while the ozone depletion crisis—and especially the discovery of the Antarctic ozone "hole"—showed how easily human activities could affect the atmosphere.

It was in this context that during the sweltering Washington summer of 1988, pioneering NASA climatologist James Hansen famously told Congress that he believed with "99 percent confidence" that a long-term warming trend had begun, probably caused by the greenhouse effect. By 1989, Hansen, a government employee, found his scientific testimony to Congress edited by the first Bush administration's Office of Management and Budget to emphasize scientific uncertainties in a way that Hansen did not deem warranted.

As environmentalists and some members of Congress began to call for reduced carbon dioxide emissions from the burning of fossil fuels, industry fought back, employing an approach familiar from earlier debates over acid rain and ozone depletion. In 1989 the electric power, petroleum, and other industries forged the Global Climate Coalition to ensure that "good science must be the foundation for policy." By 1992 the group could be found drawing on contrarian scientists (including Dixy Lee Ray and S. Fred Singer) to emphasize the uncertainties of climate science and climate modeling. Ray flatly denied that global warming was under way, claiming that humans could not alter "the enormous phenomena that take place in nature." Conservative think tanks also got in the game. Perhaps most prominently, the George C. Marshall Institute released several skeptical reports on the science of climate between 1989 and 1992.

Yet even as industry mobilized the forces of "skepticism," an unprecedented international scientific collaboration emerged that would change the terms of debate forever. In 1988, under the auspices of the United Nations, scientists and government officials inaugurated the Intergovernmental Panel on Climate Change, a global scientific body that would eventually pull together thousands of experts to evaluate the issue, providing the gold standard of climate science. The IPCC's first assessment report, published in 1990, debated whether the evidence showed that human activity had caused actual global warming. The science remained open to reasonable doubt.

But then came the IPCC's second assessment report, completed in late 1995. The report concluded for the first time that amid purely natural factors shaping the climate, humankind's distinctive fingerprint stuck out. "The balance of evidence suggests that there is a discernible human influence on global climate" thanks to emissions of carbon dioxide and other greenhouse gases, the IPCC famously stated.

Since then, the impact of fossil-fuel burning on global climate has only grown more apparent. But the 1995 findings sparked massive debate, with so-called skeptics and their industry allies predictably arguing that human factors affecting the climate were not so easily separated from all the noise. In addition, the "skeptics" mercilessly attacked the integrity of the IPCC process, even bringing unsubstantiated charges of scientific wrongdoing against Lawrence Livermore National Laboratory climatologist Benjamin Santer, lead author of the IPCC report's crucial eighth chapter.

The Gingrich Republicans provided a soapbox for the "skeptics," most notably University of Virginia climatologist Patrick Michaels, a recipient of substantial energy industry funding in the past. On November 16, 1995, the Republicans held their second "scientific integrity" hearing on climate change, with Michaels on tap. Once again, the hearing took the form of a "science court," with the mainstream on trial and Republicans aligned with the outliers. "Our '96 budget does not operate on the assumption that global warming is a proven phenomenon," Rohrabacher had declared earlier in the year. "In fact, it is assumed . . . at best to be unproven and at worst to be liberal claptrap; trendy, but soon to go out of style in our Newt Congress."

Michaels provided the alleged basis for such statements at the hearing. Most prominently, he criticized the reliability of computer models that climatologists use to project future climate change, arguing that an emphasis on "observed data" suggested that the models exaggerate warming. The criticism echoed the Gingrich Republicans' broader argument that "empirical" evidence, rather than theoretical tools (such as models), should serve as the chief basis for policy.

In fact, this entire argument has been described as attacking a "straw man." Obviously, computer models cannot perfectly simulate the massively complex climate system. Instead, they suffer from the same basic shortcomings as any other model, whether used by economists or population biologists. All future projections suffer, by their very nature, from limitations and uncertainties, and scientists should not report their results without a candid statement of caveats.

The central question, though, is whether climate models can profitably inform both scientific understanding and policy—*shortcomings notwithstanding*. In other words, should policymakers consider the range of possibilities suggested by these highly sophisticated attempts to project future climate change? Clearly, they should (and models have only grown more sophisticated since 1995). "You can't prove what's going to happen in the future," explains Naomi Oreskes, who has written extensively on the uses of modeling. "But that doesn't mean you don't have a rational basis for action." Where Michaels questioned the models' reliability, others would merely call their results uncertain—but no reputable scientist ever claimed otherwise.

At the House hearing and in submissions afterward, two highly distinguished climatologists defended the use of models: Robert Watson, who would later go on to chair the Intergovernmental Panel on Climate Change, and Jerry Mahlman, then director of the Geophysical Fluid Dynamics Laboratory at the National Oceanic and Atmospheric Administration. As Watson noted of the models, "The fact that they do not simulate every bump and wiggle in the observational record is not surprising given that they do not attempt to simulate every natural phenomenon that affects the earth's climate on short time scales." Mahlman, meanwhile, noted that models may well *underestimate* future warming rather than overstating it: "When you have made your best estimate of the way things are, it automatically says that you do not know whether you are wrong on the high side or the low side." In other words, uncertainty cuts both ways.

The attempt to draw lines between models and empirical or observed data also creates a false dichotomy, since climate modelers use empirical observations to calibrate their models. As a 2001 National Academy of Sciences report on the subject noted, "To evaluate model realism, model outputs are compared to each other and to environmental observations. The results of these comparisons form the basis for changes to model code, which improve the mathematical representation of physical processes."

In essence, models emerge from a creative interaction between empirical observations and theoretical considerations, themselves based on physical laws. Scientific *understanding* relies on both—not one or the other in isolation. Yet to this day, opponents of mandatory actions to address global climate change follow Michaels's lead in criticizing climate models and suggesting that they're somehow undercut by observations.

Conservative climate change "skepticism"—and the political misuse of science to support it—did not begin with the 104th Congress. Still, the Gingrich Republicans' actions represented a new level of abuse. These new kids on the political block lent Congress's official endorsement to climate change skepticism and denial, encouraging what has become a massive cottage industry of "skeptics" tied to conservative think tanks. "The Gingrich Congress allowed conservative Republicans to hear from contrarian scientists what they wanted to hear, and to make sure those people became witnesses at hearings," says biogeochemist William Schlesinger, head of the Nicholas School of the Environment and Earth Sciences at Duke University. "And they basically politicized the issue so that both sides now go to the table with their hired guns, and they beat on each other for a while, and there's never agreement, so nothing happens—and CO_2 continues to rise."

CHAPTER 6

JUNKING "SOUND SCIENCE"

THE LACK OF SHAME with which the Gingrich Congress carried out its assault on scientific expertise had little precedent. In fact, one of the most telling signs that the conservative war on science had attained a new level of intensity was that it now had an official slogan.

During the "scientific integrity" hearings and more generally, the Gingrich Republicans uttered a pervasive rallying cry—the demand for the use of "sound science" in the making of public policy. Much of the conservative science agenda today goes under the same enigmatic heading. On the front page of its website, the George C. Marshall Institute declares, "Our mission is to encourage the use of sound science in making public policy about important issues for which science and technology are major considerations." President George W. Bush has invoked "sound science" on issues ranging from climate change to arsenic in drinking water.

Understanding what the Gingrich Republicans meant by "sound science" exposes the ideological core of the modern conservative science agenda, as well as the strong continuities between the Gingrich era and the administration of George W. Bush. As with many phrases co-opted for PR purposes, however, to determine the true meaning of "sound science" we must first take a close look at the term's spoken and written uses.

It is hard to pinpoint exactly when "sound science" emerged as a conservative watchword. Strategic uses by the business community trace back at least to a 1981 report to the membership of the American Industrial Health Council, an industry group, describing the organization's quest to ensure the "adequacy and quality" of science used to support government action and its support for "regulatory reform" legislation. The report also celebrated the Reagan administration's "recognition of the role of sound science in the development of federal policies concerning chronic health hazards," and applauded the president's move to empower the White House budget office to vet regulatory decisions made by federal agencies, which sometimes involved challenging agency experts on specifically scientific grounds.

In the early 1990s, President George Herbert Walker Bush used the phrase "sound science" in relation to climate change. "Sound science" also became a talking point for Bush's Environmental Protection Agency administrator William Reilly. But at that time, the term's precise meaning remained fairly amorphous.

A key development, however, came in late 1992, when an Environmental Protection Agency report (technically referred to as a "risk assessment") estimated that secondhand smoke causes some three thousand lung cancer deaths each year—a much smaller number than that of deaths attributed to active cigarette smoking, but certainly nothing to be ignored. The EPA classified secondhand smoke as a human lung carcinogen. Numerous high-caliber scientific reports had already reached a similar conclusion, but as a government agency, the EPA stood in a position to take action, and secondhand smoke regulations, including potential bans on smoking in public places, posed a serious risk to tobacco industry profits.

The Tobacco Institute promptly labeled the EPA's conclusions "another step in a long process characterized by a preference for political correctness over sound science." And as we now know from tobacco documents made available as a consequence of litigation, the industry decided to do something about it.

On the issue of secondhand smoke, Big Tobacco would turn to the playbook it had first developed in the 1950s and 1960s, as evidence mounted linking active cigarette smoking to lung cancer and other illnesses. Especially in the wake of a famous 1964 U.S. surgeon general's report that tied smoking to a range of conditions including lung cancer, chronic bronchitis, heart disease, and emphysema—although the basic strategy had emerged even earlier—the industry mobilized to undermine the science. With its lawyers often funding and directing research, Big Tobacco pumped considerable resources into sowing public doubts about scientific studies showing risks from active smoking. The technique, which has been dubbed "manufacturing uncertainty," finds perhaps its best articulation in this oft-quoted passage from a circa 1969 Brown & Williamson document: "Doubt is our product, since it is the best means of competing with the 'body of fact' that exists in the mind of the general public. It is also the means of establishing a controversy."

Even as Big Tobacco worked to obscure the risks of active cigarette smoking, the industry realized it would have to fight a rearguard action as well. In the early 1980s, epidemiological reports began to appear suggesting that wives of smokers had increased cancer risks due to secondhand smoke, often called environmental tobacco smoke, or ETS. The tobacco industry quickly sprang into action to fight the new science, once again seeking to sow doubt and stoke controversy about the risks of passive smoking. This occurred both through attacks on the armada of studies demonstrating secondhand smoke risks, and—following the 1992 EPA report—through the support of a "sound science" advocacy group that would challenge the credibility of government research.

In late 1993, the Advancement of Sound Science Coalition (TASSC) burst on the scene, describing itself as a "grassroots-based, not-for-profit watchdog group of scientists and representatives from universities, independent organizations, and industry that advocates the use of sound science in the public policy arena." But all was not as it seemed. Internal documents show that following the EPA's report on secondhand smoke, Philip Morris had sought to form such a group to serve its interests.

"Our overriding objective is to discredit the EPA report and to get the EPA to establish a standard for risk assessment for all products," wrote Ellen Merlo, a Philip Morris vice president, in a February 17, 1993, memo to Philip Morris USA president William Campbell. She added that Philip Morris had hired the PR firm APCO Associates "to form local coalitions to help us educate . . . about the dangers of 'junk science.'"

Even as the industry attacked the EPA's study directly, it would also draw upon "non-industry messengers" and associate the EPA's work with "broader questions about agency research and government regulations."

By the autumn of 1993, TASSC's public launch was imminent. In a September 23, 1993, letter, APCO informed Philip Morris that "the groundwork we conduct to complete the launch will enable TASSC to expand and assist Philip Morris in its efforts with issues in targeted states in 1994." The cigarette giant didn't want to be publicly tied to the "sound science" group, however. Shortly before the launch, a company memorandum noted that "we thought it best to remove any possible link to PM [Philip Morris]," and added, "With regard to media inquiries to PM about TASSC, I am putting together some Q & A. We will not deny being a corporate member/sponsor, will not specify dollars, and will refer them to the TASSC '800-' number, being manned by [APCO]."

Headed by former New Mexico governor Garrey Carruthers, TASSC claimed a wide diversity of members. But according to a February 22, 1994, memo, Philip Morris had budgeted an impressive $500,000 for the group for 1994. Nevertheless, TASSC rarely issued direct challenges to the science linking secondhand smoke to health risks, focusing instead on other issues at the intersection of science and regulatory policy such as the use of ethanol in fuels and, revealingly, "the incomplete science used to create ozone repair policy."

Still, in September 1994, TASSC released a poll of scientists suggesting that politicians were abusing science on issues such as "asbestos, pesticides, dioxin, environmental tobacco smoke, or water quality." And one of its newsletters directly criticized the EPA secondhand smoke report. The group also questioned the scientific credibility of the EPA more generally, thereby fostering a broader climate of skepticism about government science that surely worked in the tobacco industry's favor.

At around the same time, coincidentally or otherwise, the pro-tobacco Republican Congress adopted "sound science" as a mantra. Shortly after the November 1994 elections, Gingrich and company had set the tone. "Property rights" and "sound science" had become "the environmental buzzwords of the new Republican Congress," a Knight-Ridder news story noted. The perceptive report also included a definition of "sound science," suggesting that it meant a lot more than simply "good science." Instead, the point was deregulation: "'Sound science' is shorthand for the notion that anti-pollution laws have gone to extremes, spending huge amounts of money to protect people from minuscule risks."

The politicians using tobacco-friendly terminology also received the embattled industry's money. In the first half of 1995, the Republican Party netted more than $1.5 million in tobacco donations, "five times as much as in the same period" the previous year, according to the *New York Times*. The Democratic Party received only about one-tenth that amount during the same time period.

Not only did the "sound science" movement seek to discredit past government "junk science"; it also lobbied for a crackdown on such work in the future. Calls for "sound science" closely accompanied the push to enact a key plank of the Republican "Contract with America": regulatory reform, an industry-backed gambit that would have used science itself to thwart future environmental, health, and safety regulations. Such reform, in various guises, had been a dream of industry at least since the early 1980s, and it had been supported by the Reagan administration. Now, with a Republican-controlled Congress, it suddenly seemed within reach.

By the mid-1990s, a good case existed for moderate reforms of the federal regulatory state, so as to increase flexibility and draw more extensively on market forces and individual initiative in solving problems. However, the Gingrich Congress had nothing of the sort in mind. Their "regulatory reform" package had been cunningly misnamed; in actuality, the "reformers" sought to raise steep hurdles to future regulation.

To bolster their case, the Republicans repeatedly cited dubious anecdotal examples of outrageous government overreach. To address this supposed crisis of "regulatory overkill," reform bills sponsored by the House Republicans and Senate majority leader Bob Dole would have required a broad range of government agencies to apply the same "one-size-fits-all" standards for determining whether a particular danger should be regulated.

While seemingly innocuous, the remedy is in fact perfectly tailored for tobacco and other industries aiming to tilt the rules under which regulations are scientifically adjudicated. Agencies would have been required to conduct formal "risk assessments" before undertaking most actions, and to follow a stringent set of rules laid down by the would-be science cops of the Gingrich Congress. Under the guise of "sound science," federal scientific agencies would have been hamstrung by burdensome new obligations and effectively prevented from doing their jobs.

Indeed, the leading regulatory reform proposals would have inappropriately legislated the very nature of science itself. Not only did the bills prescribe rigid rules for the assessment of risk, potentially stifling scientific adaptability; in many cases, agencies would also have had to submit their studies to "peer review" panels, which could potentially be dominated by industry scientists. Furthermore, regulatory reform would have created new opportunities for federal court challenges over the validity of the scientific studies required under the new law, an ideal opportunity for business interests to deploy their lawyers and engage in scientific warfare over analyses they didn't like (the EPA secondhand smoke report, for instance). The whole process, Georgetown University law professor David Vladeck wrote at the time, smacked of an attempt to achieve "paralysis by analysis."

"Reformers" didn't describe it that way, of course. In a *Washington Post* commentary, Dole defended his bill as an attempt to ensure that agencies used "the best information and sound science available." Yet the notion that Republican reformers merely sought better science in the abstract—instead of a dramatically higher burden of proof and substantial time delays before regulation could proceed—is hard to swallow. At the same time that they pushed for regulatory reform, the Gingrich Republicans sought to slash funding for government scientific research and even to do away with a crucial government scientific agency, the U.S. Geological Survey, which, among other tasks, tracks earthquakes and volcanic eruptions (and truly proved its worth in the wake of the Asian tsunami catastrophe of late 2004). And, of course, they successfully strangled the Office of Technology Assessment.

The Advancement of Sound Science Coalition cheered the regulatory reformers on, focusing much of its energy on the issue in early 1995. In an October 1994 speech, TASSC chairman Garrey Carruthers endorsed a regulatory reform proposal by conservative Louisiana Democratic senator J. Bennett Johnston (cosponsor of the Dole bill). Then in 1995, the group released a study protesting negative media coverage of regulatory reform, which Dole, a tobacco ally, subsequently cited in a statement. Carruthers heralded the study, without mentioning tobacco. "We want to offer information on how scientific issues are communicated to the public as another means of ensuring that only sound science is used in making public policy decisions," he stated.

Regulatory reform ultimately failed to pass the Senate. But as we will see in Chapter 8, many elements of the Gingrich-era package have since been implemented by the administration of George W. Bush—still, in many cases, under the aegis of "sound science."

The ideological origins of the "sound science" movement—and the term's close connection both to the science-abusing regulatory reform agenda and to Big Tobacco—seem fairly clear. However, this information appears to remain unknown to most Democrats, liberals, journalists, and advocacy groups, who unwittingly continue to use the phrase "sound science," apparently convinced that it simply means "good science." Conservatives clearly mean something more by "sound science," and have thoroughly co-opted the phrase for strategic reasons. It can be used naively, perhaps, but not innocently.

To the Right, "sound science" means requiring a higher burden of proof before action can be taken to protect public health and the environment. In other words, "sound science" isn't really a scientific position at all. Indeed, in an online discussion of "sound science and public policy," the Western Caucus of the U.S. House of Representatives, chaired by the aforementioned Rep. Chris Cannon, notes that "environmental laws should be made with great caution and demand a high degree of scientific certainty"—a prescriptive policy statement having nothing to do with what constitutes good science.

Similarly, Thomas Roskelly, of the Annapolis Center for Science-Based Public Policy, which celebrated Cannon in 2003 for his "sound science" initiative, describes the concept as a "counterpoint" to the so-called precautionary principle, a regulatory philosophy embraced by many environmental groups that recommends erring on the side of caution when faced with an unknown or inadequately quantified risk. If "sound science" stands in contrast to the "precautionary principle," then that contrast, by definition, must occur within the sphere of policy, not science. After all, no serious environmentalist would call the precautionary principle "scientific." Whatever the merits of that principle, at least its proponents realize that making policy in areas involving risk relies on a complex interaction between scientific and political or moral considerations.

Other conservatives define "sound science" with more nuance, but in a way that is, finally, consistent with the Western Caucus's approach. For instance, I asked William Kovacs, a vice president at the U.S. Chamber of Commerce and a longtime player in the regulatory battles between industry and environmentalists, what the term meant. "The environmentalists try to paint that as, 'if you can't show it to a scientific certainty, it's not sound science,'" Kovacs replied. "And we never meant the phrase that way. When we were using the phrase, what we meant is, 'you put all your science on the table, including an identification of the uncertainties that are there.'" In another interview about "sound science," Marshall Institute CEO William O'Keefe added, "What I have said all along, and the people I know have said, is, 'Look, you need to take account of uncertainty.'"

There is nothing wrong with recognizing uncertainty, at least in principle. And O'Keefe admits that uncertainty is "not an excuse for not acting." But under the guise of "sound science," the Gingrich-era regulatory reformers would clearly have made government action vastly more difficult, and that appears to have been the objective all along. "I believe that's a good thing, to have a fair degree of certainty if what you're going to do is impose massive costs on the economy as a whole," says Robert Walker, who shepherded the "sound science" bill through the House Science Committee.

Now that we fully understand both the history and the meaning of "sound science," we can see why the terminology is so objectionable. First, this Orwellian phrase masks an agenda that has little to do with the responsible use of science in formulating public policy. Indeed, the record makes clear that the "sound science" movement sought to smear a wide range of scientific activities conducted by taxpayer-funded government experts, greatly undermining the public's trust in science in the process. Rather offensively, "sound science" proponents aimed to discredit the agencies charged with protecting us and in many cases to substitute for their scientific judgments the views of self-interested companies.

Furthermore, the "sound science" movement also confuses the *quality* of scientific analysis with the degree of scientific certainty that has been achieved on a given question. This fosters an entirely inaccurate picture of how government agencies use science, a process that has very little to do with the pursuit of certain knowledge. True, agencies should do the best research they can, and interpret existing information carefully. But the goal of certainty in science is as elusive as certainty in most spheres of human endeavor, making a "weight of the evidence" standard much more appropriate for the work of federal agencies. After all, regulatory delays can have serious consequences.

If "sound science" proponents want to reverse the current precaution-ary orientation of our federal government's regulatory bodies, they should say so openly, rather than seeking to confuse the public about the proper role of science in protecting public health and the environment.

The Gingrich Congress's dismantling of the Office of Technology Assess-ment, its attacks on climate science and the theory of ozone depletion, its advocacy of "regulatory reform," and its demands for "sound science" are all of a piece. Fundamentally, the Gingrich Republicans wanted to dra-matically raise the burden of scientific proof required before the govern-ment could take action to protect human health or the environment. Moreover, they wanted to claim the mantle of science while doing so, even though their goal was not to acquire the best scientific research, but rather to cripple the development of government regulation. Finally, when their handpicked experts raised doubts about scientific assessments that could lead to action, the Gingrich Republicans didn't want to find these experts contradicted by Congress's very own scientific branch, the Office of Technology Assessment.

The "sound science" program set in motion during the Gingrich era remains a core component of the conservative science agenda today. In

2003, for instance, the Environmental Working Group exposed a strategic memo on the environment by Republican tactician Frank Luntz, the "pollster of record" for the Gingrich "Contract with America." Written prior to the 2002 congressional elections, the memo advised Republicans that "the most important principle in any discussion of global warming is your commitment to sound science."

But consider what "sound science" actually meant to Luntz on the issue of climate change. "The scientific debate is closing [against us] but not yet closed," he wrote. "There is still a window of opportunity to challenge the science." Luntz then provided a blueprint for the politicization of scientific expertise: "You need to be even more active in recruiting experts who are sympathetic to your view, and much more active in making them part of your message." As we shall see in the next chapter, Luntz's public relations jujitsu—dubbed "Luntzspeak" by the National Environmental Trust—seems to have been quite influential among modern-day foes of mainstream climate science, and particularly Senate Environment and Public Works Committee chairman James Inhofe.

To be sure, the Right employs other Orwellian phrases besides "sound science" on science policy issues. We must not forget "junk science," its frequent antonym, which broadens the whole lexicon.

Just as "sound science" doesn't really describe good science, "junk science" doesn't actually describe inappropriately conducted or fraudulent work in most cases. Instead, the label frequently attaches to research that fails to mesh with the laissez-faire viewpoint of regulated companies. According to a 2001 media analysis of uses of the phrase "junk science" between 1995 and 2000, most charges weren't backed up by substantive evidence of "questionable or unacceptable scientific activities." Instead, an overwhelming number of uses occurred in the context of an "antiregulatory message or admonition." That is certainly how conservative commentator Steven Milloy, publisher of the website JunkScience.com, often seems to use the term. It will come as no surprise that Milloy served as executive director of The Advancement of Sound Science Coalition in the late 1990s.

In a 2004 lecture, University of Texas law professor Thomas McGarity, president of the Center for Progressive Reform, amusingly summed up the conservative worldview with the following slogan: "Our science is sound science and their science is junk science." Conservatives are wrong on both charges, but they have been vastly successful in getting the rest of the political world to adopt their loaded terminology. These cynical slogans, "sound science" and "junk science"—reminiscent of *Fox News*'s laughable "fair and balanced" mantra—have for too long masked the true intentions of a political movement whose basic mode of operation relies on the manipulation and abuse of science.

And even as they continue to call for "sound science," today's conservatives, keepers of the Gingrich flame, have embarked on an unprecedented campaign of science politicization on behalf of business and other economic interests. The next four chapters detail how the modern Right— and especially the administration of George W. Bush—has carried forward its grand ideological tradition of pro-industry science politicization:

- On global warming, perhaps the most pressing science-based issue today, conservatives follow the Frank Luntz "sound science" strategy to a tee. Senate Environment and Public Works Committee chairman James Inhofe (R-OK) counts as the worst offender, but George W. Bush's administration comes in a close second.

- With their below-the-radar creation of, and willingness to employ, the so-called Data Quality Act, conservatives carry forward the misleading legacy of the Gingrich-era regulatory reformers, seeking to throw hurdles in the way of the government's use of scientific information to support regulatory action. As it happens, the Data Quality Act is associated with the efforts of Philip Morris via a classic Washington character named Jim Tozzi, its unofficial creator.

- When it comes to the relationship between various unhealthy foods and the global rise of obesity, conservatives and food industry interests have led an all-out attack on a 2003 World Health Organization report that dared to suggest a connection between consuming sugar-sweetened foods and poor health. On the issue of mercury contamination of fish, meanwhile, conservatives similarly hype uncertainty and cherry-pick their science.

- Finally, under the mantle of "sound science," conservatives want to reform the Endangered Species Act (ESA) to make it more difficult than ever to protect wildlife in peril of extinction. In July 2004, I traveled to southern Oregon's Klamath River Basin to witness how a battle over allocating scarce water resources among farmers and endangered fish had transmogrified into a full-fledged science fight. Klamath irrigation farmers have broadly denounced a 2001 decision to protect several species of fish as based on "junk science" and called for "sound science" reforms, yet their true sentiments came across clearly as they carried signs bearing slogans like "tame the raging bull . . . castrate the ESA."

The Klamath farmers didn't really care about improving wildlife science; they simply wanted government off their backs. Yet science has become the battlefield of choice for conservatives seeking to block environmental protections, and the Klamath irrigators were simply revealing themselves as true to type. The type itself, of course, is the real problem. It represents an alarming development both for science itself and for a Republican Party that once counted conservationist Theodore Roosevelt among its heroes.

PART TWO

THE BUSINESS OF SCIENCE

Everything's going in the science direction. All of the environmental issues, the regulatory issues, are really becoming more and more based on either science or economics. And what we're finding is that none of the information out there is really very accurate.

WILLIAM KOVACS, U.S. CHAMBER OF COMMERCE

CHAPTER 7

"THE GREATEST HOAX"

AMONG TODAY'S CONSERVATIVE "Luntzspeak" practitioners, few have the lingo down better than the honorable James Inhofe, senator from Oklahoma. Recall from the previous chapter Luntz's advice for dealing with the issue of global warming, which includes the following precepts: (1) emphasize your commitment to "sound science"; (2) seize the remaining "window of opportunity" to challenge and dispute the scientific consensus; and (3) find experts "sympathetic to your view" and make them "part of your message." It's a cunning strategy, provided that you are not ashamed of following in the footsteps of the tobacco industry, and Inhofe doesn't appear to have much shame. From his perch as chairman of the Committee on Environment and Public Works, he has followed the Luntz memo's recommendation flawlessly. In a written response to questions, Inhofe's staff fully admitted he'd read it.

A former Tulsa mayor and small-businessman, Inhofe was first elected to the Senate in 1994, the year of the Gingrich Revolution. Of average height, ruddy-faced, and conservative to the core, he has a history of extreme statements, having once dubbed the Environmental Protection Agency a "gestapo bureaucracy." For his 2002 Senate reelection race, Inhofe got more campaign donations from oil, gas, and electric companies—who surely appreciate his repeated challenges to the scientific basis of virtually every environmental problem—than from any other industries. No wonder he has received consistent zero ratings from the League of Conservation Voters (though he improved his score to 4 percent in the 108th Congress).

Following the 2002 congressional elections, which brought Republicans control of the Senate, Inhofe ascended to the chairmanship of the Environment and Public Works Committee, replacing northeastern moderate (and Republican apostate) Jim Jeffords, of Vermont. Few had any doubt that Inhofe would take a far more conservative tack than either Jeffords or his Republican predecessor, Bob Smith, of New Hampshire. "You think I'm bad—try Inhofe," Smith had warned environmentalists upon taking control of the committee in 1999.

Since becoming chairman, Inhofe has rhetorically stressed his commitment to scientific rigor. He has regularly invoked the concept of "sound science" and pledged that on his watch, the committee will "improve the way in which science is used." In practice, however, Inhofe has adopted Gingrich-style science abuse tactics that come off as even less defensible today than they did ten years ago. And that is nowhere more true than on the pressing issue of global climate change, the likely flashpoint for battles over science and politics in the coming years.

Inhofe has served as a true general in the modern Right's war on modern climate science. In 2003, he pulled out all the stops to defeat Senators John McCain and Joe Lieberman's Climate Stewardship Act, which would have created the first caps on greenhouse gas emissions ever agreed to by the U.S. government (the bill failed by a 43–55 vote). Inhofe's sallies included stacking a Senate panel with climate science contrarians and delivering a twelve-thousand-word Senate floor speech entitled "The Science of Climate Change," outlining conclusions that he said he had reached after several years of studying the issue. The talk ended with the suggestion that global warming caused by human activity might be "the greatest hoax ever perpetrated on the American people."

With lines like that, it may seem tempting to dismiss Inhofe as a fringe wacko. In fact, he is probably the Republican Party's leading environmental spokesman. And though Inhofe frequently winds up in a more extreme position than those of his colleagues, his actions reflect the modern Right's broader strategy on global warming. In June 2003, for instance, the U.S. Senate Republican Policy Committee released a report challenging mainstream climate science that in many ways anticipated Inhofe's own forays.

As a science abuser, Inhofe may be out of control, but he is not at all out of the Republican mainstream. By examining his activities, we simultaneously witness how the modern Right has carried the Gingrich legacy on climate change forward to the present. We will also see how Inhofe's "bad cop" antics tend to conveniently overshadow the Bush administration's own abuses and distortions of climate science, which have been subtler, perhaps, but no less serious.

The Luntz approach of invoking contrarians and "skeptics" to challenge the global warming theory may have constituted a defensible political strategy in 1990, when the science linking human activities to actual observed climate changes remained relatively tentative. But when Inhofe and his ilk take the same tack today, they simply cannot avoid ignoring or misrepresenting a highly robust scientific conclusion.

In the decade since the Intergovernmental Panel on Climate Change's (IPCC) cautious 1995 statement that "the balance of evidence suggests that there is a discernible human influence on global climate," the science of climate change has strengthened considerably. Even as industry-friendly "skeptics" questioned the scientific underpinnings of the Kyoto Protocol on greenhouse gas emissions in the late 1990s, the field continued to mature, while climate models increased in sophistication. With the release of the IPCC's third assessment in 2001, a strong consensus position emerged: Notwithstanding some role for natural variability, humans are almost certainly heating the planet *now* through greenhouse gas emissions, and could ramp up global average temperatures by several degrees Celsius by the year 2100. "Consensus as strong as the one that has developed around this topic is rare in science," wrote *Science* executive editor-in-chief Donald Kennedy in a 2001 editorial.

As editor of one of the scientific community's flagship journals, an outlet that publishes papers on climate regularly, Kennedy ought to know. And in a 2004 interview, he told me something quite striking: *Science* does not exclude contrary arguments about the role of humans in causing climate change; rather, there simply isn't anything to consider publishing. "There is no example of a paper that disagreed strongly with the general consensus that has been peer reviewed and failed," Kennedy told me. "The fact is, they're just not being sent. And of course, then our opponents would say, 'Well, that's because they know you'll never take them.' Well, hell, if there's been no case, how could they know that?"

Kennedy's statement draws strength from a paper published in *Science* in late 2004 that reported on a literature review of scientific papers listed with the keywords "global climate change" between 1993 and 2003. After examining 928 papers, the author, Naomi Oreskes, found that "Remarkably, none of the papers disagreed with the consensus position." As Oreskes hastens to add in an interview, this hardly means that no truly contrarian or dissenting papers exist. But given the size of her sample, she suggests that such literature must be "vanishingly small."

That is quite a consensus, and in many cases, even the so-called skeptics have found themselves subtly influenced by it. They have gradually changed the tune of their arguments, notes Harvard biological oceanographer James McCarthy, who cochaired the working group on the impacts of climate change for the third IPCC assessment. "In the late '80s, early '90s, it was, 'Nothing is changing,'" says McCarthy of the contrarian viewpoint. "And by the mid-'90s, it's, 'Well, things are changing, but just a little bit, and by the way, humans aren't causing it.' By 2000, it's, 'Well, things are changing a little bit, humans are causing it, and there may be some impacts, but you know what, it won't matter.'"

The strengthening of the consensus even converted some leading corporations that had previously supported "skepticism." Major oil companies such as Shell, Texaco, and British Petroleum, as well as automobile manufacturers such as Ford, General Motors, and Daimler-Chrysler, left the industry backed Global Climate Coalition, which itself became inactive after 2002.

Yet some forces of skepticism—including ExxonMobil, the world's largest nongovernmental oil company—have persisted. Inhofe and the Bush administration alike carry their water, as do a range of conservative think tanks, which often provide Inhofe and his confreres with arguments and talking points. These groups may no longer reflect the collective views of the fossil fuel industry, but they certainly represent the views of some industry powerhouses.

By 2002, ExxonMobil was donating over a million dollars annually to policy groups and think tanks involved in battling against the scientific mainstream on global warming, including the George C. Marshall Institute, the Competitive Enterprise Institute, Frontiers of Freedom, the Heartland Institute, the website TechCentralStation.com, and many others. Whenever a major new development occurs in climate science, these groups kick into high dudgeon, nitpicking and debunking state-of-the-art science in online commentaries, reports, press releases, and newspaper op-ed pieces. As a case in point, consider the late 2004 release of the *Arctic Climate Impact Assessment*, a scientific report showing that human-fueled global warming has already had alarming impacts on the Arctic region, such as the melting of glaciers and sea ice. The Marshall Institute promptly challenged the report's science; and then Inhofe, in issuing his own challenge, cited the Marshall Institute.

The evidence suggests that these think tanks, and politicians such as Inhofe who draw upon them, are engaged in a very familiar strategy: "manufacturing uncertainty" about global warming. When it comes to attempts by fossil fuel interests to sow doubt and create controversy, we lack the same documentation that we have on Big Tobacco's notorious scientific activities. But what does exist suggests an industry on the defensive, generating its own bogus science to prevent the advent of mandatory limits on greenhouse gas emissions.

In 1998, for instance, the *New York Times* exposed an internal American Petroleum Institute memo outlining a strategy to invest millions to "maximize the impact of scientific views consistent with ours with Congress, the media, and other key audiences." "Victory will be achieved," the document stated, when "recognition of uncertainties becomes part of the 'conventional wisdom.'" Perhaps most startlingly, the memo cited a need to "recruit and train" scientists with "new faces," "individuals who *do not* have a long history of visibility and/or participation in the climate change debate," to make the contrarian case. This seems to signal an awareness that after a time, journalists catch on to the connections between contrarian scientists and industry.

According to the memo, a representative of the George C. Marshall Institute helped develop the plan, as did a representative of the Advancement of Sound Science Coalition ("junk science" meister Steven Milloy, nowadays a regular debunker of mainstream climate science in his online column for *FoxNews.com*). In an interview in March 2004, the Marshall Institute's William O'Keefe, who was an executive at the American Petroleum Institute when the memo emerged, told me that the agenda it outlined was never implemented. But its very existence raises suspicions about a cynical industry plot to raise doubts about mainstream climate science.

Also involved in developing the American Petroleum Institute plans was ExxonMobil lobbyist Randy Randol, who has since retired but appears to have plied his trade quite effectively during much of George W. Bush's first term. When Bush took office, Randol left his calling card in the form of a February 6, 2001, memo to the White House Council on Environmental Quality (CEQ). The memo—when its existence was reported, ExxonMobil took the position that Randol had indeed forwarded the document to the CEQ, but neither he nor anyone else at the company had written it—denounced then-IPCC chairman Robert Watson, a leading atmospheric scientist, as someone "handpicked by Al Gore," who had used his position to "get media coverage for his views." "Can Watson be replaced now at the request of the U.S.?" the memo asked. It went on to single out several other Clinton administration climate experts, asking whether they had been "removed from their positions of influence."

It was, in short, a hit list. Whether influenced by Randol's memo or on its own initiative, a year later, the Bush administration campaigned against Watson's reelection for the post of IPCC chairman. Watson lost to a State Department–supported Indian engineer, Rajendra Pachauri.

It is in this context—of hardening scientific consensus and increasingly desperate conservative attempts to undermine it—that we must consider both James Inhofe's abuses of climate science and the Bush administration's more below-the-radar activities. As previous chapters have noted, the current administration has continually distorted science, exaggerated uncertainty, and interfered with the activities of expert agencies on this issue. But while the White House subtly misrepresents and undermines the scientific consensus, James Inhofe has basically thrown himself in front of the scientific equivalent of a moving train.

On July 28, 2003, as the McCain–Lieberman legislation to reduce carbon emissions gathered momentum, Inhofe delivered a massively detailed Senate floor speech, well over an hour in length and complete with a moving picture show of colorful charts and diagrams, directly challenging the conclusions of mainstream climate science. The address began, like the hodgepodge that it was, with a brief allusion to ozone depletion contrarianism; a harsh attack on science journalist David Appell; and the questionable invocation of former CIA director and Carter administration energy secretary James Schlesinger—a board member of Peabody Energy, the world's largest coal company, though Inhofe didn't mention that—as an expert on climate. Inhofe then proceeded to dismiss concerns about climate change as little more than alarmism, stating that "after studying the issue over the last several years, I believe that the balance of the evidence offers strong proof that natural variability is the overwhelming factor influencing climate."

Major scientific bodies with expertise in the area have reached a very different conclusion, however. The IPCC, the National Academy of Sciences, the American Meteorological Society, the American Geophysical Union—all agree that human activity is causing climate change. Moreover, nature itself keeps providing more evidence: According to NASA's Goddard Institute for Space Studies, the years 1998, 2002, 2003, and 2004 were the four warmest years in the temperature record since the 1890s. But never mind: "The claim that global warming is caused by manmade emissions is simply untrue and not based on sound science," Inhofe concluded, and then delivered the following line as a parting shot: "With all of the hysteria, all of the fear, all of the phony science, could it be that manmade global warming is the greatest hoax ever perpetrated on the American people? It sure sounds like it."

One leading climate scientist—former Geophysical Fluid Dynamics Laboratory director Jerry D. Mahlman, whom we last encountered defending climate modeling against the criticisms of Patrick Michaels—provides the following reflection on Inhofe's "scientific" speech: "It's the kind of thing you write Monty Python skits about."

Suppose we take Inhofe's words at face value, and assume that he means that there is an actual global warming *hoax*. This raises the question, Who are the hoaxers? For global warming truly to qualify as a hoax, a dramatic level of coordinated action would have to exist on the part of scientists. This would make past scientific hoaxes seem like minor peccadilloes.

Perhaps Inhofe believes in a vast conspiracy tying together federal regulators, government-funded scientists, and environmental groups. But due to the nature of the climate science field, the conspiracy couldn't exist just within the United States. It would have to have a global scope, which makes the notion dramatically more implausible. As Donald Kennedy put it to me, "It would be astonishing if the Intergovernmental Panel on Climate Change, the vast majority of American scientists, the government of the United Kingdom, and all those people were consensually involved in a climate hoax."

Besides conspiracy mongering, Inhofe's Senate speech contained predictable attacks on climate modeling and the IPCC. Most audaciously— and echoing the Luntz memo's admonition to find experts "who are sympathetic to your view"—the senator engaged in a scientific land grab. Inhofe listed a slew of scientific authorities who supposedly back his view on climate change, naming a number of well-known contrarians—Sallie Baliunas, S. Fred Singer, Patrick Michaels, MIT's Richard Lindzen—as well as several more mainstream scientists. What happened next was extremely revealing: Two of the mainstream scientists publicly protested that their work had been severely misrepresented.

Perhaps the most outspoken response came from Tom Wigley, a distinguished meteorologist who serves as senior scientist with the National Center for Atmospheric Research, in Boulder, Colorado. Inhofe drew upon a 1998 paper by Wigley in the journal *Geophysical Research Letters* in order to argue that the Kyoto Protocol would result in only a small reduction in temperatures by 2050 or 2100. But in a letter to Senators Tom Daschle and Bill Frist, Wigley objected to Inhofe's "selective use of scientific information, his misrepresentation of much of the scientific evidence on global warming, and his misrepresentation of my own published work in particular." Wigley's paper relied on the *hypothetical* assumption that no further policies to deal with climate change would follow the Kyoto

Protocol, which expires in 2012—a rather huge qualification that Inhofe failed to note. The whole point of the paper, Wigley pointed out, was to show that Kyoto would be only "the first step in a long and complex process of reducing our dependence on fossil fuels."

Inhofe also cited a 2001 *Nature* commentary by the distinguished Stanford climatologist Stephen Schneider, painting Schneider as a critic of the IPCC's 2001 projections of possible warming by 2100 (between 1.4 and 5.8 degrees Celsius, which Inhofe dubbed an "extreme prediction"). The same pattern ensued. In a written response to questions from Senator McCain, Schneider protested being cast as an IPCC critic: "It is misrepresenting my views to characterize them as even implying that IPCC has exaggerated or failed to describe the state of the science fairly," he wrote.

Schneider notes that he never claimed, in *Nature,* that the IPCC had "overestimated the likelihood of future temperature rises." Instead, he had merely suggested that in order to help policymakers, the IPCC should have attempted to provide a probability estimate to go along with each of its projections. Thus using Schneider's work to question the IPCC's projections seems a particularly stunning abuse. Despite Wigley's and Schneider's protestations that they had been misrepresented, Inhofe has continued to cite their work, instead of apologizing for his error.

And the day after his "hoax" speech—sure to live on in infamy even after the United States and other holdout nations finally lurch into action to deal with global climate change—Inhofe compounded the offense. He used his Senate chairmanship to provide a forum for two climate science contrarians, both linked to energy interests, to challenge a researcher (Michael Mann, now at Penn State University) whose work has contributed to the powerful global scientific consensus on this issue. In the process, Inhofe carried forward into the twenty-first century the Gingrich Republicans' strategy on climate change: using scientific outliers to sow doubt about mainstream findings in public rather than scientific forums.

Inhofe opened the hearing by swearing fealty to "sound science." He then lavished praise on a highly controversial paper, heavily criticized at the time by mainstream climate scientists and since largely discredited, by two scientists at the Harvard-Smithsonian Center for Astrophysics. "In many important ways," Inhofe declared in portentous history-of-science speak, the study "shifts the paradigm" away from the generally accepted view, defended by Mann's work and numerous other scientific papers as well as the IPCC, that the late twentieth century saw unprecedented temperature spikes (presumably as a result of greenhouse gas emissions). Historically, scientific paradigm shifts have rarely occurred before the U.S. Senate—but that didn't seem to faze Inhofe.

And so the cacophony began. If Inhofe's goal was to create confusion and hence the appearance of scientific uncertainty, he succeeded brilliantly. The two sides testifying disagreed about virtually everything except basic physics (and even that took them a while to come to terms on). Harvard-Smithsonian astrophysicist Willie Soon, one of the paper's authors, plugged his recent work, claiming that his review of other scientists' "proxy records" for past temperatures—themselves based on measurements from tree rings, ice cores, corals, and other sources—suggested that twentieth-century warmth did not, in fact, constitute an anomaly. "The climate of the 20th century is neither unusual nor the most extreme," Soon testified. Soon drew support at the hearing from David Legates, of the University of Delaware, another contrarian scientist and coauthor on a later version of the paper.

Since these scientists failed to disclose their industry ties before the Senate, let us do it for them. First, the paper coauthored by Soon was partly funded by the American Petroleum Institute. Furthermore, Soon has served in the past as a "senior scientist" with the George C. Marshall Institute, which received $110,000 in total contributions from Exxon-Mobil in 2003, the year of the hearing. Soon also serves as "science director" for TechCentralStation.com ($95,000 from Exxon Mobil in 2003 for "climate change support"), and is listed as "science director" with the Center for Science and Public Policy at Frontiers of Freedom ($195,000 in 2003). Legates seems unable to match Soon in the area of corporate ties, but he puts up a respectable fight. Having collaborated with the Marshall Institute in the past, as of 2004, Legates also served as an "adjunct scholar" with the National Center for Policy Analysis, in Dallas, Texas, which received $75,000 from ExxonMobil in 2003.

Pitted against these two doubters stood Michael Mann, a lone representative of the overwhelming consensus view among climate scientists. Much of Mann's work has focused on what we can learn about climate history from tree rings and other "proxy" indicators of what temperatures were prior to actual thermometer-based observations; his university webpage shows him posing with the cross section of a tree trunk. Mann and his colleagues have reconstructed climate records suggesting that recent temperatures represent an anomaly in the context of the past millennium, a conclusion embodied in an iconic "hockey stick" graph showing relatively moderate oscillations until temperatures spike upward at the end of the twentieth century (the "blade" of the stick).

Though multiple studies confirm the basic thrust of this work, Mann's association with the "hockey stick" graph has made him a target of various climate science "skeptics" and contrarians, including (but not limited to) Soon and Legates. Just as the "skeptic" camp attacked Benjamin Santer following the 1995 IPCC assessment, with the apparent objective of discrediting the U.N. organization's larger body of work, so they have attacked Mann after the 2001 IPCC report. After all, the report had prominently displayed a graph based on one of the Mann team's "hockey stick" reconstructions in its opening "Summary for Policymakers" section, which stated that "the increase in temperature in the 20th century is likely to have been the largest of any century during the past 1,000 years."

At Inhofe's hearing, Mann defended both his own work and the conclusions of the IPCC, which channels the work of hundreds of experts. But for those keeping count in the Senate that day, the intellectual ticker showed a score of two to one, not a handful versus a horde. Such was Inhofe's conception of "balance." At one point, for example, the senator asked the panelists whether they agreed or disagreed that rising carbon dioxide levels can "produce many beneficial effects upon the natural plant and animal environments of the earth." Here were the results:

Dr. Soon: I agree.
Dr. Mann: I find little in there to agree with.
Dr. Legates: I would tend to agree.

A similar hung jury occurred on other scientific questions, such as whether our pumping of carbon dioxide into the atmosphere has caused glacial retreats, and how to interpret the cooling trend that occurred between 1940 and 1970, prior to the current temperature spike. All the disagreement led Senator Craig Thomas, a Wyoming Republican, to throw up his hands and state, "We are expected to make some policy decisions based on what we ought to be doing with regard to these kinds of things, but yet there does not seem to be a basis for that kind of decision."

By now, the problems with Inhofe's attempt to turn Congress into a science court should be apparent. The validity of Michael Mann's particular "hockey stick" analysis remains open to debate among experts, and has in fact been prominently challenged in the peer-reviewed literature. But holding a heated public hearing between mainstream scientists and contrarians will hardly help determine its merits. "That's why the federal government turns to the National Academy of Sciences for advice, or the governments of the world turn to the Intergovernmental Panel on Climate Change," explains Princeton University climate expert Michael Oppenheimer.

Moreover, although it might create good publicity, the Right's selective attack on Mann's work ultimately presents a huge diversion for policymakers trying to decide what to do about global warming. Mann points out that he's hardly the only scientist to produce a "hockey stick" graph—other teams of scientists have come up with similar reconstructions of past temperatures. And even if Mann's work *and* all of the other studies that served as the basis for the IPCC's statement on the historical temperature record are wrong, that would not in any way invalidate the conclusion that humans are *currently* causing rising temperatures. "There's a whole independent line of evidence, some of it very basic physics," explains Mann.

And on top of that, Willie Soon's rebuttal to Mann's work, coauthored with fellow Harvard-Smithsonian astrophysicist and longtime climate science contrarian Sallie Baliunas, turns out to be highly dubious science. Indeed, Inhofe's decision to highlight this work suffered from extremely unfortunate timing. The very day before Inhofe's hearing, the editor in chief of *Climate Research,* the small journal where the Soon and Baliunas paper originally appeared, had resigned to protest deficiencies in the review process leading up to the paper's publication. Several other editors also subsequently resigned. Some "paradigm shift"!

The first to resign, Hans von Storch (who has himself since criticized Mann's work), later complained to the *Chronicle of Higher Education* that climate science contrarians "had identified *Climate Research* as a journal where some editors were not as rigorous in the review process as is otherwise common." Soon, an article by Mann, Oppenheimer, Wigley, and ten other leading climate scientists, harshly critical of the Soon and Baliunas paper, appeared in the American Geophysical Union publication *EOS*. Among other flaws, the authors objected that Soon and Baliunas use historic drought and precipitation records to infer information about past temperatures—which Mann called a "fundamental error" in his Senate testimony. In a subsequent editorial, *Climate Research* founder Otto Kinne agreed with critics that Soon et al.'s published findings "cannot be concluded convincingly from the evidence provided in the paper."

Inhofe's reliance on this questionable study—nay, his willingness to devote half of a Senate hearing to it—may seem a tad embarrassing. But the Bush administration has followed suit, seizing on the questionable Soon and Baliunas study in an attempt to undermine the scientific consensus on global warming. In early 2003, when the White House controversially sought to edit the climate change section of a draft EPA report (an action exposed by the *New York Times*), it removed a diagram of the Mann group's "hockey stick" and inserted a reference to the Soon and Baliunas work. In response, a dissenting EPA memo called the Soon and Baliunas paper "a recent, limited analysis [that] supports the Administration's favored message."

Inhofe may have rough edges, and may make the Bush White House's stance on climate change seem moderate by comparison. But in reality, the two aren't so far apart.

Indeed, this comes across clearly when you consider how Inhofe and the Bush administration have treated a true thorn in the side of the climate change "skeptics," and therefore of industry: a 2001 National Academy of Sciences (NAS) report on climate science, requested by the Bush White

House, that confirms the basic 2001 findings of the Intergovernmental Panel on Climate Change. In the opening sentence of its summary, the NAS report states point blank that "greenhouse gases are accumulating in Earth's atmosphere as a result of human activities, causing surface air temperatures and subsurface ocean temperatures to rise. Temperatures are, in fact, rising."

In Inhofe's lengthy Senate climate science tirade, he took an interesting approach to this key study: He ignored it entirely. But in ducking findings by the United States' leading scientific body, Inhofe isn't alone. Even though the Bush administration asked the NAS to review the IPCC's 2001 conclusions—an action that some have likened to asking a district court to review a Supreme Court decision—it too has run away from the end product. Attempted White House edits to the climate change section of the 2003 EPA report, mentioned above, also sought to remove references to the NAS's findings.

In written questions, I asked Inhofe how he could give a major speech on climate, criticize the IPCC, and yet ignore that the NAS had very good things to say about the IPCC's work. In response, the senator's committee staff proceeded to misrepresent the NAS report egregiously, a tack the Bush administration has also taken on occasion. Much of the misrepresentation exploits a classic strategy for abusing science: magnifying uncertainty.

In their response, Inhofe's committee staff highlighted the many references to scientific uncertainty in the NAS report. They went on to suggest that these passages somehow reconcile Inhofe's position—that "natural variability" is the main factor influencing the climate—with the mainstream. Yet interviews with a number of climate scientists, including some NAS report authors, confirm that the uncertainties described in the report do not undermine the view that human activity continues to fuel global warming. Therefore, they hardly back Inhofe's view of manmade global warming as a "hoax . . . perpetrated on the American people."

The NAS report discusses uncertainty at length because the Bush administration expressly (and perhaps strategically) asked the panel to define areas in climate science "where there are the greatest certainties and uncertainties." It doesn't follow, though, that the NAS report can rescue Inhofe from the scientific hall of shame. Inhofe accepts the current uncertainties in climate science—the things scientists aren't sure about yet—but refuses to accept what they actually *do know*: that uncertainties notwithstanding, human activities are almost certainly causing the current warming trend. But Inhofe should not be allowed to pick and choose which parts of the NAS report he likes or doesn't like; the *entire report* reflects the scientific consensus.

In the process of hyping uncertainty, Inhofe's staff also highlighted a single sentence from the NAS report, one that conservatives love to cite. It reads in full as follows: "Because of the large and still uncertain level of natural variability inherent in the climate record and the uncertainties in the time histories of the various forcing agents (and particularly aerosols), *a causal linkage between the buildup of greenhouse gases in the atmosphere and the observed climate changes during the 20th century cannot be unequivocally established*" [Italics added]. George W. Bush's 2004 presidential campaign cherry-picked precisely this sentence when asked about its climate change policies by *Science* magazine. Based on passages like this, the Bush campaign asserted that "the nation's most respected scientific body found that key uncertainties remain concerning the underlying causes and nature of climate change."

The above sentence might indeed give one pause about the alleged scientific consensus on human-caused climate change, at least initially. But you have to understand what it actually means. Very little in science has been "unequivocally established." In general, science doesn't deal in unequivocal knowledge. It seeks to determine what we can and can't be reasonably sure of in a complex world.

With that in mind, I asked several members of the 2001 NAS panel how to interpret this seemingly damning sentence. "I think that's true only because there is no way of establishing causation absolutely in a field that is not an experimental field," Edward Sarachik, an atmospheric scientist at the University of Washington, in Seattle, remarked of the statement. In other words, you can't create another Earth, change the CO_2 levels, and then measure temperatures to prove a causal link. Sarachik's University of Washington colleague John M. Wallace, also an NAS panelist, even said he wished the sentence had been written differently: "We didn't think about how it would sound as an isolated quote, separated

from the rest of the report." Finally, I also spoke with NAS panelist Eric
J. Barron, a geoscientist at Penn State University, about whether this sentence undermines the notion that humans have caused recently observed
climate changes. "There are a lot of things that cause climate variability,"
Barron explained. "So, are you unequivocally certain? No. Is every finger
pointing at it? Yes."

Obviously, then, the NAS did not mean for this sentence to contradict
the notion that the best evidence suggests human-caused climate change
to be underway. Yet conservatives have seized on the passage, either without understanding or without caring what it means in context, and
quoted it to back up their policy agenda. On this point, yet again, Inhofe
and the Bush administration seem to be sharing the same cheat sheet.

And Inhofe's staff misrepresented the NAS report in yet another way. They
questioned whether the NAS had actually "confirmed" the IPCC 2001 findings, even though the NAS report plainly states, "The IPCC's conclusion
that most of the observed warming of the last 50 years is likely to have been
due to the increase in greenhouse gas concentrations accurately reflects the
current thinking of the scientific community on this issue."

In suggesting that the NAS somehow criticized the IPCC instead of
basically agreeing with it, Inhofe's staff and other conservatives love to
cite statements by one NAS panel member in particular, the famed climate science contrarian Richard Lindzen, of MIT. A respected meteorologist and National Academy of Sciences member who has advised the
Bush administration on climate science, Lindzen has earned his reputation as the most scientifically renowned of the various "skeptics." "I think
he's offered a healthy level of caution to the scientific community that's
made people think," says Duke's William Schlesinger. At the same time,
however, Lindzen remains a controversial figure known for his highly
contrary bent. A smoker, he has reportedly even questioned how strong
the link is between cigarette smoking and lung cancer.

Shortly after the NAS report emerged, Lindzen wrote a *Wall Street
Journal* op-ed piece putting his own particular spin on its findings. He
took the media to task for claiming that scientific unanimity existed on
the climate change question, and argued that "we are not in a position to

confidently attribute past climate change to carbon dioxide or to forecast what the climate will be in the future." Lindzen also lit into the 2001 IPCC report's opening section—known as the Summary for Policymakers—charging that document with "a strong tendency to disguise uncertainty" and noting that "within the confines of professional courtesy, [the NAS] essentially concluded that the IPCC's Summary for Policymakers does not provide suitable guidance for the U.S. government."

Given these strong words, it is no wonder that Inhofe's staff and other conservatives rely on Lindzen so heavily, and use him to claim that the NAS somehow did not ratify the IPCC's findings. But that is a misinterpretation. In an interview that I tape-recorded and provided him a copy of (at his request), I spoke with Lindzen about whether, in his various writings, he had really questioned whether the NAS had confirmed the IPCC's work. Lindzen made clear that he took issue more with the IPCC's Summary for Policymakers (SPM) than with the report itself, and agreed that the NAS had endorsed the credibility of the broader document, which (despite some misgivings) he called "an okay summary of what's gone on in the field, read in total."

In fact, leading climate scientists, including NAS report members, hold a range of views concerning whether the SPM inappropriately disguises uncertainty, with Lindzen occupying probably the strongest anti-SPM position. Stanford's Stephen Schneider, in contrast, takes a very different stance. "I would say there is virtually no difference between the [IPCC] summaries and the rest on fundamental matters," he told me.

But once again, this is all a diversion. Climate science's robust consensus position on human-caused global warming hardly relies on the IPCC Summary for Policymakers—a hand-holding document designed for elected representatives and the media—for its validity. In arguing over discrepancies between a report's summary and its actual body (something they also do with respect to the NAS report itself), conservatives once again seek to distract attention from the big picture.

Despite the Right's repeated hyping of uncertainty and attempts to cherry-pick particular passages of reports, it has grown all but impossible to ignore the science community's central conclusion on climate change. Ralph Cicerone, former chancellor of the University of California at

Irvine, recently elected president of the National Academy of Sciences, and chairman of the 2001 NAS climate science panel, may have put it best. Appearing on the *NewsHour with Jim Lehrer* on June 7, 2001, to describe the NAS's findings, Cicerone stated, "We agreed with the previous findings that the weight of the evidence, the weight of the scientific opinion, is that most of that warming of the past twenty years is caused by human activities."

But when they don't like what science has to tell them, those wielding political power often have another option besides running away from the truth: suppression. Perhaps the most antiscientific part of Inhofe's agenda thus far has been his involvement in a legal push, largely centered at the conservative Competitive Enterprise Institute (which received a whopping $465,000 from ExxonMobil in 2003), to stop government "dissemination" of a pathbreaking Clinton-era study that focused on the possible impact of climate change on different regions of the United States—officially entitled *Climate Change Impacts on the United States: The Potential Consequences of Climate Variability and Change*, but often simply referred to as the "National Assessment." Once again, Inhofe's activities in this regard merely put him at the forefront of the broader conservative movement's tactics. The Bush administration, too, has cooperated with the war on the National Assessment.

In 2002, a huge contretemps arose over the U.S. *Climate Action Report*, a State Department document required by the 1992 United Nations Framework Convention on Climate Change, an international treaty. Long in the works, the report was quietly submitted to the U.N. in late May 2002. But it didn't go unnoticed, as the Bush administration appears to have hoped. Instead, Andrew Revkin, of the *New York Times*, wrote a front-page story declaring that the report represented a "stark shift for the Bush administration" because it took the consequences of global warming seriously and accepted the notion of human-caused climate change. That, in turn, prompted Bush to distance himself from the document, which he said had been "put out by the bureaucracy"—yet another case in which the Bush administration ran away from the scientific consensus on climate change.

The *Climate Action Report* did indeed take climate change seriously. In part, that is because it reiterated the conclusions of the 2001 National Academy of Sciences report. And in part, it is also because the *Climate Action Report* relied on the aforementioned National Assessment, a first-ever attempt to inform different U.S. states and regions of the risks they could face if, say, sea levels were to rise or snowpacks to melt. Because the National Assessment makes the possible consequences of climate change very clear to stakeholders in vulnerable communities across the United States, it has been ferociously attacked by those hoping to stop preventive action.

Led by the Competitive Enterprise Institute, conservatives first brought a lawsuit over the National Assessment in late 2000, with Clinton soon to leave office. The suit, with Inhofe as a coplaintiff, named Clinton as a defendant but refused even to admit that he was president, hilariously describing him as "a citizen of the State of New York residing in Washington, D.C., who serves as chairman of the National Science and Technology Council." It alleged various procedural deficiencies in the process by which the National Assessment Synthesis Team, a federal advisory committee, had worked to develop the report, and then, stunningly, demanded a block on the report's production or utilization in either draft or final form—in other words, the suppression of scientific information. Such a request found its apparent justification in claims like those of coplaintiff Jo Ann Emerson, a Republican congresswoman from Missouri, who charged, in characteristic conservative lingo, that "the administration is rushing to release a junk science report in violation of current law to try to lend support to its flawed Kyoto Protocol negotiations."

After Clinton left the White House, the Bush administration settled the lawsuit with an admission that the National Assessment, merely a government report, did not represent official policy. But meanwhile, Emerson had sponsored a brief appropriations bill rider that would become known as the Data Quality Act. As subsequently interpreted by Bush's Office of Management and Budget, the act—a godchild of the Gingrich-era regulatory reform efforts—provides a new means for parties to submit complaints, and potentially lawsuits, over the scientific quality of government information. In his push to improve the use of "sound science" by his committee, Inhofe has welcomed the Data Quality Act.

After the 2002 flap over the *U.S. Climate Action Report*, the Competitive Enterprise Institute (CEI) pursued various "data quality" complaints about the National Assessment (and the *Climate Action Report*'s reliance on it) at various government agencies. Ultimately, in August 2003, CEI launched the very first lawsuit under the act, demanding a halt to the report's dissemination. The lawsuit objects that the National Assessment disseminates data from "demonstrably inaccurate computer models," complaining that the two models relied on by the study have been "proven invalid."

In a February 2002 petition that preceded the CEI lawsuit, another industry-supported group—Jim J. Tozzi's Center for Regulatory Effectiveness, about which much more in the next chapter—had also challenged the computer models used by the National Assessment. Tozzi's group relied on a scientific critique from Patrick Michaels, who argued, "The basic rule of science is that hypotheses must be verified by observed data before they can be regarded as facts. Science that does not do this is 'junk science,' and at minimum is precisely what the [Data Quality Act] is designed to bar from the policymaking process."

This "junk science" claim relies on a by-now familiar misconception. The National Assessment necessarily relied on climate models to project possible climate change consequences for the United States. These models vary and don't necessarily always agree with one another, and none provides a foolproof *prediction*. But introductory text from the National Assessment's "Overview" report provides a fully appropriate statement concerning such models' limitations: "Scenarios are plausible alternative futures—each an example of what might happen under particular assumptions. Scenarios are not specific predictions or forecasts. Rather, scenarios provide a starting point for examining questions about an uncertain future and can help us to visualize alternative futures in concrete and human terms."

Once again, Michaels's criticism attacks a straw man. How exactly could such a "scenario"—a future projection—be "verified by observed data"? "Basically, their argument has been that since we ran two scenarios and they were different, therefore one must be wrong," explains Michael MacCracken, a climate scientist who coordinated the National Assessment process during the Clinton administration. "Well, as scientists what we would say is, 'all scenarios are wrong.' We don't know what's going to happen in a hundred years. The question is not, 'are they wrong?' the question is, 'are they plausible and do they provide insight?'"

James Inhofe didn't participate in the second National Assessment lawsuit directly. But in a flap reported by the *Washington Post*, his committee invited the lead attorney on both cases, CEI's Christopher Horner, to attend a State Department sponsored briefing for congressional staff on international climate change issues and policies—including the controversial *Climate Action Report*. This outraged many Democrats present, who described the appearance as "highly unusual and a breach of congressional protocol," according to the *Post*. Horner says he was "not there in pursuit of information relating to any pending lawsuits."

For his part, Horner speaks with utter contempt for the National Assessment, which he calls an "absolutely absurd document." But bear in mind that the National Academy of Sciences has both relied on, and heaped praise on, the report. The 2001 NAS report on climate change science actually based an entire section on the National Assessment, which, it said, "provides a basis for summarizing the potential consequences of climate change." (For obvious reasons, conservatives don't cite this part of the report.) A National Academy of Science panel charged with reviewing the Bush administration's ten-year strategic climate-change research plan, meanwhile, observed in 2004 that the National Assessment made "important contributions to understanding the possible consequences of climate variability and change," and also commended the process by which it had been created and reviewed.

Clearly, the NAS does not regard the study as "junk science," no matter what conservative think tanks may say. Moreover, though the study certainly wasn't perfect, lawsuits are not generally a means of resolving scientific questions. "I think those lawsuits are an embarrassment. That's just not the way the science community works," says Penn State's Eric Barron, a member of the National Assessment synthesis team.

Despite the National Assessment's value and utility, the suits have had a chilling effect. After the White House settled with CEI the second time, for example, the government website displaying the National Assessment was amended to include a prominent disclaimer saying that the report had not been "subjected" to Data Quality Act guidelines. That is technically true, but very misleading, since the Data Quality Act wasn't even in effect when the report was prepared. Moreover, given the extensive review process to which the National Assessment had been subjected, it would likely have passed any data quality scrutiny with flying colors.

Yet the crusade against the National Assessment provided a pretext for the Bush administration to discard the study and create a taboo against any further use of it in relevant planning and policymaking. The administration's ten year strategic plan on climate change research contains just a single mention of the National Assessment—something for which it was taken to task not once but *twice* by the aforementioned National Academy of Sciences panel that reviewed the plan. "All the way through the climate change science plan, they clearly had distanced themselves, in a not very shy way, from the U.S. National Assessment," says Jerry Mahlman, who sat on the NAS review panel.

For Michael MacCracken, who coordinated the National Assessment process (and for his labors found himself singled out in the notorious 2001 Randy Randol–forwarded memo to the Bush administration), the result has been thoroughly discouraging. MacCracken admits that the massive report may have had its flaws. After all, such a study had never been undertaken before. But he simply doesn't understand why anyone would want to prevent the government from using or citing it. "What you want to do is give people a chance to plan ahead" about climate change, says MacCracken. "Because if you plan ahead you can normally make better decisions than if you don't plan ahead. I mean, information is normally of value."

At the end of a long interview, MacCracken related a story about the National Assessment process that carries a heavy irony given the subsequent lawsuits, and the Bush administration's seeming capitulation to them. When the National Assessment first emerged in 2000, MacCracken sent three copies of the document to every U.S. state governor, along with a letter. After all, he surmised, the potential impact of climate change on their states might be of interest, at least to some governors.

"Darned if I didn't get a response," says MacCracken, laughing. Most governors did not acknowledge receipt of the study, but one did write back: George W. Bush, then the governor of Texas. The letter, which

MacCracken has kept, came from Bush's office with his authentic signature. It was dated December 15, 2000, a time when Bush undoubtedly had other things on his mind (like the Florida recount). "Thank you for your letter and enclosed copies of your assessment about the potential consequences to the United States of a climate change," it read. "I appreciate the work that went into preparing this information."

If you want to know what conservative think tanks have on their collective minds, take a look at whom they invite to their awards dinners. At its 2003 annual dinner—an event that received funding from ExxonMobil—the Annapolis Center for Science-Based Public Policy celebrated Utah Republican congressman Chris Cannon for his initiative to create a "sound science" caucus. The next year, Annapolis followed up with a celebration of none other than James Inhofe—for his support of "rational, science-based thinking and policy-making." The award may make your jaw drop, but it also demonstrates how committed the Right has become to claiming that it goes into political battle with science on its side. So central has this strategy become to conservatives, apparently, that they will present an absolute know-nothing like James Inhofe with a science award.

I tried to attend the Inhofe dinner, but the Annapolis Center said that the event was full. So I don't know precisely what Inhofe said, or how *Fox News's* Tony Snow introduced him. Given the contents of this chapter, though, one can certainly imagine.

Inhofe's "hoax" line about global warming may have historical infamy written all over it, but the senator won't be remembered as the only villain. Contrast Inhofe's claims about a manmade global warming "hoax" with the Bush White House's far more subtle argument, made from the beginning of the administration, that climate science remains "incomplete" and that we need more research to determine policy directions. In effect, Inhofe's outrageous statements conveniently allow the White House to split the difference between his view and the mainstream scientific position (which acknowledges remaining uncertainties but asserts that we are reasonably certain that human activity is heating the globe and know enough to justify action). From a political standpoint you can't blame the Bush administration for using Inhofe as a foil. But this clever positioning has had the ultimate effect of massively misleading Americans about the true findings of modern climate science.

The harm of such behavior may be quite literally incalculable, and it is global in scope. We have already pumped massive amounts of CO_2 into the atmosphere in an ongoing experiment with the planet's thermostat. We cannot take any of that CO_2 back; we can only act to prevent the potential consequences of our actions—a rise in sea level, the extinction of species, more frequent extreme weather events, and much else—from getting any worse.

To be sure, it remains up to policymakers to decide whether the economic costs of such preventive measures outweigh the benefits. But that key question isn't even being properly debated. Instead, climate change has become an issue on which conservatives have elected to fight over science at least as much as over economics, relying on stunning distortions and a shocking disregard for both expertise and the most reputable sources of scientific assessment and analysis.

If this situation is maddening, it is also tragic. There may be no other issue today where a corruption of the necessary relationship between science and political decision-making has more potentially disastrous consequences. And together, James Inhofe and the Bush administration have made that corruption systematic and complete. Not only do they strive to prevent the public from understanding the gravity of the climate situation, but in sowing confusion and uncertainty, they help prevent us from doing anything about it. And this—*this*—is what the Right calls "rational, science-based thinking and policy-making."

Update

Since this chapter was written, global average temperatures have continued to rise. According to NASA, 2005 was the hottest year in more than a century—slightly hotter even than 1998, the record El Niño year that triggered anomalous warmth. Meanwhile, ExxonMobil has continued to support think tanks involved in battling over climate change science and policy. According to the company's 2004 giving report, in that year it gave $270,000 to the Competitive Enterprise Institute, $250,000 to Frontiers of Freedom, and $170,000 to the George C. Marshall Institute, just to name a few examples. These groups have not only continued to put out contrarian takes on the science but have followed the twists and turns of the debate in the media, predictably challenging the notion that global warming might be causing stronger hurricanes in the wake of the disastrous 2005 storm season.

Another trend has continued as well: Repeated scandals over political interference with the science of climate change have emerged from the Bush administration and the Republican Congress. In the past year, several government scientists have denounced either political control over their ability to speak and release accurate information to the public about their findings, or political interference with the content of government scientific reports or press releases that deal with issues of climate.

In mid–2005 Rick Piltz, formerly an employee of the U.S. government's Climate Change Science Program, exposed documents showing that a White House official, once employed by the American Petroleum Institute, had edited the text of program reports. The changes generally had the effect of casting doubt on the increasingly strong scientific conclusion that human beings are causing global warming and its attendant environmental effects. In a complaint memo Piltz went further still, noting that references to the U.S. National Assessment of climate change's impact on the country had been "systematically [deleted]" from program documents. Then in early 2006, NASA's famed climate expert James Hansen went public with charges that a political appointee in the agency's government affairs office had sought to restrict Hansen's communications with the media and public about global warming. Another round of media convulsions over the Bush administration and science ensued, followed by needed reforms at NASA, for which the agency deserves much credit. But as of this writing, problems with scientists' freedom of speech persist at another agency involved in climate research, the Environmental Protection Agency.

Such science scandals in the federal government are hardly surprising when the president himself does not accept the science of global warming. In March 2006, President Bush responded to a reporter's question on climate science by acknowledging that "the globe is warming" but then stated that a "fundamental debate" exists over whether such warming is "manmade or natural." In fact, there remains no substantial scientific debate on this question. Bush had been told as much in 2001, in a report from the National Academy of Sciences (NAS) that he himself had requested. To be sure, the NAS may not be where the president prefers to get his science advice. According to journalist Fred Barnes, who originally broke the news, in early 2005 Bush met with the novelist Michael Crichton, and their conversation reaffirmed Bush's "dissenter" views on global warming. It's hard to imagine a more complete breakdown of the scientific advisory process than this.

But if the Bush administration has rejected much of climate science and interfered with climate scientists, perhaps the most egregious abuse nevertheless came from a member of Congress. And believe it or not, it wasn't from James Inhofe, who delivered a series of contrarian speeches on global warming but was upstaged in the role of climate bad guy by the extraordinary behavior of Joe Barton, chair of the Energy and Commerce Committee of the House of Representatives. In a move that drew denunciations from the National Academy of Sciences, the American Association for the Advancement of Science, and the American Geophysical Union, Barton launched an attack on the original authors of the "hockey stick" study, demanding access to their data and information about their sources of funding arcing back over their entire careers. It was an extraordinarily intimidating and burdensome demand for a prominent politician to direct toward a group of scientists, and it represented yet another attempt to block action on global warming through scientific disputation. The protests of the scientific organizations were soon echoed by a number of moderate Republicans, including Senator John McCain and Science Committee chairman Sherwood Boehlert.

In 2005, like James Inhofe before him, Joe Barton won the Annapolis Center for Science-Based Public Policy's "rational, science-based thinking and policy-making" award. Meanwhile, in 2006, the National Academy of Sciences put together an expert panel to assess the current state of understanding on the historic temperature record—in other words, to straighten out the debate over whether or not the hockey stick's handle ought to be more curved. Even as members of Congress put on show hearings and target individual teams of scientists, the scientific process continues—as does the warming of our planet.

CHAPTER 8

WINE, JAZZ, AND
"DATA QUALITY"

O F ALL THE SCIENTISTS and other characters who populate this book, Jim J. Tozzi has at least this distinction: He is the only one with whom I have done a "Dirty Girl Scout" shot.* That's the thing about Tozzi; he loves partying with the enemy. On numerous occasions, I've spotted this sixty-seven-year-old creator of the despised Data Quality Act hanging around at events sponsored by liberal-leaning groups like the Center for American Progress and the Center for Science in the Public Interest, busy making connections and cracking jokes. When the speaker wraps up, Tozzi always rises to ask a question. His word-slurring musician's jive can be disarming; certainly, it is not what you would expect from a well-heeled lobbyist wearing "JJT"-monogrammed shirt cuffs.

A former Office of Management and Budget (OMB) official and corporate consultant who now runs the industry-funded Center for Regulatory Effectiveness (CRE), Tozzi clearly relishes his notoriety on the left. But in person, his leathery smoker's face grinning madly, he manages to get under the skin of his liberal critics. Rena Steinzor, a member scholar at the Center for Progressive Reform, calls Tozzi her "favorite bad guy."

---

*Vodka, Kahlua, Bailey's Irish Cream, and white crème de menthe (although the shot we did used green crème de menthe).

Sidney Shapiro, another Center for Progressive Reform scholar, recalled to me how once, on a stalled conference call, Tozzi entertained everyone with a tune on his office piano. "It's hard not to like the guy; he's just utterly charming," said Shapiro.

He's also a little odd. Tozzi describes himself as a regulatory policy "nerd," but his self-confessed voyeuristic streak seems closer to something out of *Revenge of the Nerds*. After seeing his piano and—if you're lucky—his cache of alcohol (for a while, Tozzi sold wine under his own label, Villa Tozzi), a visit to Tozzi's office culminates in a look through his brass telescope, poised in a seventh-story window overlooking Washington, D.C.'s Dupont Circle and plainly visible from the ground. Apparently the telescope—which Tozzi calls a "landmark"—comes in particularly handy after dark. "That can get you in a lot of trouble," he told me. "I'm a dirty old man. I love it."

Extracurricular uses aside, Tozzi's telescope provides an apt metaphor for the brand of lobbying and pro-corporate activism that has become his specialty. Out of a broad landscape of scientific information used by government agencies to safeguard public health and the environment, he has helped industry groups focus in on particular studies to challenge or dispute. Tozzi says that his overarching goal is to "regulate the regulators," preventing agencies from arbitrarily releasing or relying on bad information that can needlessly hurt companies' profits. But others consider the strategy little more than an extension of the Gingrich-era push to achieve "paralysis by analysis" under the guise of regulatory reform.

During George W. Bush's administration, Tozzi has managed to change the very rules of the regulatory game itself. Granted, he has had some help. In 2000, Tozzi drafted two sentences of legalese that Rep. Jo Ann Emerson then tucked into a massive appropriations bill signed into law late in the Clinton administration. As subsequently interpreted by the Bush administration—and, as we have seen, wielded against the National Assessment on climate change impacts—the so-called Data Quality Act creates an unprecedented and cumbersome process by which government agencies must field complaints over the data, studies, and reports they release to the public. It is a science abuser's dream come true.

On its face, the act merely seeks to ensure the "quality, objectivity, utility, and integrity" of government information. In practice, it saddles agencies with a new workload while empowering businesses to challenge not just government regulations—something they could do anyway—but scientific information that could potentially lead to regulation somewhere down the road. "Anyone who is involved in the regulatory process knows it begins ten years or so before you ever see a rule," says William Kovacs, a Chamber of Commerce vice president who says he has met with Tozzi "a hundred times" to plot strategy. The Data Quality Act "allows you to begin inputting, and access the process [from] the very beginning."

As we saw from the assault on the National Assessment, the act may also enable companies to sue agencies that reject their data quality complaints, thereby dragging individual studies into the courtroom (ironic given that conservatives generally rail against frivolous litigation). Furthermore, the Bush administration has used the act's thin language as an excuse for implementing a government-wide "peer review" system for scientific information, an industry-friendly superstructure that once again calls to mind the Gingrich Republicans' regulatory reform agenda.

Given industry groups' determination to use arguments over science to block regulations, the Data Quality Act clearly gives them an edge. Sure enough, mid-2004 analyses by the *Washington Post* and the liberal-leaning OMB Watch showed that the act, in effect for almost two years by then, had been overwhelmingly used by industry. This was no big shock: Tozzi first came up with "data quality" while helping Philip Morris with strategies to block regulations on secondhand smoke. No wonder conservatives call the Data Quality Act a "sound science" law.

The Data Quality Act represents the culmination of Jim Tozzi's decades-long career as a master of the obscure intricacies of the government regulatory process, a man once described, in a 1998 Philip Morris memo on whether to retain his lobbying services, as someone whose "contacts at OMB are second to none." That career is worth surveying in some detail, since it illustrates a revealing progression in which industry groups and their allies have increasingly turned to disputing scientific information itself to generate regulatory paralysis, seeking to upend the process before regulators can even get their shoes on. This strategy, of course, has brought with it an epidemic of science abuse.

An economics Ph.D. who'd hung around New Orleans until he realized he'd never make it playing jazz cornet, Tozzi began his Washington days in 1964 working for the secretary of the army's office. Charged with overseeing the Army Corps of Engineers, he gradually expanded his scope to include reviewing not only Corps projects but Corps *regulations* (or "rules"), something largely unheard of at the time. Having established a reputation as "that nerd over there in the Pentagon that really likes to review rules," Tozzi then moved to OMB in the Nixon White House to scrutinize regulations churned out by the newly created Environmental Protection Agency. There he became wildly controversial for second-guessing attempts to enforce the agency's mandate. Environmentalists would ask, "Christ, who's running EPA—Tozzi?" he recalls with a cackle.

Tozzi stayed at OMB through several administrations, biding his time and trying to further establish a precedent for government-wide review of agency regulations. Signed into law late in Carter's term, the Paperwork Reduction Act established a new branch of OMB: the Office of Information and Regulatory Affairs, or OIRA. Reagan then swept in and began to implement his pro-industry deregulatory agenda. He appointed Tozzi deputy administrator of the new office while signing an executive order that centralized review of government regulations at OMB.

This controversial step represented a landmark for industry groups. Suddenly, a White House political branch would take an active role in curbing and controlling the actions taken by federal agencies. Soon Tozzi's office became known as the "black hole" of the regulatory process for sucking in rules proposed by agencies and never letting them see light again. Tozzi himself garnered the nickname "Stealth." "I don't want to leave fingerprints," he once memorably told the *Washington Post*.

But the establishment of regulatory review at OMB would not ultimately satisfy industry interests, who wanted still more bites at the apple. Corporations—including tobacco—would increasingly clamor for a way of challenging the scientific studies that can set the regulatory process rolling to begin with. Moving to the private sector in the middle of the Reagan years and launching a consulting shop called Multinational Business

Services, Tozzi became instrumental in this push as well. By the 1990s, he had become a key player in the tobacco industry's campaign to battle scientific information suggesting that cigarettes endanger not only smokers themselves, but innocent bystanders as well.

We have already seen how tobacco interests supported the Advancement of Sound Science Coalition as part of a broader push against secondhand smoke regulation. A related strategy involved lobbying for the adoption of scientific standards for "good epidemiological practices" that would have made it difficult or impossible to prove the dangers of secondhand smoke. As one 1994 Philip Morris memo explained, "Within the sound science concept, we are uniquely interested in criteria for good epidemiological practices (GEP). And, narrowing the scope somewhat further, we must ensure that GEP apply specifically to epidemiology applied in ETS [environmental tobacco smoke] research." In some cases, the proposed epidemiology standards included the suggestion that unless studies demonstrated a doubling of risk from a particular substance (i.e., an increase of one hundred percent or more), the finding was questionable. Tobacco interests were trying to define away secondhand smoke risks by changing the rules of science.

Epidemiologists, however, rejected the suggestion that they should view lower relative risks skeptically. "Many studies demonstrate that effects of this magnitude can be studied, and on a biological basis we anticipate that many effects relevant to public health will fall into this range," wrote Johns Hopkins University epidemiologists Jonathan M. Samet and Thomas A. Burke in a 2001 *American Journal of Public Health* editorial denouncing the tobacco industry's tactics. Indeed, notes tobacco researcher Stanton Glantz, of the University of California, San Francisco, tobacco's proposed standards would bar regulation of nearly any environmental toxin, including secondhand smoke.

Tozzi played an instrumental role in the tobacco industry's push for new epidemiology standards. In the mid-1990s, while receiving donations from Philip Morris, Tozzi's nonprofit organization Federal Focus convened scientists to draft "Principles for Evaluating Epidemiologic Data in Regulatory Risk Assessment" (which came to be called the "London Principles"). The London Principles did not suggest that reliable epidemiological studies should demonstrate an increased risk of 100 percent or more from a substance or pollutant. But a set of Philip Morris "Criteria for Epidemiology," apparently agreed on at a meeting attended by Tozzi, argued that risks below that level "fall into the realm of 'weak association.'"

Meanwhile, under the auspices of Multinational Business Services, Tozzi worked to support "legislative mandates on epidemiological standards" and to increase "debate on ETS risk assessment within EPA." A signed 1994 Philip Morris contract set a cap on his payments for such work at $610,000 for that year.

And not only did tobacco interests seek to influence the rules of epidemiology. Cigarette companies and other industry groups also sought other devices that would empower them to slow down the regulatory process by disputing the scientific basis for government action. Once again, they drew on Tozzi's services.

In 1998, in a dress rehearsal for the Data Quality Act, Tozzi worked on the so-called Shelby amendment—named after its official "author," Alabama Republican senator Richard Shelby—for Philip Morris. Also a brief insert to an appropriations bill, and equally loathed by the scientific community, the one-sentence amendment allows for the use of the Freedom of Information Act to obtain "all data produced" by any publicly funded scientific study.

Though subsequently limited in scope by a wary Clinton administration, the Shelby amendment helps companies conduct their own audits of studies they don't like, as many firms, including tobacco, have often sought to do. In the process, they can reanalyze the data and put a different spin on it. In short, Shelby provided yet another tool for industry's scientific battles against regulation. Tozzi takes part of the credit for conferring this new power (though he clearly was not the only proponent). "We proposed the Shelby; we were the first ones on the Shelby amendment," he says proudly.

The Data Quality Act has been called the "daughter" of Shelby. Sure enough, Philip Morris documents show that Tozzi circulated a "data quality" proposal to the company and received comments in response. "Yeah, there was a dozen or twenty companies that saw our papers. Yeah.

And they were one of them. Yes. But they did not come up to me and say, 'pass Data Quality,'" he says. While Tozzi agrees there was "definitely" a connection between the tobacco industry and the Data Quality Act, he insists that it was "not a cause and effect." But he adds that if tobacco companies had "continued challenging stuff, data quality would have been a big help to them."

With the Data Quality Act, industry had finally found a means of disputing agency science in the earliest stages of the regulatory process. At the merest public mention that a government agency might be looking at a particular study and wondering whether it could compel regulatory action, industry interests could file a complaint challenging the agency's "dissemination" of the information. "Industry tends to think there's some magic bullet somewhere that's going to protect them from regulation," says Gary Bass, executive director of OMB Watch and a kind of counterpart to Tozzi on the Left. "If they can just get there earlier. And now it's in the science, it's in the data."

In fact, both the Shelby amendment and the Data Quality Act represent successful under-the-radar attempts to pass bite-sized pieces of legislation highly reminiscent of the failed Gingrich-era "regulatory reform" bill. That is not by accident. As it gradually became clear that no comprehensive "regulatory reform" law would pass, "Jim and I split up the issues," recalls the Chamber of Commerce's William Kovacs. The Data Quality Act represents the crowning achievement of this piecemeal approach. "In the end," says Kovacs, "what we're going to get is far more than we could have ever gotten by having a comprehensive regulatory reform law passed."

As soon as the Data Quality Act came into effect in October 2002, corporate interests took it for a test drive. Tozzi's Center for Regulatory Effectiveness teamed up with the Kansas Corn Growers Association and a group called the Triazine Network to challenge an EPA risk assessment document that had discussed evidence suggesting that atrazine, a herbicide in widespread use in U.S. cornfields and on other crops, causes gonadal abnormalities in male frogs. The groups' petition objected that "there are no validated test methods for determining whether atrazine causes endocrine effects." It also alleged that findings of endocrine effects from the chemical had not been reproduced in other studies. The EPA's claims about atrazine's hormonal interferences were "not based on sound science," Tozzi and company concluded.

Ostensibly filed on behalf of agricultural groups that use atrazine on their crops, the petition omitted a rather relevant detail. In 2002, the year the document appeared, Jim Tozzi's Multinational Business Services received $20,000 in lobbying fees from Syngenta, the Swiss-based corporation that manufactures and sells atrazine. The Triazine Network, a group supporting the use of atrazine-like chemicals, has also received some financial support from Syngenta in the past for events to inform members about the EPA regulatory process. The petition itself, however, did not disclose any connection to Syngenta. Instead, Tozzi's group claimed "affected person" status with respect to the EPA's atrazine deliberations because of a "long-standing and active interest in ensuring the quality of information disseminated by federal agencies."

When I asked Tozzi about the support from Syngenta, he agreed that he was registered as a lobbyist "at some point" but said, "You can't make a necessary connection" to any work on atrazine. Syngenta spokeswoman Sherry Ford added that "while it's true that the Center for Regulatory Effectiveness has filed petitions concerning atrazine, Syngenta does not provide funding to CRE specifically for this purpose." Rather, she said, Syngenta supports CRE "because of the organization's extensive knowledge of the federal government."

The atrazine story provides a perfect example of how industry-friendly groups meddle in science and how the Data Quality Act—just one tool among many, but definitely a powerful one—facilitates the process. At a time when the EPA was weighing tighter restrictions on atrazine, the act provided Tozzi and cohorts with a highly convenient device for challenging the cutting-edge work of a scientist who had found disturbing effects from the chemical at extremely low doses.

In 1997, EcoRisk, Inc., a consulting company working for the corporation that would later become Syngenta, hired a University of California, Berkeley, developmental endocrinologist named Tyrone Hayes to study atrazine. A cheerful but outspoken scientist who recently received the distinction of becoming Berkeley's youngest full professor, Hayes had a long history of studying frogs. But despite being well matched for the job, Hayes quit working for EcoRisk in late 2000, claiming that the company tried to stall his research once he began detecting troubling effects from atrazine in his animals. Hayes has subsequently charged that the company tried to buy his silence (a charge EcoRisk denies). Syngenta's Sherry Ford, for her part, provides a very different gloss on the fallout: "The [EcoRisk] atrazine science panel was not confident of the quality of the Hayes study and encouraged him to repeat it. Hayes disagreed with the panel's interpretations of his study. He instead chose to leave the panel and ultimately sought publication on his own rather than repeat the study as part of the panel."

Hayes not only sought publication, he succeeded. The big article appeared in the April 16, 2002, issue of the *Proceedings of the National Academy of Sciences*, a very highly regarded journal. In it, Hayes and Berkeley colleagues sought to test the hypothesis that atrazine may have contributed to the decline of amphibian populations worldwide by disrupting the animals' endocrine systems, and thus interfering with reproduction. Sure enough, they reported that after exposing tadpoles to different levels of atrazine in water in the laboratory, "gonadal abnormalities," such as multiple gonads or hermaphroditism, developed in up to 20 percent of the animals at all but the tiniest concentrations. Male frogs exposed to atrazine also had much smaller larynges (a demasculinizing trait that could interfere with mating calls). These effects probably occur, according to Hayes, because the weed-killing chemical induces the expression of a gene whose action creates estrogen hormones. In an interview, Hayes observed that atrazine has a "chemical castration effect."

Previously, atrazine had been deemed safe at low concentrations. But Hayes's group found negative effects at levels as low as one-tenth of a part per billion, a level far lower than the Environmental Protection Agency's tolerated limit. The study noted that "in total, atrazine exposure at these levels has been replicated 51 times by our laboratory with similar results," an astonishing number of replications, although by only one research group. But Hayes explains that he wanted to do a large number of tests because it was "surprising" to find an effect at such a low dose.

Hayes's work suggests that atrazine's use may have contributed to a marked decline in amphibian populations worldwide. Moreover, because the chemical has been detected in water supplies, his research also raises the concern that troubling effects might occur in humans exposed to high levels of atrazine, such as farm workers. "People who actually apply atrazine have levels that are a hundred times what we're looking at in the laboratory," Hayes says.

Hayes had launched a frontal attack on an extremely popular and widely used chemical, reminiscent of Rachel Carson's broadside against DDT. As in Carson's case, the pro-industry backlash came quickly, and it got very nasty. For his labors, Hayes was labeled a "junk scientist," "determined to scare the public about atrazine," guilty of producing research "more akin to a Brothers Grimm fairy tale than science," and someone backed by "mindless atrazine-hating, eco-extremist supporters," all by former Advancement of Sound Science Coalition executive director Steven Milloy in a series of *FoxNews.com* columns. While not directly accusing Hayes of research fraud, one of Milloy's columns bore the title "Freaky-Frog Fraud," clearly insinuating as much. Yet Milloy never even bothered to contact Hayes for comment before attacking him, according to Hayes. Syngenta's atrazine website provides a link to one of Milloy's columns. (Milloy did not respond to requests for comment for this book.)

Meanwhile, Hayes kept publishing more research, attempting to go beyond laboratory data to gauge atrazine's effects in the wild. After his new work emerged in *Nature* and *Environmental Health Perspectives*, an even bigger (and cannier) fish got into the game on the side of industry. In late 2002, Jim Tozzi, who once employed Milloy at Multinational Business Services, weighed in with his data quality challenge, the first of several he would file relating to atrazine. The timing couldn't have been better, coming as the EPA neared the tail end of determining whether to regulate atrazine more stringently. The Natural Resources Defense Council, a leading environmental group, had petitioned the EPA for a ban on atrazine partly based on Hayes's research.

Tozzi's petition repeated the substance of one of Milloy's charges, claiming that Hayes's former colleagues at EcoRisk had been unable to reproduce his test results. These scientists, however, may have a conflict of interest, given that their work was conducted "on behalf of" Syngenta, according to an EcoRisk press release. Moreover, Hayes has scathingly critiqued the Syngenta-supported work in an article in the journal *Bioscience*. Analyzing a selection of sixteen studies on the effects of atrazine on frog gonads, Hayes detected a strong correlation between industry sponsorship and negative results. "100 percent of the negative studies were funded by Syngenta," Hayes reported. (Asked to comment on the study, Syngenta's Ford countered that "other studies by independent scientists find no effect of atrazine on frogs," citing two examples.)

And if the evidence allegedly exonerating atrazine seems questionable, Hayes also noted in *Bioscience* that several independent research teams have supported the notion that atrazine affects the gonads of frogs. Indeed, the European Union has moved to phase out use of the herbicide by 2007.

Nevertheless, the U.S. EPA concluded that hormone disruption did not constitute a good reason to restrict atrazine's use because of the lack of an accepted test for measuring endocrine effects. "Basically, what they're saying is, you can't use new science to make new policy," says Hayes incredulously. Tozzi's petition may not have been the only cause of this determination, but it probably had an effect, notes Sean Moulton, of OMB Watch, which has monitored the Data Quality Act closely: "You put the sequence together and I think it's common sense to say it had influence." Syngenta's Ford disagrees that Tozzi's petition changed the outcome of the EPA's atrazine deliberations, but she says that it "was properly used to characterize the uncertainty surrounding atrazine's potential effect on frogs." In short, the Data Quality Act had served its role as a device to raise doubts about scientific findings with the potential to trigger government regulation.

Meanwhile, in mid-2004, Tozzi filed his second and third atrazine-related Data Quality challenges. Once again cosigned by the Kansas Corn Growers Association (as well as other sponsors), both targeted the National Toxicology Program at the National Institutes of Health, which had begun the process of reviewing whether to list atrazine as a "known" or "reasonably anticipated" human carcinogen. Neither petition mentioned Syngenta in any way.

The atrazine challenges represent just one of many uses of the Data Quality Act. In addition to atrazine and the previously chronicled attack on the National Assessment of climate-change impacts, the act has also been employed in a wide variety of lesser-known industry-friendly campaigns: by a chemical company challenging EPA information on barium, its product; by a law firm that has reportedly represented companies involved in asbestos lawsuits, objecting to an EPA document on asbestos risks from work on vehicle brakes; and by paint manufacturers questioning the basis for regulations on emissions from their products, to name just a few examples.

All three of the above-cited data quality challenges targeted the Environmental Protection Agency. But other agencies have also seen an influx of complaints. For instance, logging interests challenged several U.S. Forest Service studies and documents relating to habitat protection for the northern goshawk, submitting a massive 281-page petition outlining their objections.

Some liberals and environmental groups—including the Environmental Working Group, which famously exposed the Frank Luntz "sound science" memo—have also attempted to employ the Data Quality Act. And then there is Americans for Free Access, which used the act to dispute government claims that marijuana has no medical uses. "My musician friends will finally appreciate my work," cracked Tozzi in an e-mail alerting me to the pot petition.

But unsurprisingly, industry uses of the Data Quality Act greatly outnumber public interest–oriented ones. In any all-out war over regulatory science, corporations will ultimately have the financial edge. That goes a long way toward explaining why they love the Data Quality Act so much, and why the law represents a legislative abuse of science. In effect, it shifts the regulatory playing field still further in industry's favor while innocently claiming to improve "science."

Moreover, while the atrazine and National Assessment stories demonstrate how the Data Quality Act allows pro-industry groups to disrupt the regulatory process through scientific disputation, they also describe simpler and more orthodox uses of the act. Since Tozzi slipped this lilliputian piece of legislation through Congress, however, both he and the Bush administration have sought to dramatically expand its scope, setting the stage for a far more diverse array of uses—and, once again, seeking to provide self-interested industry groups with still more bites at the regulatory apple than they have already.

Tozzi has consciously sought to increase the Data Quality Act's potency in a variety of ways. He has drafted sample legislation for states. He has written letters both to the World Health Organization and to universities warning that if they want their science to influence the U.S. government, it had better meet "data quality" standards. His group has even tried using the act preemptively to rule out government use of *comments* submitted to the EPA by the Natural Resources Defense Council, a foray that led to an outraged counterblast from the liberal-leaning Center for Progressive Reform, which called Tozzi's group's letter "an attempt to quell scientific debate."

Though subtle and arcane, these attempts at Data Quality Act aggrandizement could have substantial long-term effects. More troubling still is industry's strategy to ensure a right to judicial review under the act, allowing for federal lawsuits in cases where agencies reject data quality complaints. Slowly, Tozzi and his allies are laying the groundwork for a broader assault on the regulatory state, one in which industry lawyers as well as industry scientists can get in on the action.

The lawsuit over the National Assessment, discussed in the previous chapters, did not result in a legal ruling; instead, the parties settled out of court. But in early 2004, the Chamber of Commerce teamed up with the Salt Institute, an industry organization, in a data quality lawsuit challenging a National Institutes of Health study showing that reduced salt intake lowers blood pressure—a strategic test case to establish judicial review under the act. In a clear setback for Data Quality Act fans, in November 2004 the judge in the case rejected the complaint and denied a right to judicial review.

In an early 2006 ruling, the case fared no better on appeal. Still it will only be a matter of time before some judge writes a precedent-setting opinion in a Data Quality Act lawsuit. If that happened and higher courts concurred, a brand new body of law would emerge, consisting largely of corporate lawsuits against scientific analyses.

In fact, drawing on the U.S. Supreme Court's controversial 1993 decision in *Daubert v. Merrell Dow Pharmaceuticals, Inc.*—which tasked federal judges with determining the reliability of expert scientific testimony—conservative legal scholars now want judges to play a similar role in assessing the validity of government regulatory science. Some see the Data Quality Act as a potential springboard into this brave new world of science litigation, one in which industry lawyers would have a field day getting judges to invalidate study after study based on their alleged "flaws."

This narrow debunkers' approach to scientific information has grown so prevalent among industry defenders that University of Texas law professor Thomas McGarity has come up with a term to describe it: "corpuscular." In McGarity's analysis, companies love to nitpick individual studies they don't like (the corpuscles). Then they charge that the results of those studies can no longer be relied on. The effect is to raise doubts about specific analyses in a courtroom-style proceeding, while disregarding the cumulative weight of available scientific evidence.

The atrazine example epitomizes the "corpuscular" approach: Tyrone Hayes's work came under ferocious attack even though other scientific groups had reached similar conclusions. "We were not the first to show the effects of atrazine on gonadal development . . . in the laboratory, nor were we the first to show the effects of atrazine on gonadal development in wild amphibians," Hayes has noted. You would hardly know it from all the attacks on him.

But the biggest Data Quality Act aggrandizer hasn't been Jim Tozzi or the Chamber of Commerce. Instead, we must reserve that distinction for the controversial John Graham, who for five years headed Tozzi's old haunt—the OMB Office of Information and Regulatory Affairs—in the Bush administration.

Formerly head of the Harvard Center for Risk Analysis, a group funded in part by industry, Graham also served on the policy advisory board of the Advancement of Sound Science Coalition. A strong backer of "regulatory reform," he once even commented, in reference to desires on the part of some right-wingers to abolish the Environmental Protection Agency outright, "Maybe we're going to get to that some day, but I don't think that's a helpful way to talk about these issues." Decide for yourself whether Graham's statement indicates outright extremism, or merely moderation in the company of extremists. But his nomination proved such a flashpoint that thirty-seven senators ultimately voted against his confirmation.

Without Graham, Tozzi's data quality efforts might not have gone very far. Instead, Graham quickly seized on the Data Quality Act—which had sailed through Congress with little debate—and instructed federal agencies to draw up their own data quality guidelines by October 2002, when the law would go into effect. But that wasn't enough, apparently. Claiming that the Data Quality Act gave his office a newfound role in improving the quality of government science, in September 2003 Graham proposed using the act's thin language to justify an unprecedented government-wide "peer review" system for agency research. Once again, the step recalled the Gingrich Republicans' regulatory reform agenda.

Without a doubt, peer review—the rigorous vetting of unpublished research by independent, qualified experts—represents a cornerstone of modern science. Academics' reputations hinge on the ability to get their work through peer review and into leading journals; university presses employ peer review to decide which books to publish; and federal agencies like the National Institutes of Health use peer review to weigh the merits of applications for federal research grants. When members of Congress make an end run around this vetting process and pump R & D cash directly into their home districts, they are widely disparaged for supporting a particularly odious and antiscientific version of pork-barrel politics.

Yet peer review varies widely in form and purpose, and suits certain situations better than others. Many of the studies conducted to determine the appropriateness of government regulatory action cannot proceed under the same circumstances that govern curiosity-driven university-based research intended for publication in a scientific journal. Time and resource constraints—as well as the difficulties of conducting science at the edges of what's known or with the goal of projecting possible future risks—often mean that policy-oriented scientific research is of a different nature, and must be evaluated based on different criteria. That's not to say that policy-oriented research is less rigorous or credible than traditional scientific studies, but rather that it is conducted with different goals and constraints in mind.

Similarly, while academic peer review chiefly affects individual reputations, the peer review of government science implicates the economic fortunes of major corporations, especially big polluters who want environmental and health rules softened. Such companies have a strong incentive to influence the scientific determinations made by the agencies that regulate them, and plenty of money available for lobbying and litigation. Any attempt to impose a new regulatory peer review superstructure thus has major economic and political implications, and serious consequences may result from hampering the ability of federal agencies to make science-based (but not *only* science-based) regulatory decisions.

The burdensome set of suggested protocols contained in Graham's initial proposal provoked a huge hue and cry from the nation's science community, which saw little in the plan resembling standard academic peer review. Scientific heavyweights like the American Public Health Association, the Association of American Medical Colleges, and the Federation of American Societies for Experimental Biology issued scathing critiques of the idea (the latter two jointly), as did a range of other organizations and experts. The hallowed American Association for the Advancement of Science—which publishes the preeminent peer-reviewed journal *Science*—also announced worries. Science watchdog Henry Waxman and an accompanying group of Democratic members of Congress even dubbed the proposal a "wolf in sheep's clothing."

Far beyond merely ensuring good "science," Graham's "peer review" system would have dramatically slowed down the regulatory process. Under it, agencies would have had to ensure that all "significant regulatory information" intended for public release be peer reviewed. And for "especially significant regulatory information"—defined as data "with a possible impact of more than $100 million in any year"—additional hurdles would exist. In a laborious-sounding "formal" process, carefully chosen reviewers would have had to sift through public comments before submitting final peer review reports, to which the regulating agency would, in turn, have to respond. The head of OIRA—Graham or his successor—would have the power to determine which information counts as "especially significant"; in such cases, agencies would have to consult with OIRA, a political office, to ensure that they were conducting adequate peer review.

Not only did this represent an overbroad and expensive plan likely to stymie agency efforts to protect public health and the environment, but most pointedly, critics charged that Graham's proposal would block academic scientists from serving as reviewers if they had obtained or were seeking "substantial" research funding from the government agency in question—a condition likely to exclude leading academic experts in the field!—yet showed little concern about the participation of industry scientists (who are, if anything, far more conflicted). In short, the proposal seemed to flagrantly privilege industry science over university-based or publicly funded science—just like the Gingrich-era regulatory reform bills that preceded it.

Later, Graham released a revised "peer review" bulletin that responded to many of the scientists' critiques, and boasted that it had been "substantially revised" in reaction to the volley of complaints. Admittedly a far more moderate document, the new bulletin emphasized that federal agencies would have discretion in determining the type of peer review they wished to conduct in many cases, and no longer barred anyone who had ever received a government grant from participating. It also provided for exemptions from peer review in cases in which agencies might need to take emergency action.

Yet criticisms leveled against the original peer review proposal by the American Association for the Advancement of Science still stand. Most importantly, no convincing explanation has been provided for why this "peer review" system is needed in the first place. While no government-wide standard for peer review existed in the past, that may have been a good thing: After all, different agencies have different needs, just as different scientific disciplines employ different methodologies and standards of proof. Moreover, as legal scholar Wendy Wagner, of the University of Texas, pointed out in a recent article entitled "The 'Bad Science' Fiction," there is little real evidence to support the notion that government agencies churn out "junk science" or that their existing peer review protocols are inadequate.

Furthermore, the concern about onerous and unnecessary intrusions into the regulatory science process remains warranted, especially when it comes to OMB's standards for peer review of "highly influential scientific assessments" (studies whose "dissemination" could have a financial impact of more than $500 million in a year). The process outlined in the new bulletin for vetting such information seems quite burdensome, encompassing not only public participation, but also the preparation of a peer review report that must be made public (along with the identities of the reviewers) and a written response to that report from the agency (which must also be made public). Such procedures will only further ossify an already sluggish regulatory process. And as "peer review" critic Sidney Shapiro, of the Wake Forest University School of Law, observes, they are required even for "routine information" that is not "complex, controversial, or novel."

All in all, then, the basic criticism of the peer review proposal still seems to hold. It is an industry-friendly solution in search of a problem, much like the Data Quality Act that has been used to justify it. Moreover, the proposal brings conservatives even closer to achieving, through non-legislative means, the objectives that appeared dead with the failure of "regulatory reform" in 1995. The Dole regulatory reform bill, too, would have required new levels of "peer review" for government science. "I think if you take a step back, and look at initiatives that have been imposed by the Bush administration, one could say that they've achieved through the back door that which Dole could not achieve through the front door," says David Vladeck, an associate professor of law at Georgetown University who has specialized in the regulatory arena.

Nevertheless, in December 2004, shortly after George W. Bush's reelection, Graham's office finalized the peer review plan with only "minor revisions." Its provisions for "highly influential scientific assessments" took effect on June 16, 2005. The media hardly even noticed.

You have to admire conservatives for what they have managed to make out of the brief, covertly created Data Quality Act. Unable to obtain passage of a regulatory reform bill that would have slowed government agencies and provided a new means of challenging agency science, conservatives instead found a flamboyant lobbyist (Tozzi), a legislator (Jo Ann Emerson), and a White House administrator (Graham) to do the job.

In regard to his success in obtaining passage of the Data Quality Act, Tozzi has claimed (with characteristic flair) that "sometimes you get the monkey, and sometimes the monkey gets you." He almost makes it sound like dumb luck. In fact, Tozzi's statement downplays the strategic achievement that the act embodies.

As past chapters have shown, industry groups have continually felt it necessary to battle against scientific information that could potentially bolster the case for regulation. It is hard to think of a better solution to this problem than the Data Quality Act, which—to extend the military metaphor—greatly improves industry's battlefield position by drawing both independent scientists and government agencies within the range of industry's snipers.

But none of this has anything to do with science. Rather, the measures taken under the Data Quality Act—including the related peer review initiative—simply give industry groups, who already have a right to challenge final regulations in court, multiple additional opportunities to attack the regulatory process in its earlier stages.

In the three decades since the regulatory revolution that created the Environmental Protection Agency and other fixtures of the federal bureaucracy, industry has become increasingly adept both at weighing down the rulemaking process with years of preliminaries and at challenging regulations once promulgated. Now, each regulatory decision seems to descend into a "science" fight, not because we don't have enough qualified scientists in federal agencies, or because they are doing a poor job, but rather because those seeking to avoid regulation constantly try to raise the burden of proof required for action. Such is the charade perpetrated by "sound science," "data quality," and "peer review" proponents, and it is fair to ask what the next step will be to improve the role of "science" in the regulatory process. Peer review of the peer reviewers, perhaps?

Update

On the "data quality" front, the most important news comes in the form of an early 2006 decision from the U.S. Court of Appeals for the Fourth Circuit, denying that Jim Tozzi's creation enables federal lawsuits over the validity of government scientific research. The act "creates no legal rights in any third parties," ruled the court. "Instead, it orders the Office of Management and Budget (OMB) to draft guidelines concerning information quality and specifies what those guidelines should contain."

This strong and unambiguous precedent represents a serious setback for those, like Tozzi and the U.S. Chamber of Commerce, who had sought to grow the Data Quality Act into an even more powerful device to challenge the use of inconvenient science in the government regulatory process. But "data quality" proponents aren't giving up. Tozzi thinks a different case could achieve a more positive result, perhaps a potential lawsuit by Americans for Safe Access, a group trying to use the Data Quality Act to compel the government to cease disseminating what they claim is misinformation about medical marijuana. Meanwhile, "data quality" fans in Congress, like Rep. Candice Miller (R-MI), have suggested they might try amending the act to include an explicit right of judicial review, an action that could present significant legislative hurdles but that would certainly render the current legal issues moot and finally give industry interests what they want.

At the same time, recent events have shown that in at least a few cases, the Data Quality Act can be used constructively. Public Employees for Environmental Responsibility successfully invoked the act to correct questionable Fish and Wildlife Service information about the endangered Florida panther. And two sexual health groups have used the act to challenge scientific misinformation promulgated by federally funded abstinence education programs (for more on the problems with these programs see Chapter 13). Especially when wielded against the Bush administration, it's no surprise that the Data Quality Act might occasionally help correct real misinformation. Nevertheless, it remains a tool that inherently favors and enables the well-known scientific warfare tactics of private industry.

Attempted misuses of the act remain frequent. Its potential for mischief is perhaps best epitomized by a sweeping petition from two conservative groups—the Washington Legal Foundation and the American Council on Science and Health—seeking to prevent the Environmental Protection Agency from classifying chemicals as "likely" human carcinogens based solely or centrally on rodent tests (a practice the groups label "junk science"). "Change your evil ways and start adhering to sound science," wrote the American Council on Science and Health's Gilbert Ross in an online commentary addressed to EPA.

Yet in an imperfect world like our own, it is perfectly reasonable for scientists to use evidence from animal studies to infer risks to humans. After all, scientists can't test chemicals directly on people to look for cancerous effects—that's why they turn to animal studies in the first place. There are many uncertainties associated with such studies, but that hardly makes them equivalent to "junk science." Still, the American Council on Science and Health petition perfectly demonstrates how defenders of industry hope to use the Data Quality Act: to attack good research that can cost companies money. In truth, the petition doesn't challenge bad science at all; rather, it attacks what EPA calls its "public health-protective" policy, under which the agency would rather use imperfect science to prevent risks to the public than take no action at all.

Finally, John D. Graham stepped down from his post as administrator of OMB's Office of Information and Regulatory Affairs in early 2006. After five years on the job, he had dramatically transformed the government regulatory process—creating multiple new avenues by which special interests can attack scientific information that they dislike for some reason. Graham didn't leave quietly. In his final days he set in motion a new policy to require strict and uniform standards for risk assessment studies across U.S. governmental agencies, even though these agencies often have very different missions. Such uniformity for its own sake will surely stifle flexibility and adaptability—as well as cause substantial delays. Once again, Graham's maneuver echoed an earlier Republican "regulatory reform" agenda, and specifically the Dole bill (see Chapter 6). And once again, Graham's office cited the Data Quality Act as one justification for the new policy.

CHAPTER 9

EATING AWAY AT SCIENCE

FOR SHIRIKI KUMANYIKA, the attacks came almost out of nowhere. An epidemiologist and associate dean at the University of Pennsylvania School of Medicine, Kumanyika served in early 2002 as vice-chair of a United Nations panel on diet, nutrition, and the prevention of chronic diseases. At the event and several subsequent meetings in Geneva, Kumanyika and other international experts reviewed the evidence linking poor diet to chronic conditions such as obesity, type 2 diabetes, and cancer. Poring over comments from scientists, government officials, nongovernmental organizations (NGOs), and industry groups, they compiled a thick World Health Organization and Food and Agriculture Organization (WHO/FAO) report documenting their conclusions.

The whole process sounds unexceptional, technocratic, perhaps even a bit dull. The report's published recommendations—eat more fruits and vegetables, cut back on fats and sugars—are common sense. But when the document emerged in early 2003, Kumanyika suddenly found her integrity challenged. For participating in a scientific process that produced an official recommendation to "eat less"—a concept that is anathema to many food companies—she became a victim in our nation's politicized battle over the science of obesity and its causes.

In early 2003, the U.S. Sugar Association launched a blistering attack on the report, enraged by its recommendation to limit so-called free sugars to ten percent of daily calories.* In an April 14 letter, the trade group warned the WHO that it would "exercise every avenue available" to challenge the study, including asking congressional allies to block the United States' funding of the global health body. Sugar manufacturers also joined other food industry groups, including the Corn Refiners Association (whose members make high-fructose corn syrup, a leading soft-drink sweetener) in requesting that Health and Human Services Secretary Tommy Thompson personally intervene to have a draft version of the report removed from the WHO's website. And sugar industry political allies such as Idaho Republican Larry Craig and conservative Louisiana Democrat John Breaux, both of the "U.S. Senate Sweetener Caucus," also contacted Thompson, calling on him to instruct WHO to "cease further promotion" of the report.

Many food industry groups criticized the WHO/FAO report, but none fought as hard as the U.S. Sugar Association. On April 23, association president and CEO Andrew Briscoe held a press event to criticize the report's scientific basis, during which he cited Kumanyika's connection to a study by the U.S. Institute of Medicine—a branch of the National Academy of Sciences (NAS)—that had reached a different conclusion on sugar. He accused her of "speaking out of both sides of her mouth," according to journalists participating in the press conference.

Two days later, a sharper version of the same attack appeared in the media. In his "Junk Science" column published by the conservative *Fox News.com,* former Advancement of Sound Science Coalition executive director Steven Milloy wrote, "Last fall, WHO panel member Shiriki Kumanyika oversaw a U.S. Institute of Medicine panel concluding that diet quality was unaffected until added sugars exceeded 25 percent of daily calories. Now months later, Kumanyika's WHO panel recommends a 10 percent limit. No new science supports such a drastic change—it's simply arbitrary and capricious." Milloy also lists Kumanyika as a "junk scientist" on his website, JunkScience.com.

---

*The WHO defines "free sugars" as those added to food products by "the manufacturer, cook, or consumer," as well as concentrated sugars found in "honey, syrups, and fruit juices" (which have the same nutritional drawbacks).

In fact, both the Sugar Association and especially Milloy were wrong in implying that Kumanyika had any direct responsibility for the content of the U.S. Institute of Medicine (IOM) study. And on top of that, they also misrepresented the IOM study itself. Nevertheless, the sugar defenders had employed a politically powerful tactic: If you don't like the science, attack the scientists. (Milloy did not respond to interview requests for this book; the Sugar Association did not respond to written questions.)

A distinguished expert on diet, Shiriki Kumanyika sits on the Food and Nutrition Board at the Institute of Medicine. As with all Food and Nutrition Board reports, her name therefore appeared in the introduction to the report in question. But Kumanyika hardly "oversaw" the expert panel that produced the document. She merely reviewed a draft and submitted comments. "I did not determine any of the final statements in the report," she explains.

Moreover, the Institute of Medicine report did not actually conclude that people could get 25 percent of their calories from sugar without harm, or issue any recommendation that contradicted the WHO/FAO report. After sugar interests and some reporters wrongly read the IOM study in this way, the institute made sure to set the record straight. In an April 15, 2003, letter to Tommy Thompson, IOM president Harvey Fineberg wrote that "interpretations suggesting that a sugar intake of 25% of total calories is endorsed by the Institute's report are incorrect." The 25 percent figure was not intended as a dietary recommendation, Fineberg explained, but rather as a "maximum intake level" to prevent nutrient losses.

The hit on Kumanyika represented just one of many offenses as food interests—backed by conservative allies in think tanks, the media, and the Bush administration—sought to undermine the WHO/FAO report. Ultimately, the document found itself the target of a debunking campaign at least as determined as the conservative war on the National Assessment on climate change, or on Tyrone Hayes's atrazine studies. The effort provides a case study in the wide array of techniques used by conservatives to undermine scientific findings perceived to conflict with industry profits.

Welcome to the nasty, highly politicized field of dietary science. Welcome to the obesity wars.

The WHO's concern about obesity and global health had been building for some time. With diet-related chronic conditions like obesity and cardiovascular disease rapidly overtaking infectious diseases as the leading cause of death worldwide, public health experts have turned their attention to the obesity epidemic. This new emphasis has, in turn, put the food industry on the defensive. And in battling back, food interests have predictably turned to their political allies for help.

Obesity rates have risen steadily in the United States over the past several decades. Roughly 31 percent of Americans were obese in 2000, according to the Centers for Disease Control and Prevention (CDC), compared with 14.5 percent in 1971. The same study found that in 1971, American women on average consumed 1,542 calories daily, and men 2,450; by 2000, the averages were 1,877 and 2,618. The obesity epidemic has taken a horrible toll on America's kids. Roughly fifteen percent of adolescents and children are obese, according to the American Obesity Association.

From a scientific standpoint, there is no doubt that obesity presents a serious health threat, upping the risk of heart disease and type 2 diabetes as well as some types of cancer and other medical problems. Consider the condition's impact in the U.S. alone (many other nations present a similar picture). According to a January 2004 study by the CDC and RTI International, a nonprofit research institute, medical costs related to obesity reached $75 billion in 2003.

Few dispute the rise of obesity or the attendant health costs. But laying blame is another matter, and one about which the American public remains ambivalent. Have food corporations created a "toxic food environment," much in the way that tobacco companies addicted consumers to their products? Or do individuals bear sole responsibility for the foods they eat? The science that could influence our conclusions on this question has been fiercely contested.

It might seem obvious that factors such as a boom in the fast food industry and an increasing trend toward eating out would at least partially account for the larding of America. But scientists have had to strive to demonstrate this connection. Just as it is impossible to rerun the history of our planet to prove that human CO_2 emissions contribute to global warming, researchers can't easily disentangle the role of food consumption from other lifestyle factors, like exercise, that also contribute to weight and health (a fact skeptics exploit to their advantage). "You cannot shut down restaurants for a year and then measure the population's weight afterwards," explains Kumanyika.

It's even tough to determine what people are eating in the first place. Scientists generally have to rely on individual reporting to establish eating patterns, and this can be unreliable. Still, increasingly strong evidence indicates that Americans' eating habits—and particularly widespread consumption of fast food and soft drinks—have lined them with excess fat.

Consider a January 2004 study published in the journal *Pediatrics* by scientists at the U.S. Department of Agriculture (USDA) and Harvard Medical School. The researchers analyzed dietary data compiled by the USDA on 6,212 children and adolescents. Nearly a third ate fast food on the average day, and these super-sizing kids consumed more fat, saturated fat, added sugars, and other not-so-healthy stuff than their peers, while eating far fewer healthy foods like fruits and vegetables. On average, they took in 187 calories more per day than other kids—which, spread across the population, amounted to a theoretical "6 pounds of weight gain per child per year." Fast food consumption, the authors concluded, "seems to have an adverse effect on dietary quality in ways that plausibly could increase risk for obesity."

That's cautious scientist-speak, but the conclusion becomes far more powerful when viewed in the context of other studies with similar findings. When it comes to sugar-sweetened soft drinks—which provide loads of calories without any particular nutritional benefit or providing a feeling of satiety that would lead to less additional consumption of calories—the data present an especially troubling picture. In terms of the obesity epidemic in the United States, notes Yale global health expert Derek Yach (who oversaw the WHO/FAO diet report process), "I think the one thing which is emerging stronger and stronger is the role of soft drinks."

Just consider: A 2001 study of 548 Massachusetts schoolchildren, published in the *Lancet,* found that the risk of becoming obese increased 60 percent with each additional soft drink downed on the average day. "Consumption of sugar-sweetened drinks is associated with obesity in children," the authors concluded. More recently, a much larger study of fifty thousand nurses, published in the *Journal of the American Medical Association,* found a strong link between the consumption of sugar-sweetened soft drinks and weight gain. Furthermore, just one drink per day (over the course of four years) was linked to an 83 percent increased risk of type 2 diabetes.

These studies, and others like them, paint a consistent picture that supports basic common sense: Our nation's unhealthy eating habits are making us obese. "The bulk of evidence in fast food and obesity is pretty clear, that there's a relationship," says psychologist Kelly Brownell, director of the Yale Center for Eating and Weight Disorders. "There's a strong and growing case that can be made that consumption of fast foods and soft drinks are important contributors to the childhood obesity epidemic," adds Harvard obesity researcher David Ludwig.

Still, the evidence has hardly stopped the denials from food companies, which generally oppose policy solutions to combat obesity such as taxes on unhealthy products, restrictions on the advertising of junk food to children, or bans on the sale of soft drinks in high schools. "Obesity is a complex problem with many causes and no single, easy solution," proclaims Dr. Richard Adamson, of the American Beverage Association (formerly the National Soft Drink Association). "All foods can be a part of a healthy diet," adds the National Restaurant Association.

The food industry's political allies—often the same conservative Republicans who have misrepresented science on issues like climate change—have echoed these misleading arguments. On March 10, 2004, the Republican-controlled House of Representatives passed the so-called "cheeseburger bill," legislation exempting food companies from tobacco-style lawsuits charging that their products cause obesity. During the debate, Utah's Chris Cannon, "sound science" supporter extraordinaire, parroted the industry line. "It is not junk food that is making teenagers overweight, but rather a lack of activity," he declared.

Cannon would have had a hard time reaching this conclusion by read-ing the relevant scientific literature, which implicates both specific foods and lack of activity in the obesity problem. But in an age of think tanks and advocacy groups, he didn't have to. Instead, industry "science" on obesity emerges in reports like the U.S. Chamber of Commerce's 2003 "Burgers, Fries, and Lawyers: The Beef Behind Obesity Lawsuits," a flip and chatty rebuttal to the tide of class action suits seeking to hold fast food companies accountable for making customers fat. The Chamber of Commerce, it should be noted, has the president and CEO of the Na-tional Restaurant Association on its board of directors.

In questioning the science linking specific foods to obesity, food compa-nies have merely employed the tactics used by any number of American corporations, a trend going at least back to the lead industry. Although food products hardly pose the same health threat as cigarettes, the general approach also recalls the tobacco industry's challenge to the link between smoking and disease.

But industry isn't the only problem. The brass knuckles tactics used by the Sugar Association to challenge the WHO/FAO report certainly do the organization no credit. But when the administration of George W. Bush proceeded to parrot industry's critiques of the document, the abuse of science in the service of economic self-interest morphed into the abuse of science by our government.

The WHO/FAO report had considerable policy relevance. That is what made it so controversial. Over the course of 2003, the WHO had prepared a global strategy on "Diet, Physical Activity and Health," calling on governments worldwide to battle excessive weight and obesity through means long loathed by many food companies: controls on advertising to kids, taxes on junk foods, and subsidies to promote healthier eating. A draft version of the global strategy, released in late 2003, specifically cited the controversial WHO/FAO report, which, in turn, gave those dis-pleased by the strategy a strong incentive to attack the science.

That included the U.S. government. On January 5, 2004, William R. Steiger, director of the Office of Global Health Affairs at the Bush Department of Health and Human Services (and, as it happens, George

H. W. Bush's godson), sent a missive to WHO director-general Lee Jong-wook in which he invoked "sound science" and included an exhaustive critique of the already finalized technical report from the group of which Kumanyika had been a part. Stating that the United States had a different "interpretation of the science," Steiger's document challenged the link between specific foods (including those containing high levels of added sugars) and obesity, and further questioned the WHO's lack of emphasis on the role of "personal responsibility" in choosing a healthy and balanced diet.

These talking points closely mirror those of food industry groups like the Grocery Manufacturers of America and the U.S. Sugar Association, both of which had previously challenged the WHO/FAO expert report. The Bush administration's food industry ties are, of course, extensive. As George W. Bush ran for reelection in 2004, his campaign listed among its "Rangers"—fundraisers who raise at least $200,000 in donations—the sugar magnate Pepe Fanjul. Meanwhile, in the campaign's class of "Pioneers"—raisers of a minimum of $100,000—was Robert E. Coker, a United States Sugar Corporation senior vice president. Other "Rangers" and "Pioneers" included executives from Coca Cola and Nestlé USA.

Nevertheless, the administration claimed that industry played no role in determining its "scientific" stance. "If they're mirroring us, you'd have to talk to them, but we're not taking guidance," Department of Health and Human Services spokesman Bill Pierce told me in February 2004. In an interview in early 2005, meanwhile, William Steiger described the submission as representing "the collective thinking of a group of government experts . . . who reviewed in depth the report and the scientific citations."

But do U.S. government experts really question the existence of a link between specific foods and obesity? If the burden of scientific proof is set high enough, of course, a scientist will question anything, and one major thrust of the U.S. government's submission was to demand a higher burden of proof in WHO documents. Yet even one scientist cited by Steiger and Pierce as a contributor to the U.S. critique, CDC Division of Nutrition and Physical Activity director Dr. William Dietz, has made public statements quite consistent with the WHO/FAO position on the connection between unhealthy foods and obesity.

Whereas the U.S. officially disputed the claim that "specific foods are linked to noncommunicable diseases and obesity," Dietz partly attributed the national obesity epidemic to growing fast food and soft drink consumption in May 21, 2002, testimony before the Senate. "The rapidity with which obesity has increased can only be explained by changes in the environment that have modified calorie intake and energy expenditure," Dietz stated. "Fast food consumption now accounts for more than 40 percent of a family's budget spent on food. Soft drink consumption supplies the average teenager with over 10 percent of their daily caloric intake." That's pretty food-specific. It also accurately represents the consensus of diet researchers.

That the U.S. Department of Health and Human Services chose to challenge this position in a highly detailed, line-by-line critique of the WHO/FAO report turns science on its head, says Kelly Brownell. "They went through and just picked and picked and picked," he says. "If you used that criterion in making public policy, you wouldn't be able to do anything about anything. You could question the laws of gravity."

Brownell calls the specific dietary recommendations made by Kumanyika's panel "ho-hum—they're consistent with dozens of reports from many countries." Derek Yach agrees: "The actual sugar recommendation is not dramatically different from that which is operative in twenty, twenty-five countries around the world." In fact, though food interests may have found it convenient to attack, the WHO/FAO report hardly represented the sole scientific foundation of the WHO's global initiative to combat obesity. "There's lots of science behind the global strategy, not just this technical document," says Dr. Kaare Norum, past president of the University of Oslo and chair of the scientific panel that advised WHO during the global strategy process. But by singling out one report, and one individual associated with it, for criticism, the food industry and its political allies strategically ignored the big picture.

Sure enough, industry groups also wheeled out the Data Quality Act in an attempt to undercut any influence the WHO/FAO report might have within the United States. A September 2003 "data quality" petition from Jim Tozzi's Center for Regulatory Effectiveness—about which Tozzi would say only that it was filed on behalf of "somebody in the food business"—challenged whether the U.S. Departments of Agriculture and Health and Human Services could draw on the WHO/FAO report in assembling the 2005 U.S. dietary guidelines (the basis for the famous "Food Guide Pyramid"). Tozzi's group even e-mailed every member of the U.S. Dietary Guidelines Advisory Committee announcing its challenge to federal use of the WHO/FAO report. It was, in a sense, a preemptive scientific strike, and one with clear economic motivation.

Even though the Institute of Medicine's president Harvey Fineberg had clearly set the record straight, Tozzi's submission incorrectly suggested that the institute's diet report contradicted the WHO/FAO study on the question of sugar intake. Far from being contradictory or out of the mainstream, the 10 percent free sugars recommendation in fact comports with both the findings of a previous WHO panel that released a technical report in 1990 and even the Department of Agriculture's food pyramid.

Granted, dietary recommendations can never achieve the status of scientific "fact." By their very nature, they are recommendations *based* on nutritional science. Nevertheless, such recommendations have been broadly consistent on sugar intake. "You have committees all over the world, in Norway, in Sweden, in Europe, in England, United States, saying that 10 percent energy from sugar is, from a scientific judgment point of view, a good figure," says Norum.

Still, Tozzi's petition closely mirrored the U.S. government's own line on the WHO/FAO report. William Steiger's missive to the WHO, submitted several months after Tozzi's challenge, prominently cited the Data Quality Act in objecting to the scientific methodology used by the world health body's group of experts. The Steiger document even went so far as to provide the WHO with a full-length copy of the Department of Health and Human Services' "data quality" guidelines. In effect, this reiterated the basic content of an earlier letter from Tozzi to the WHO, which had advised the global health body that it must follow the Data Quality Act in order for U.S. agencies to be able to rely on its work.

It's no surprise to find Tozzi and food interests seeking to corral the WHO into obeying U.S. "data quality" standards. As implemented by the U.S. Department of Health and Human Services, the Data Quality Act sets a far higher bar for scientific proof in government research than had previously existed, calling for time-consuming and costly procedural safeguards like "external" peer review. And Tozzi's goal of demanding tighter scientific vetting dovetailed with Steiger's own approach. "It occurred to me that perhaps we wanted to point out to the director general that he might want to raise the bar for the burden of proof in his own agency," says Steiger, who also criticizes the WHO/FAO report as "not a peer-reviewed document" and for having undergone "no kind of international peer review."

It is not as though the WHO/FAO report was not peer reviewed *at all*, of course. It grew out of a number of background papers that had been requested by the WHO, and these were sent out for review. When Kumanyika and the rest of the expert group met in early 2002, they took into account both the papers and reviewers' comments on them. As for a more comprehensive review—for example, sending out the entire draft report for an external critique—Kumanyika says that "WHO doesn't have those resources."

Still, the WHO sought comments on a draft of the report, both from industry and nongovernmental organizations. In total, the expert panel received over a hundred written responses, and revised the draft to take into account those that they deemed relevant. Given this process—and the fact that the WHO's conclusions broadly track dietary recommendations that have prevailed as common wisdom for decades—the report hardly seems to represent any violation of "data quality." True, it used a different peer review protocol than the Bush administration or industry defenders may have liked. But peer review comes in many different forms, some more appropriate than others for different situations.

Nevertheless, by calling for unnecessary levels of scientific review, nitpicking over well-founded conclusions, attacking individual scientists, and even employing political threats, food interests did whatever they could to discredit the WHO/FAO report. They had political help in this process from the Bush administration, and in a sense, the strategy paid off.

So outraged were some nations, and particularly sugar producers, over the WHO/FAO report that when the World Health Assembly (WHA) adopted a final global strategy on diet and health in May 2004, it omitted, as a political compromise, any reference to the contested document.

Perfectly "sound" dietary science had been effectively censored thanks to a concerted political campaign to discredit it.

Within the United States, meanwhile, Tozzi's "data quality" attempt met with more mixed success—thanks in part, perhaps, to the growing body of scientific evidence linking sugar to obesity.

Tozzi's complaint originally came in response to an August 2003 press release from the Departments of Health and Human Services and Agriculture, which announced an intention to draw on the WHO's work to assemble the 2005 U.S. Dietary Guidelines. For obvious reasons, this must have alarmed critics of the WHO/FAO report. When the Dietary Guidelines Advisory Committee completed its background scientific report in late August 2004, however, it merely issued the tepid recommendation that the guidelines should advise Americans to "choose carbohydrates wisely for good health." This appeared to step back from previous guidelines that had told Americans to "moderate your intake of sugar." It also seemed a major victory for sugar interests.

Deep in the fine print, however, the advisory committee report included one important advance: a statement actually linking intake of sugar-sweetened beverages to obesity. After much debate, a majority of committee members had decided to include this simple language despite the resistance of a few members, who believed that every study showing a link between sugar and obesity was scientifically flawed, according to committee member Carlos Camargo, a Harvard epidemiologist. "Fortunately, there comes a point in every debate where there's sufficient evidence for most scientists to reach a conclusion, and even the cynics can't stonewall forever," says Camargo.

Nevertheless, the Dietary Guidelines Advisory Committee report did not rely on the much-attacked WHO/FAO report to reach this conclusion. While the committee's final report does cite the WHO/FAO study in its section on "Fats," it ignores it entirely in the section relevant to sugar: "Carbohydrates."

And at least if you believe the investigative reporting of the *Washington Post*, the political dispute over the WHO's attempt to combat global obesity may also have left a structurally different, but equally troubling, instance of political interference with science in its wake.

Shiriki Kumanyika wasn't the only American scientist to work on the heavily attacked WHO/FAO report. Members of the "expert consultation" that produced the document also included two U.S. government employees, scientists at the CDC and NIH. "At every stage, we tried to get the very best U.S. brains that we could on board," says Derek Yach, who oversaw the development of the WHO/FAO report while working with the WHO. "And I think we were successful."

Perhaps too successful. In 2004, the Bush administration informed the WHO that it would adopt a controversial new policy regulating which U.S. health experts would participate in WHO scientific deliberations, a policy that the *Washington Post*, quoting "several government scientists" speaking on "condition of anonymity," reported was partly linked to the flare-up over the WHO/FAO report. "Two CDC researchers and an NIH economist were involved in writing or reviewing scientific material that undergirded [the WHO's global strategy]," noted the paper.

Once again the communiqué, dated April 15, 2004, came from the Department of Health and Human Services' William Steiger, whose office works on developing international health policy but also has responsibility for the activity of department personnel overseas. "Effective immediately," Steiger declared, the WHO could no longer ask specific U.S. experts to serve as scientific or technical advisers. Instead, the WHO would have to submit requests to Steiger's office, which would then identify an "appropriate expert" to participate. Steiger's letter also stated that U.S. government scientists must represent the U.S. "at all times and advocate U.S. Government policies." The WHO, in contrast, requests that its scientific experts serve independently of the governments for which they work, and not go into deliberations representing official policy positions.

Not surprisingly, Steiger disputes the *Post*'s suggestion that the new policy links back to the controversy over the WHO/FAO report. Instead, he says, the shift came about because the WHO "was continuing to invite a rather small slice of our workforce," and he wanted to see the diversity of department employees better included. Steiger also states that with regard to scientists representing the United States abroad, "if you're a federal employee, you're a federal employee, and there are rules and regulations about your dealings with the outside world that apply to me just as much as they apply to anyone across the department." Later in our interview, he disputed the notion that "somehow if you're a scientist, you get a pass, because you're engaged in some activity that is inherently more noble than anything else."

This control-oriented approach has made Steiger an extremely controversial character in the scientific community; in September 2004, *Science* magazine even profiled him as "the man behind the memos." The article noted that Steiger—a Bush family friend with a Ph.D. in Latin American history—had also placed tight restrictions on government scientists accustomed to frequent international travel to scientific meetings. Indeed, under Steiger, U.S. scientists were forced to clear visits to the Washington, D.C., offices of the WHO and other United Nations bodies as "foreign travel." Steiger counters that he got "many more positive phone calls about that article than negative ones," and defends the travel rules. "These are international organizations," he says. "People go representing the U.S. government, and we'd like to know about that, so we can keep track of the collaboration that's going on."

Yet it's not hard to see how explicitly vetting which scientists attend international meetings, and warning them that they must represent U.S. positions, could have a chilling effect on the free exchange of opinions, especially in controversial areas. And as soon as the new HHS policy on scientists' collaborations with the WHO emerged, a backlash followed. "This is a raw attempt to exert political control over scientists and scientific evidence in the area of international health," declared Rep. Henry Waxman in an outraged letter to Health and Human Services' Tommy Thompson.

The WHO balked at the vetting process outlined by Steiger. Meanwhile, additional denunciations came from such luminaries as D. A. Henderson, the famed epidemiologist credited with the eradication of smallpox when he worked for the WHO, and Jeffrey Koplan, former director of the CDC. Roger Pielke, Jr., director of the Center for Science and Technology Policy Research at the University of Colorado at Boulder, even depicted the new Health and Human Services policy as a case study in the misuse of science. "It seems abundantly clear that the HHS decision, while apparently not illegal, does serve to delegitimize science in the sense that it puts a political filter between the WHO and U.S. government scientists," Pielke wrote.

Indeed, the new policy suggests an attempt to control the input into a scientific deliberation so as to control the output: an assault on the integrity of the scientific process. At least as far as international consultations go, the policy also inappropriately attempts to limit and control professional interactions among individual scientists. This is nothing less than interference with the free exchange of information and knowledge that has made science such a successful force in modern society.

Steiger is certainly correct in stating that government scientists are federal employees. But the government shouldn't employ scientists merely to tell policymakers what they want to hear, whether at home or internationally. In this light, the WHO's insistence that scientists advising it have no policy commitments makes eminent sense. And despite Steiger's defense of his moves to control scientists' international interactions, these are precisely the sort of distrusting, control-minded activities that have contributed to such a stark estrangement between the scientific community and the modern political Right.

The war over the WHO/FAO report on diet and health shows just how fiercely business interests will fight against scientific research that threatens their bottom line, how readily today's conservative politicians and policymakers take up the scientific cause of industry, and how thoroughly such tendencies have infiltrated the Bush administration. And these tendencies are not limited to issues arising from the levels of fats and sugars in Americans' diets. We see a very similar pattern on mercury emissions, an issue that merges problems of food consumption with concerns over environmental pollution.

Mercury is a naturally occurring heavy metal released through a number of processes, most prominently coal burning. Emitted into the atmosphere from electric utilities and other sources, mercury eventually winds up being deposited back on land or in bodies of water, where microorganisms convert it into methylmercury, a strong neurotoxin. Methylmercury makes its way up the aquatic food chain and bioaccumulates in predatory fish; humans are then exposed through fish consumption. Pregnant mothers run the gravest risk, since their consumption of mercury-tainted fish can lead to neurological damage in their children. According to the Environmental Protection Agency, approximately 630,000 newborn children in the United States had dangerous blood mercury levels in 1999–2000, while nearly all of the nation's rivers and lakes suffer from mercury contamination.

Mercury from power plants has long gone unregulated by the Environmental Protection Agency, but the Clinton administration took steps toward issuing tough new rules requiring offending plants to use the "maximum achievable control technology" to cut pollution. In a late 2003 about-face, however, the Bush administration's EPA proposed far weaker regulations, with goals that now would be achieved through a market-based "cap and trade" system—an extremely controversial plan that the agency finalized in March 2005.

This laxer approach finds part of its alleged justification in a classic case of conservative science abuse. Political allies of the electric power industry have selectively highlighted the single large epidemiological study suggesting that chronic exposure of pregnant mothers to mercury *doesn't* pose risks to fetuses, even though two other large studies, as well as a comprehensive assessment of the existing literature by the National Academy of Sciences, say that it does.

Not surprisingly, one of the power industry's most dedicated allies on mercury pollution has been global warming denier James Inhofe. A July 2003 hearing on mercury pollution before Inhofe's committee shows the Right's two-pronged strategy for twisting the science on this issue, both by challenging epidemiological findings concerning mercury's health risks and by questioning the significance of U.S. industrial emissions in contributing to domestic mercury pollution.

Inhofe's hearing presented the testimony of three scientists. The first, from the industry-supported Electric Power Research Institute, or EPRI, predictably argued that much mercury deposited in the U.S. comes from other countries and that emissions reductions would do little to reduce that pollution. No counterpoint was offered on this controversial question. The rest of the hearing then pitted Dr. Deborah Rice, a former EPA toxicologist defending the case for concern about mercury's health risks, against Dr. Gary Myers, of the University of Rochester, a coauthor of the only large epidemiological study that did not find harmful effects from high levels of fish consumption (conducted in the Seychelles, an island nation off the eastern coast of Africa where the population eats large amounts of fish).

In her testimony, Rice described the full range of evidence on the question of whether low-dose exposure of pregnant mothers to mercury through fish consumption posed a risk to fetuses. "At least eight studies have found an association between methylmercury levels and impaired neuropsychological performance in the child," she noted (two of these were large epidemiological studies). "The Seychelles Islands study is anomalous in not finding associations between methylmercury exposure and adverse effects." Using a similar approach in its own mercury study, the National Academy of Sciences also considered the weight of the existing evidence, and concluded that it could not, and should not, rely exclusively on the one study, from the Seychelles, that did not find a harmful effect. Myers, however, implicitly disparaged *all* the other studies as well as the NAS in his testimony. "We do not believe that there is presently good scientific evidence that moderate fish consumption is harmful to the fetus," he testified.

Inhofe leaned toward the industry-friendly position. His questioning emphasized that fish are part of a healthy diet, and challenged the prominent Faroe Islands mercury study, which reached a different conclusion than the Seychelles study. "By bringing Dr. Myers in, it gave an awful lot of emphasis on the one negative study," notes Rice of the hearing. Industry groups have done likewise. The Chamber of Commerce, for example, released a report suggesting, based on the Seychelles study, that "the current levels at which Americans consume fish are not harmful." It highlighted Myers's research and testimony before Inhofe's committee.

Unmentioned at the hearing, however, was the following: Through the Food and Drug Administration and University of Maryland's Joint Institute for Food Safety and Applied Nutrition, EPRI has contributed $486,000 in funding to Seychelles-related mercury research conducted by Myers's group. In an interview, Myers said that EPRI did not support "the main study in the Seychelles," but rather "some related studies." The main study, he said, has been funded by the NIH's National Institute of Environmental Health Sciences.

The National Academy of Sciences did not find "any serious flaws" in the Seychelles study; industry funding certainly doesn't overturn that assessment. As a general rule, we should never consider the funding source of a study as prima facie evidence either of its validity or otherwise. Yet such funding may not be entirely irrelevant, either, especially given industry's well-known proclivity for underwriting sympathetic scientific research. Rice, for example, calls industry funding "something that needs to be part of the equation when you think about the way in which investigators are interpreting results."

Other conservative groups have similarly lionized the Seychelles study. As a case in point, consider the Center for Science and Public Policy (CSPP) at the conservative Frontiers of Freedom, which also challenges mainstream climate science and received $195,000 in total funding from ExxonMobil in 2003. In 2003, the first year of its existence, CSPP came out with a report claiming that the EPA's initial push to regulate mercury stringently, later watered down dramatically by the Bush administration, was "not justified by science." The report treated the Seychelles work of Gary Myers as definitive. Moreover, it filtered directly into the sympathetic conservative media.

In an April 8, 2004, editorial entitled "The Mercury Scare," the *Wall Street Journal* blithely pronounced that there is "no credible science showing America faces any health threat at all from current fish consumption." The editorial called Myers's Seychelles work the "gold standard in mercury research" and levied dubious criticisms against the rival Faroe Islands study, the true "gold standard" (clearly identified as such by the NAS, although not precisely in that language). The *Journal's* editors appear to have based their scientific statements on the CSPP. But when citing the organization, they did not identify it as an industry-funded group.

The *Journal's* editorial prompted a letter from Faroe Islands study senior investigator Philippe Grandjean, of the Harvard School of Public Health. The Center for Science and Public Policy, Grandjean complained, hardly counted as a reliable scientific source. Debunking CSPP's "claim to enlist the expertise of world-renowned scientists," Grandjean noted that in fact, the group provides only a "unilateral" view of mercury's health risks.

The *Wall Street Journal's* folly didn't surprise Grandjean. As the mercury pollution issue rose in prominence, he witnessed multiple attacks on the Faroe Islands study, frequently accompanied by uncritical praise of the Seychelles study. In an interview, Grandjean told me that his Faroe Islands work gets "most of the critique" on this issue because "people are trying to find reasons to explain away the mercury." But singling out a single study for criticism—the "corpuscular" approach described in the previous chapter—is not how science is supposed to work. "You have to look at the total evidence," Grandjean says.

The mercury issue links back to the politicization of climate science in a surprising way. One of the authors of the Center for Science and Public Policy's report on mecury was none other than Dr. Willie Soon, described in that document as "science director" of the Center for Science and Public Policy.

In recent years, Soon has made industry-friendly scientific arguments on several environmental topics. Besides his work on climate change, Soon has also reportedly argued, on behalf of the CSSP and while sharing the stage with a former executive from Peabody Energy—a major coal company—that U.S. power plants provide only a tiny fraction of the global mercury burden. The implication is that mercury restrictions in the U.S. are not justified.

This argument should sound familiar by now. As with other atmospheric pollution issues—acid rain, ozone depletion, climate change—industry interests hope to deny responsibility for mercury pollution and blame it on natural sources or other countries. That would mean, of course, disregarding the problem. Yet once again, the science doesn't come out in their favor.

No one disputes that a substantial amount of mercury has natural origins (such as volcanic eruptions). But the best evidence suggests that human sources have increasingly driven the so-called global mercury cycle. U.S. Geological Survey scientists studying ice cores from the Upper Fremont Glacier in Wyoming attribute 70 percent of mercury inputs over the last hundred years to human activity. Similarly, a global assessment report on mercury pollution by the United Nations Environment Programme (UNEP) completed in late 2002 found that "available information indicates that natural sources account for less than 50 percent of the total releases" of mercury.

It's true that U.S. industrial sources account for only a relatively small percentage of total global emissions and that mercury pollution can circulate worldwide. "China now accounts for nearly forty-five percent of the global anthropogenic mercury release to the atmosphere," explains

Nicola Pirrone, an Italian atmospheric scientist who chaired the European Commission's working group on mercury.

Nevertheless, in late 2000, the Environmental Protection Agency estimated that 60 percent of mercury deposited in the United States comes from sources within this country. Other scientists put the estimate somewhat lower, but still in the neighborhood of 40 percent or more. But the take-home point is this: No matter what other countries may be doing, and in spite of natural mercury sources, our domestic industries contribute very significantly to the problem. This fact, combined with the dangerous threat that mercury poses to unborn children, explains perfectly well why the Clinton administration had sought to regulate the substance stringently in the first place.

That hasn't stopped the denials from conservative Republicans. In February 2005, Reps. Richard Pombo and Jim Gibbons of the GOP-controlled House Committee on Resources released an astonishing contrarian report on mercury pollution claiming (among other things) that "current, peer-reviewed scientific literature does not show any link between U.S. power plant emissions and mercury in fish." The report had the gall to cherry-pick language from an EPA *Federal Register* notice suggesting scientific uncertainty on this matter. Yet it failed to note that in the very same paragraph, the EPA had stated that "there is a plausible link between emissions of mercury from anthropogenic sources (including coal-fired electric utility steam generating units) and methylmercury in fish." Indeed, while scientists may not be able to quantify exactly how much U.S. power plant mercury ultimately winds up in fish that reach the dinner table, that hardly means they are unjustified in assuming that a very significant amount does.

To be fair, U.S. environmental groups have not always spoken forthrightly about the natural and overseas sources of mercury pollution, preferring at times to selectively emphasize domestic industries' contributions without fully describing the complexities of the global mercury cycle. For instance, it is much easier to link domestic mercury pollution to contaminated lake and river fish than to contaminated ocean fish (like tuna). While lake and river fish likely get their mercury from nearby power plants, ocean fish could be getting it from one of several continents. Nevertheless, we have often seen all kinds of fish lumped together in discussions of the mercury issue.

But while environmentalists may have shaded the evidence their way, they have hardly concocted a problem from whole cloth. Mercury pollution *does* present a real and severe threat, especially to infants and children. Compare environmentalists' crimes to the right-wing *Wall Street Journal*'s declaration, apparently on the basis of information from one industry-funded think tank, that there is "no credible science showing America faces any health threat at all from current fish consumption." Greens may twist information now and again, but they can hardly match this blatant denial of any problem whatsoever—a denial that flies in the face of virtually all of the relevant science.

Update

Much of this chapter presents a case study demonstrating how food industry interests and their political allies employed a variety of science-based tactics to attack a single report from the World Health Organization—a report whose conclusions and recommendations perfectly tracked a large body of prior scientific work. This particular skirmish has now run its course, but it points to a larger trend. Attacks on scientific information—often accompanied by considerable distortion and misinformation—have become a leading technique by which some food interests have sought to stave off policy solutions to the problem of epidemic obesity in the United States.

How appropriate, then, that as the hardcover edition of this book went to press, media attention focused on a particularly noteworthy collaboration between food interests and a nonprofit organization dedicated to denouncing obesity "junk science." In a 2005 exposé, the *Washington Post* examined the activities of a group called the Center for Consumer Freedom, which sharply attacks what it calls the "food police" and seeks to debunk "obesity myths." The center receives funding from a number of restaurants and food companies, though its executive director, Richard Berman, wouldn't tell the *Post* which ones.

But that's not all the *Post* revealed. According to the paper, the Center for Consumer Freedom was originally founded in the mid–1990s with support from Philip Morris. Then it had a different name—the Guest Choice Network—and opposed restrictions on smoking in restaurants. The coincidences continue: In late 2005, the Center for Consumer Freedom launched the website FishScam.com, which debunks the risks of mercury contamination of fish. In a typically misleading rhetorical maneuver, the site paints the Seychelles Islands mercury study (which found no health risks from fish consumption) as reliable and criticizes the Faroe Islands study, in the process ignoring the weight-of-the-evidence assessment of these and other mercury studies by the National Academy of Sciences.

So not only do the defenders of industry employ similar tactics and strategies when it comes to fighting over scientific information and science policy. Sometimes the same organization can be found participating in multiple such campaigns—underscoring the extent to which fighting over scientific information has become a tried and true political strategy. If the tactic succeeds, it's partly because it is inherently easy to sow doubt about accepted scientific conclusions. The very nature of science, which privileges skepticism and counterargument and fully admits uncertainty, leaves it ripe for exploitation.

CHAPTER 10

FISHY SCIENCE

ARLY ON A SATURDAY MORNING in July, a double rainbow spans the sky above the drowsy southern Oregon town of Klamath Falls. With a population of less than twenty thousand lodged between the pelican-dotted Upper Klamath Lake to the north, the bald-eagle-patrolled Klamath River to the south, and the eastern Cascades, Klamath Falls can seem a rather meager outpost of humanity in a far more overpowering natural landscape. But not today. Riders dressed in cavalry garb have gathered near the mouth of Lake Ewauna, which narrows into the Klamath River as it begins its 250-mile flow through lower Oregon and northern California to the Pacific. A crowd of more than a hundred irrigation farmers and their families have arrived to join the cavalcade. The demonstrators tote signs and posters—"ESA = Flawed Science," a typical one reads, in disparagement of the Endangered Species Act. Bush-Cheney 2004 campaign stickers adorn their truck bumpers.

It is just seven o'clock, but the farmers have gathered to relive a dramatic event of the year 2001. Faced with drought conditions and a court order, the U.S. Bureau of Reclamation shut off irrigation water—channeled from Upper Klamath Lake and other sources to more than two hundred thousand acres of agricultural land in the Klamath basin—to protect a white-bellied bottom-dwelling fish called the shortnose sucker, the related Lost River sucker, and a genetically distinct variety of the coho salmon. Long in coming, the crisis flowed inevitably from the fact that a wide range of interests in the Klamath region—farmers, Native American tribes, wildlife refuges, and finally, fish—all find themselves fighting over too little water.

The shutoff marked the first time that the Endangered Species Act had forced a dramatic withholding of water from a federal reclamation project (the Klamath project dates back to the early 1900s). It triggered immense anger over the government's decision, which was perceived to elevate the concerns of fish over those of farmers. "We lost our way of living," declares Al King, a Republican Senate candidate dressed in a tie, blue jeans, and cowboy boots who has come to address the assembled marchers. In 2001, the Klamath irrigators even engaged in civil disobedience, opening the Bureau of Reclamation's "A" Canal headgates and releasing water downstream. The government spent much of the summer—and an estimated $750,000—guarding the canal. In today's procession, a woman's T-shirt, reading "Headgate Stand, July 4, 2001," recalls the drama of those days.

But the mood this morning seems more subdued, perhaps because after 2001, the Bush administration has striven to ensure that Klamath farmers continue to receive their water. Following King's speech, the irrigators listen to a prayer, sing "God Bless America," and begin their march through the city center. Their destination: the Ross Ragland Theater, where the Republican-controlled House of Representatives' Committee on Resources has staged a field hearing on the Endangered Species Act and the Klamath irrigation project. The event will galvanize political support for a bill introduced by Greg Walden, the Republican congressman who represents Klamath county, called the Sound Science for Endangered Species Act Planning Act of 2003. Just days after the hearing, the Resources Committee will pass Walden's bill and send it to the House floor, in the process renaming it the Endangered Species Data Quality Act of 2004.

As this language hints, the Klamath dispute is no typical Endangered Species Act fight pitting environmentalists and uncharismatic species against angry landowners. Since 2001, the battle has taken a new turn, with an emphasis on science. In late 2003, the National Academy of Sciences completed a two-part review of the scientific underpinnings of the water shutoff, concluding that there wasn't strong scientific evidence supporting some of the more controversial actions proposed to help the fish. Agricultural interests and their conservative allies, who had been shouting "junk science" from the beginning of the controversy, now hold up the Klamath debacle as proof that many ESA decisions lack scientific justification and that the act itself needs science-based reform.

Yet the anti-ESA rhetoric emanating from the Klamath farmers—one memorable sign that I saw read, "Tame the raging bull . . . castrate the ESA"—suggests a very different agenda. In fact, hundreds of wildlife scientists have pointedly criticized the ESA "sound science" reform campaign on specifically scientific grounds. And in response to allegations that the Bureau of Reclamation's decision to withhold water from farmers had a foundation in "junk science," two scientists from the NAS's Klamath committee have countered that "we credited federal biologists for using the best information they had available at the time." In other words, the NAS's work had been misrepresented.

The Endangered Species Act "sound science" push represents just one more front on which conservatives have sought to interfere with the processes and results of science, in this case to block species protections unpopular with their libertarian-leaning Western constituents (and the farming, logging, development, and other industries and interests they represent). On ESA issues ranging from the protection of Pacific Northwestern salmon to the role of human activity in causing species extinction, conservatives have yet again twisted and misrepresented science to achieve political goals—goals they had already failed to attain through widely unpopular (but much more honest) assaults on the ESA itself.

Passed overwhelmingly by Congress in 1973 and signed by Richard Nixon, the Endangered Species Act was deliberately written to be tougher than two previous laws that had failed to curb extinctions. The new legislation embodied an activist desire to prevent the loss of *any more* species. As a report of the House Committee on Merchant Marine and Fisheries put it at the time: "Who knows, or can say, what potential cures for cancer or other scourges, present or future, may lie locked up in the structure of plants which may yet be discovered, much less analyzed? . . . Sheer self-interest impels us to be cautious." Elsewhere, the committee wrote of endangered species that the "value of their genetic heritage is, quite literally, incalculable."

Citing such comments, the U.S. Supreme Court described the ESA as "the most comprehensive legislation for the preservation of endangered species ever enacted by any nation" in the landmark 1978 case *Tennessee Valley Authority v. Hill.* In fact, the law turned out to be more comprehensive than some had apparently realized. In *TVA v. Hill,* the court notoriously affirmed a lower court's ruling that the nearly finished Tellico Dam project must be permanently halted to protect a tiny fish called the snail darter: Preserving an endangered species trumped all other priorities. Congress quickly reined in the ESA with amendments that brought other factors (especially economic ones) into play, and provided an explicit exemption so that work on the dam could continue. (Ironically, critics noted that from a cost–benefit standpoint, the dam was a highly dubious project.)

Yet despite these reforms, the ESA remains the "pit bull of environmental laws" and the weapon of choice for pro-environment litigants, who have used it for purposes extending far beyond protecting species. In the famous case of the spotted owl fifteen years ago, the bird served as a legal surrogate for another environmental aim: protecting the Pacific Northwest old-growth forests in which the owls live. "It's the most effective, substantive law that we've got for helping out the biota," says Holly Doremus, an ESA expert at the University of California, Davis, School of Law.

That may explain why since 1973, as environmental issues have grown fiercely contested, the ESA has made so many enemies. At the Klamath event in July, speaker Elliot Schwarz, of a group called the Rural Resource Alliance, even dubbed the act a "weapon of mass destruction in the hands of the eco–al Qaeda," a line that drew copious applause from the Klamath farmers. After all, the ESA ties the hands of developers, loggers, ranchers, farmers, federal agencies—pretty much anyone whose designs on a particular piece of property may be hampered by the flora and fauna residing there.

Over the years, many conservatives have attempted to dismantle or undermine the ESA. During the heady days of Newt Gingrich's tenure as Speaker of the House, Republicans crusaded for reforms that would have largely stripped the act of its potency, leading to pitched battles with environmental activists (dubbed "waffle-stomping socialists out to destroy the Constitution" by one GOP ESA reformer, Alaska's Don Young). The Gingrich Republicans' ESA reforms failed, however, and the act has gone without significant amendment since 1982.

The Right hasn't given up, though. Instead, just as they have done in other areas, conservatives have veiled their latest attacks on the ESA by claiming that they are acting on behalf of science. Instead of challenging the act itself, they have cleverly questioned its methodological underpinnings. The past decade has seen an increasing number of science-based legal challenges to ESA decisions, and explicit "sound science" laws, which have also gathered momentum over the past few years, could open up a brand new legal battleground, which is probably one objective of the proponents of such laws. The strategy, advanced by many of the same political actors who once argued for all but repealing the current ESA outright, turns on a seemingly earnest desire to improve the law's implementation, which everyone agrees would grind to a halt without good scientific analysis.

Science plays a central role in the ESA. Repeatedly, the act's text requires that actions taken—species listings, moves to protect listed species, and so on—rely on "the best scientific and commercial data available." Most evidence suggests that agencies have faithfully followed this mandate. In a 1995 report, the National Academy of Sciences failed to uncover "any major scientific issue that seriously hinders the implementation of the act." Similarly, a 1996 report by the Ecological Society of America noted that government scientists charged with ESA implementation "generally try to use the best scientific information and methods available." Failures on this front, the report continued, are "generally due to inadequate budgets and overworked staff."

Even studies requested by "sound science" proponents haven't backed up their concerns. A 2003 report from the Government Accountability Office (GAO) concluded that Fish and Wildlife Service listing decisions were "generally based on the best available science" (though problems existed when it came to the adequacy of data used to support designations of "critical habitat" for species). The GAO report had been requested by several Republican congressional leaders, including House Committee on Resources chairman Richard Pombo, a Californian who had once coauthored a book, *This Land Is Our Land*, harshly criticizing the ESA. Pombo's "property rights" screed likens the environmental movement to communism and absurdly claims that leaving "nature" free of human interference will itself trigger a decline in biodiversity because "some species tend to dominate others." Pombo has backed his colleague Greg Walden's "sound science" bill.

But if major studies fail to show a problem with the scientific basis for enforcement of the Endangered Species Act, why are conservative "sound science" proponents trying to remedy a nonexistent crisis?

A look at Walden's proposed "sound science" bill helps clear up the mystery. Since we already know that "sound science" describes a *policy* agenda to require a higher burden of proof before the government can take action, it will come as no surprise that the bill demands added vetting of the relevant science before Endangered Species Act enforcement can occur. Walden, in fact, belongs to the House Western Caucus, which (as noted previously) has explicitly stated that "environmental laws should be made with great caution and demand a high degree of scientific certainty." Although ostensibly a science-based reform, his proposed law did not arise from advocacy on the part of wildlife scientists. Rather, its champions have been libertarian-leaning Western Republicans and conservative Democrats sympathetic to farmers, ranchers, and developers who have a serious ax to grind against the ESA.

These sympathies shine through in the proposed law itself. As one of its central tenets, the Walden bill would require scientific "peer review" before virtually any ESA action could be taken. When it comes to the survival of either species or communities, Walden asked rhetorically during the 2004 congressional hearing in Klamath Falls, "Why in the devil wouldn't we ask for peer review so that we get it right?"

This argument fails for the same reason that the case for government-wide "peer review" under the Data Quality Act does. Peer review represents a hallowed scientific institution, and no one could possibly oppose it in a general sense. However, Walden's question elides a key distinction that has been highlighted by Harvard science policy scholar Sheila Jasanoff. We shouldn't confuse peer review of curiosity-driven science to be published in a journal with peer review of science used to make a government regulatory decision, Jasanoff argues. For example, in regulatory decision-making, added layers of review could create delay and red tape, effectively thwarting the ability of regulators to do their jobs (such as protect threatened wildlife).

And when it comes to the protection of endangered species, "peer review" faces another hurdle. In many cases, a very limited number of scientists actually possess the requisite expertise about a particular rare species. As a Congressional Research Service report on "sound science" and the Endangered Species Act noted, "There may be few (or no) people in the world knowledgeable about some species, and these specialists often have other duties and may not be available (or willing) to serve governmental regulators—in some cases constituting peer review panels could be difficult." The "sound science" legislation, however, proposes no way of dealing with this hurdle.

Joining in the tradition of Gingrich-era regulatory reform proposals, the "sound science" bill also attempts to legislate the definition of science itself. In order for agencies to take action under the ESA, the law wouldn't merely require them to use the best "scientific and commercial data available"; rather, the legislation would require that agencies "give greater weight to scientific or commercial data that is empirical or has been field-tested or peer-reviewed." The law also mandates that decisions to list species under the act must be supported by "field data."

Wildlife scientists read this language as a stealth attempt to undermine one of the most reliable techniques they have for understanding the vulnerability of species: population modeling, which projects species and ecosystem data into the future, and is thus neither exclusively empirical nor field-tested (though the initial data have to come from nature or the field). "When they start saying, 'You've got to give preference just to 'field-tested,' 'peer reviewed,' that is a total misrepresentation of how science goes," says Gordon Orians, a biologist at the University of Washington, Seattle, who chaired the National Academies' Board on Environmental Studies and Toxicology when it impaneled the Klamath review committee. "If you're going to say, 'We can't use models,' you might as well shut down the scientific enterprise," Orians continues.

The indirect attack on population modeling runs parallel to the attack on climate models by House Republicans during the Gingrich years. In each case, mainstream science relies on such models while fully disclosing their imperfections. Conservatives, seeking to derail inconvenient scientific findings, then cleverly target the models because of their well-known shortcomings. But unless conservatives are willing to rule out the use of *all* models in all fields (including economics), the criticism smacks of hypocrisy. Stuart Pimm, a conservation biologist at Duke University, notes that "the population models that we use, that have a lot to do with predicting whether a species will go extinct or not, are essentially the same as the models that people are using to investigate whether a small-pox outbreak would spread if some terrorist arrived in New York and tries to spread smallpox." One has heard no objection to the use of models in *that* instance.

The "sound science" bill has still other flaws. Orians adds that precisely because endangered species are so rare, it is hard to collect enough data on them to publish in the peer-reviewed literature in the first place. Walden's "sound science" proposal thus seems crafted to rule out precisely the sorts of information needed to protect a species before it is too late. The law also contains a double standard: There is no call for more stringent science in species *de*listings, though much higher hurdles are created to get species listed in the first place.

In yet another objectionable move, Walden's bill would mandate that agencies planning to list species consider data submitted by landowners (of the "I saw this bird in my field today" variety). Yet such anecdotal data would hardly have undergone scientific peer review. The shifting playing field suggests something much further from honest ESA reform and much closer to sabotage. "If you look at, in detail, what they're proposing, it can't be honest," says Orians.

Finally, adding insult to injury, the ESA reform would demand that in any case in which the current law required the use of the "best scientific data available," such data would also have to comply with the Data Quality Act. This provision appeared only in the latest version of the "sound science" bill, but the Bush administration has itself endorsed a previous incarnation of the bill. This campaign to eviscerate the ESA is not limited to Republicans in Congress; it finds sympathy in the White House.

The Walden bill clearly aims to slant the process of government science to attain a political result. And "sound science" proponents abuse science in another way in the high-profile Klamath case. They misrepresent the findings of the National Academy of Sciences panel that studied the Klamath fishes, spinning the group's reports into "proof" that the ESA's science provisions need fixing. At the 2004 Klamath Falls hearing, conservative Republican Wally Herger, who represents Klamath Basin irrigators in Northern California, claimed that the NAS's work had "vindicated" farmers, calling the science used to justify the 2001 water shutoff "fundamentally wrong." Yet as I learned from interviewing them, NAS report authors repudiate such an interpretation.

In making its controversial decision to withhold water from Klamath farmers in 2001, the Bureau of Reclamation relied on opinions about the relevant biological science from two other government agencies, the U.S. Fish and Wildlife Service and the National Marine Fisheries Service (the agencies that had listed the Klamath fish species as endangered or threatened to begin with). In late 2001, the NAS assembled an expert panel to determine, after the fact, whether the emergency actions taken during the drought had been scientifically justified. The review marked the first time ever that an outside body had rigorously examined the scientific basis for an ESA decision after the fact.

As requested, a hastily assembled 2002 interim report by the NAS committee, produced after just four months, appeared to confirm what many landowners already suspected—that the decision to maintain higher water levels in Upper Klamath Lake to protect the suckers, and higher flow levels on the Klamath River main stem to protect the coho, lacked a "sound scientific basis." The committee couldn't find a simple, clear link between lake water levels or river flow and the welfare of the fish.

These results were widely trumpeted, not least by farmers. Bush's Interior secretary, Gale Norton—who in her previous job as attorney general of Colorado once argued that the ESA was unconstitutional—dashed off a press release citing the "weaknesses" the NAS had discovered and remarking ominously that the study "will affect our decision-making process for this year and future years." Soon, the Klamath became Exhibit A in the conservative case for ESA reform. In congressional testimony defending his "sound science" bill in early 2004, Greg Walden said of his Klamath constituents, "I challenge anyone to find a group that has been more negatively affected by the inadequacy of the science used in making decisions under the Endangered Species Act."

Yet after issuing their final report in late 2003, members of the expert NAS panel lashed back at what they considered a misinterpretation of their preliminary analysis of the previous year. The committee's final report explicitly repudiates the spin put out by conservatives, noting,

> The listing agencies have been criticized for using pseudoscientific reasoning ("junk science") in justifying their requirements for the protection of species in the upper Klamath basin. The committee disagrees with this criticism. The ESA allows the agencies to use a wide array of information sources in protecting listed species. The agencies can be expected, when information is scarce, to extend their recommendations beyond rigorously tested hypotheses and into professional judgment as a means of minimizing risk to the species.

As committee member J. B. Ruhl, a legal scholar at Florida State University and ESA expert, explained to me, "A lot of people started screaming about 'junk science' after our interim report. And they just have no appreciation of what's going on."

It is important to remember, says Ruhl, that most ESA decisions will not be scrutinized as exhaustively as the Klamath decision. The review panel had a $685,000 budget and over a year to reanalyze a decision that had to be made quickly. And while the Klamath committee did find that there was "not sufficient scientific evidence to support what the agency did," Ruhl continues, "we never said what they did was a bad decision."

The distinction is crucial: In the face of scientific uncertainty and insufficient evidence, the agencies exercised their professional judgment about how best to protect endangered species. Both the interim and final Klamath reports note that there was also no good scientific evidence to support the notion that *lower* water levels in Upper Klamath Lake and on the river, as had originally been proposed by the Bureau of Reclamation to help farmers, wouldn't hurt the fish. There wasn't a lot of good evidence to go around, period, and the agencies did the best they could. They certainly didn't abuse science in any way. "The agencies are generally justified in making decisions on very limited data, and that's a far cry from saying it's 'junk science,'" says NAS committee member Peter Moyle, a fisheries expert at the University of California at Davis.

At the 2004 hearing in Klamath Falls, NAS committee chairman William Lewis, a University of Colorado limnologist, echoed Ruhl's argument about "professional judgment" in the face of inadequate evidence. Moyle has also set the record straight, clarifying that using the NAS's work as an excuse for revising the ESA has no justification. In an op-ed article published following the release of the Klamath committee's final report in 2003, Moyle and fellow committee member Jeffrey F. Mount, a University of California at Davis geologist, declared that "in the Klamath basin the Endangered Species Act (ESA) is working as intended when President Nixon signed it into law 30 years ago."

Clearly, conservatives have severely misrepresented the NAS's Klamath report and its implications. What's more, the NAS's interim report has itself come in for scientific criticism. Douglas Markle, an expert on endangered suckers at Oregon State University in Corvallis, coauthored a 2003 paper in the journal *Fisheries* suggesting that the NAS had unwisely looked for overly simple connections in a highly complex ecosystem. Markle and his coauthor called the Fish and Wildlife Service's original biological opinion on the suckers "more rigorous, thorough, and defensible" than the NAS Interim Report. Whether you side with NAS or its scientific critics, though, you still won't wind up with the conservative take on the Klamath.

Further developments also seem to have vindicated, at least to an extent, the original instincts of the federal agencies charged with protecting species. In September 2002, after the Bush administration stepped in to ensure a copious supply of water to farmers, the Klamath River witnessed a dramatic downstream fish kill in which at least thirty-three thousand salmon—largely wild Chinook but among them a number of endangered coho—died. Markle explains that the fish kill resulted from a "perfect storm" of ecological factors—including low river flow levels, warm water temperatures, and a high salmon run—with no one factor decisive on its own. Still, a 2004 analysis by the California Department of Fish and Game centrally highlighted the role of irrigation diversion, noting that "flow is the only controllable factor and tool available in the Klamath Basin" to protect against fish kills.

Following the fish kill, evidence emerged suggesting that the Bush administration had engaged in outright scientific suppression to justify its policies. Former National Marine Fisheries Service biologist Mike Kelly filed for whistleblower protection, charging that "political pressure" had corrupted the normal process whereby the fisheries agency should have assessed the potential threat to species before the Bureau of Reclamation could decide how it would regulate river flow levels. Kelly alleged that "obviously necessary analyses" went unperformed, even as he was pressured to remain "consistent" with a particularly rigid interpretation of the NAS interim report. He received the "distinct impression," he noted, that this demand came "from a very high level."

To hear Kelly talk about the fish kill is truly heartbreaking. Kelly guesses that sixty to seventy ESA-listed wild coho died (not to mention tens of thousands of dead Chinook). "But the thing to keep in mind is that they're a very rare fish," he explains. "If say fifty wild coho died, that may have been the entire run for a particular creek or subpopulation in the system. And in my mind as a biologist, losing the entire spawn in a subbasin would be a catastrophic thing." And the impact may have been worse than the number of dead fish would indicate. "While I was out there I saw a lot of stressed coho swimming around in schools that looked pretty bad," Kelly told me. "They may or may not have been able to spawn successfully."

Much like George W. Bush's decision on embryonic stem cell research, the Klamath fish kill provides a striking example of the consequences of ignoring science in political decision-making, as well a stunning indictment of the current administration's scientific stewardship. "I think it pretty clearly demonstrates that the federal agencies did a lousy job in ultimately considering the things that they need to do," says Kelly. "The fish kill is the biggest exclamation point on the end of that."

In the final analysis, the Klamath River Basin is a highly complex ecosystem that scientists don't completely understand. Forced to make a high-stakes decision in difficult circumstances in 2001, federal agency scientists charged with protecting species opted to err on the side of caution. That is true of many ESA decisions. Grappling with scientific uncertainty is the true challenge posed by the act, but "sound science" and "data quality" bills wouldn't fix this problem; they would simply gum up the process, deliberately creating paralysis instead of contributing to better and more timely analysis.

In a recently published article in *Environmental Law,* J. B. Ruhl makes a constructive suggestion. Agencies should stick with acting on the basis of "professional judgment" in most cases of scientific uncertainty, he argues, but in cases in which they decide to err strongly on the side of caution, comprehensive Klamath-style reviews could be employed to help keep government scientists honest. The proposal sounds both workable and well-intentioned—the opposite of "sound science." Once again, the Endangered Species Act "sound science" push is an excuse for inaction, not a scientific endeavor at all.

In fact, sometimes the cynicism can make your jaw drop. Consider the case of Rep. John Doolittle, the conservative California Republican we last encountered disdaining "a mumbo-jumbo of peer-reviewed documents" on the chlorofluorocarbon (CFC) ozone issue during the Gingrich years. Irony of ironies: Doolittle has signed on as a cosponsor of the Walden "peer review" bill.

At the 2004 Klamath Falls congressional hearing, Doolittle provided some comic relief. First, he tried doggedly to get NAS panel chair William Lewis to answer a *policy* question: *Should* the 2001 water shutoff have occurred? Lewis repeatedly (and properly) declined to answer the question, noting that the NAS committee had restricted itself to assessing the scientific basis for the water cutoff decision. Doolittle seemed incapable of grasping this distinction.

Later, Doolittle stumbled again, posing the following question to representatives of the Fish and Wildlife Service and the National Marine Fisheries Service: Why hadn't the agencies modified their policies following the 2001 water shutoff? "Congressman, they were [modified]," replied the National Marine Fisheries Service staffer. Apparently confused, Doolittle then launched into a tirade, instructing the agencies to "err on the side of the people" in enforcing the ESA—precisely what the law says they *must not* do—and proclaiming that "God created the earth for men and women"—an opinion that, true or false, has nothing to do with ESA decision-making. Doolittle's theological digression drew cheers from the Klamath irrigators packing the auditorium, though.

And even as the Right pushes its "sound science" reform agenda, conservatives use demonstrably *unsound* arguments to deny the need for stronger protection of species in a variety of instances. As a canonical case, consider the crusade to ease protections for more than twenty distinct populations of Pacific salmon and steelhead trout, currently defended from harm at a cost of some $700 million annually by the federal government, not to mention lost revenue claimed by landowners who have been prevented from building, farming, and logging near select streams, as well as engaging in other economic activities that could potentially harm fish habitats.

Catering to industry interests, in 2004 the Bush administration proposed counting the millions of fish churned out of hatcheries—which can actually harm wild populations through interbreeding, competition for space, and predation on smaller wild juveniles—in determining the ESA status of wild salmon stocks. The move, potentially leading to the delisting of salmon populations en masse, flew in the face of recommendations by six scientists serving on the Salmon Recovery Science Review Panel at the National Marine Fisheries Service (NMFS). These scientists, who had been approved for their roles by the National Academy of Sciences, claim that agency higher-ups instructed them to remove arguments against the counting of hatchery fish from a scientific report that they filed. The scientists later went public with their views in the journal *Science*. The hatcheries debate thus ends up involving science abuses of two types: both suppression of scientific information and the basing of policy on dubious, industry-friendly "science."

Scientists have long known that hatchery-raised salmon show strong effects of domestication. Behaviorally, hatchery-spawned salmon differ strikingly from wild fish; they also compete with them and can alter their gene pool through interbreeding. For these and other reasons, historically, NMFS has not generally counted hatchery fish for ESA purposes, and has even viewed them as detrimental to wild populations.

That began to change, however, with a 1999 lawsuit brought by the conservative Pacific Legal Foundation (PLF), a group representing industry interests that has also been a key player in support of irrigation interests in the Klamath Basin. PLF challenged the listing of the Oregon coast coho salmon (different from the Klamath population) under the ESA. The group claimed that the fisheries agency indulged in "junk science" by failing to include genetically similar hatchery fish in its count.

In a September 2001 ruling, federal district court judge Michael Hogan, of Eugene, Oregon, essentially took PLF's side. The dispute gets a bit technical, turning on how fine a distinction the government can legally make in determining which fish to protect. Essentially, Hogan ruled that because the agency already classifies hatchery fish as part of genetically distinct salmon populations—technically called "evolutionary significant units"—it must also *count* them in determining whether those distinct populations are in jeopardy. The fisheries agency had created the "unusual circumstance of two genetically identical coho salmon swimming side by side in the same stream, but only one receives ESA protection and the other does not," the judge objected.

The ruling applied to just one of the more than twenty listed Pacific salmon and steelhead populations. But the legal requirement to count hatchery fish opened the door to challenges over many other listed populations. PLF thus calls Hogan's ruling "one of the most groundbreaking environmental decisions of the last decade."

Fisheries scientists, however, maintain that even as hatcheries teem with fish, wild salmon could become extinct without protection. They even liken hatcheries to zoos, whose populations simply have no bearing on the status of wildlife populations in nature. The approach suggested by Hogan's 2001 court ruling, noted the six NMFS advisers in *Science,* "ignores important biological distinctions between wild and hatchery fish," such as genetic adaptations to domestication that make the latter less viable in the wild. "Much evidence exists," they added, "that hatcheries cannot maintain wild salmon populations indefinitely."

Nevertheless, Judge Hogan's ruling, which took effect in early 2004 after legal delays, gave the Bush administration an opportunity to advance its own industry-friendly policy—whose dark underbelly was promptly exposed by the *New York Times.* The paper revealed that Mark C. Rutzick, a lawyer previously engaged by the timber industry to battle against species protections (including for the spotted owl), had helped shape the new policy as a legal counsel to the National Marine Fisheries Service (a position he left in early 2005). According to the *Times,* Rutzick had conceived of the hatchery strategy three years earlier while employed by industry. Now the fox was guarding the scientific henhouse.

Rutzick was already on record as applauding the 2001 decision by Judge Hogan. In a November 2001 commentary posted on the website of Douglas Timber Operators, Rutzick, described as "an attorney in Portland who has represented forest product interests in natural resource disputes for 15 years," called Hogan's Pacific salmon ruling "a victory for both the coho and the rural communities in southwestern Oregon that rely on the region's natural resources for their economic base." He went on to assert that according to "experts," using hatchery fish to restore salmon runs "will bring the runs back sooner and in greater numbers."

"That, scientifically, is nonsense," observes Ransom Myers, one of the six dissenting scientific advisers to the NMFS and chair of the Department of Ocean Studies at Dalhousie University, in Nova Scotia. Myers notes that while hatcheries may help wild fish in some specific cases, Rutzick's argument is "simply not true" as a general statement.

But when Myers and colleagues made essentially that point to their superiors at NMFS—arguing that the service should not count hatchery fish as part of evolutionarily distinct salmon populations—they charge that an e-mail from an administrator labeled them "radical environmentalists." The scientists were then told to remove the recommendation from their report. "It's a clear example of policy trumping science," charges Robert Paine, an emeritus professor of biology at the University of Washington and chair of the salmon panel. When NMFS administrators try to justify their policy on hatcheries, Paine adds, "there's very little science in fact that they can offer up as saying hatcheries are good."

After Myers, Paine, and fellow panelists published their views in *Science,* NMFS officials defended themselves by arguing that the scientists had improperly strayed into making policy recommendations. But Myers dismisses the charge. The question of which fish you need to count in order to determine whether to classify a species as endangered "is certainly a purely scientific issue," he says.

The potential for public uproar appears to have dissuaded NMFS from taking its hatchery policy to its logical conclusion and delisting population after population of charismatic and symbolically potent Pacific salmon and steelhead (known for struggling powerfully against currents to return to the streams of their birth before spawning). But the policy, which was finalized in June 2005, nevertheless lumps hatchery and wild salmon together for many populations. Moreover, agency administrators have made questionable scientific statements in defending their policy, reminiscent of Rutzick's claim that hatcheries could help restore wild salmon runs. "Just as natural habitat provides a place for fish to spawn and to rear, also hatcheries can do that," NMFS regional administrator Bob Lohn told the *Oregonian* in April 2004. "You can't define a cement tank as a natural habitat," counters Paine.

The current policy also seems guaranteed to fuel lawsuits, as industry interests seek to use the foot in the door the government has provided them—namely, the introduction of hatchery fish into the ESA equation—to force the delisting of wild salmon populations. In January 2005, in another case brought by the Pacific Legal Foundation, Judge Hogan challenged the listing of the Klamath coho under the Endangered Species Act, but did not overturn the listing outright pending the final issuance of NMFS's new hatcheries policy. PLF has also filed a sweeping lawsuit to overturn a wide range of salmon protections in light of the new policy. In short, instead of adopting the clear, science-based policy recommended by its scientists, the fisheries agency may have ensured mounds of litigation. "This will be a process that will go on for years," says Myers. "It shouldn't, because the science is pretty clear on this issue."

The Bush administration stands accused of similar science games with respect to the treatment of a number of other endangered species, such as the Florida panther and sage grouse. The administration's move to avoid enforcing strong protections for the threatened marbled murrelet, a small Pacific-northwestern seabird that nests in old-growth forests, fits the same pattern. In September 2004, newspapers reported that Fish and Wildlife Service official Craig Manson had ordered the agency's scientists to change their determination that murrelets in Washington, Oregon, and California are sufficiently distinct from Canadian and Alaskan populations to warrant protection.

Supporting the picture suggested by these anecdotal case studies, in early 2005 evidence emerged suggesting that the abuse of science has become endemic within the Bush administration's Fish and Wildlife Service. The Union of Concerned Scientists and Public Employees for Environmental Responsibility (PEER) teamed up to send surveys about science abuse and politicization to more than a thousand agency scientists, and received an impressive four hundred or so back. Almost half of the respondents working on endangered species reported that they had been "directed, for nonscientific reasons, to refrain from making [findings] that are protective of species." One out of five agency employees added that they had been "directed to inappropriately exclude or alter technical information from a USFWS scientific document." Half said they were aware of cases in which "commercial interests have inappropriately induced the reversal or withdrawal of scientific conclusions or decisions through political intervention." And so on. This is damning stuff.

These abuses ultimately fall at the doorstep of Craig Manson, the Bush administration's chief Interior Department official charged with implementing the Endangered Species Act and a supporter of "sound science" reforms. As it happens, Manson is on record denying the starkly obvious link between human activities and species extinctions. In an April 2004 interview with *Grist* magazine, Manson was asked to comment on studies showing close correlations between the rise of industrialization and population growth and species extinctions. "It is a logical fallacy to suggest that because two things happen concurrently they are necessarily related, without further evidence," Manson replied.

This astonishing statement prompted a letter from Stanford conservation biology postdoctoral fellow Kai Chan, a co-creator of a website critiquing the Bush administration's science policies called ScienceInPolicy.org, noting that Manson had ignored "an overwhelming body of evidence that virtually all recent extinctions and endangerments have human-associated causes." Indeed, in its 1995 report *Science and the Endangered Species Act,* the National Academy of Sciences noted that "human activities are causing the loss of biological diversity at an increasing rate: The current rate of extinction appears to be among the highest in the entire fossil record." That quotation comes from the second sentence of the report's introductory chapter. If the government's official in charge of Endangered Species Act enforcement willfully denies something so absolutely basic to our scientific understanding of the plight of endangered species, we should hardly be surprised at calculated attempts to spin science in situations like the Klamath.

Having been sufficiently regaled by Greg Walden, Wally Herger, and especially John Doolittle at the Klamath Falls congressional hearing, I depart from the Ross Ragland Theater, happening to pause at the entrance on the way out. Makeshift metal detectors have been installed at the doorway, and peering down, I see a colorful assortment of pocket knives of various sizes lying in a box on a table alongside the machines—implements presumably confiscated earlier in the morning as Klamath irrigators poured into the building for the hearing. The image—a kaleidoscope of banned penknives—seems emblematic of the Klamath farming community, a distinct patch of red within the reliable blue state of Oregon.

Later that day, driving along the highway that conducts drivers (almost exclusively of trucks) between Klamath Falls and northern parts of Oregon, I spot a massive bald eagle perched on a telephone pole, peering down into the algae-rich waters of Upper Klamath Lake. Out over the lake, a white pelican flaps by, its black primary feathers creating a striking color contrast with the bald eagle's white-headed, black-bodied plumage. Beneath the murky waters, invisible to me but perhaps not to the hungry eagle, shortnose and Lost River suckers troll the shallow lake bottom, inhaling small crustaceans and insects.

Eagles happily dine on endangered suckers, I later learn, particularly on older or sick ones that venture near the shore or surface. "One of my field crew had a sucker dumped on them by an eagle," Doug Markle tells me. It's certainly a complicated image: one revered and majestic ESA-listed species feeding on an obscure and despised one. A species whose rescue is due in part to the act preying on one whose fate few care about.

As the contrast between bald eagle and sucker suggests, our political and aesthetic values play a large role, ultimately, in determining how much effort we pump into saving an endangered species, and which ones we will sacrifice the most to keep. Nevertheless, the Endangered Species Act remains a powerful piece of legislation, and however incompletely enforced, applies to *all species*. If political conservatives and their constituents don't like that, they should announce their earnest intention to amend the political and moral vision embodied in the ESA. "At least be honest about it and say, 'We don't really want to protect the fish,'" says Mike Kelly.

Those who feel this way have every right to push for legislation that would reverse the current requirement that government agencies take prompt action to save species. But what they shouldn't do—both because it is dishonest and because it corrupts a form of assessment that our society depends upon—is to try to blind us all with science.

Update

Since this chapter first appeared, Republicans have continued their quest to revise the science standards of the Endangered Species Act but have changed their tactics. Instead of pushing Greg Walden's narrow science-centered bill, they have instead sought to fold similar amendments into a far more sweeping reform of the entire act, centrally sponsored by House Resources Committee chairman Richard Pombo but cosponsored by

Walden and many other elected representatives, including some conservative Western Democrats.

Pombo's 88-page bill, which passed the Republican-controlled House of Representatives in September 2005, nestles science-based reforms alongside a slew of other amendments, some of them quite radical. But the proposed law once again rejects a flexible "best available science" standard for the Endangered Species Act and instead attempts to legislate the nature of science by requiring that the data used to enforce the act be either empirical, "peer reviewed," or in compliance with the already problematic Data Quality Act.

In justifying these reforms, Pombo's committee complains that "species data is, by its very nature, often vague, ambiguous, and frequently subject to best-professional judgment rather than objectively quantifiable." Precisely. That's why it makes no sense to make it even *harder* to use what little science we have to enforce the Endangered Species Act. Upping the burden of proof, requiring that every step taken be "objectively quantifiable," will only create further delays in the act's implementation. Such steps will tie the hands of the wildlife professionals who stand in the best position to know what kinds of activities must be prevented if scarce populations are to have any hope of persisting under already adverse conditions.

But of course, strong reasons exist for suspecting that the true goal here is, indeed, to hobble the scientists charged with enforcing the Endangered Species Act. That Pombo's "science" reforms rest upon a foundation of cynicism becomes more and more apparent once you consider just how energetically wildlife scientists have organized to oppose his bill. In March 2006, the Union of Concerned Scientists released a letter signed by more than five thousand biological scientists that rejected the Pombo approach and warned that "use of scientific knowledge should not be hampered by administrative requirements that overburden or slow the Act's implementation, or by limiting consideration of certain types of scientific information." It's revealing that while Pombo's reforms are couched in the language of science, the scientists who know the most about endangered species overwhelmingly oppose them.

If Republicans in Congress haven't backed down from their disturbing quest to use science to gut the Endangered Species Act, there's slightly better ground for optimism about the future of the Klamath River Basin. In a series of recent rulings, federal courts have repudiated the Bush administration's irrigation-friendly policy for managing the river's flow levels (a policy implicated in the disastrous fish kill of 2002). Most notably, in late 2005 the U.S. Court of Appeals for the Ninth Circuit vindicated the claims of National Marine Fisheries Services whistle-blower Mike Kelly, finding that the agency had allowed the river flow regime to go forward without properly considering potential risks to threatened coho salmon. According to the court, the agency hadn't even bothered to conduct the requisite scientific analyses in order to determine whether the fish would be placed in jeopardy by lower water levels, and had thus behaved in an "arbitrary and capricious" manner. "While the [National Marine Fisheries Service] can draw conclusions based on less than conclusive scientific evidence, it cannot base its conclusions on no evidence," the court noted.

Once again, work by the Union of Concerned Scientists puts this troublesome behavior at the National Marine Fisheries Service in context. In June 2005, the group teamed up with Public Employees for Environmental Responsibility to release a survey of some 460 fisheries biologists and other scientists at the agency. This produced results almost as dismal as those received when the same groups had polled the Fish and Wildlife Service: Over a third of respondents said they had been "directed, for non-scientific reasons, to refrain from making findings that are protective" of marine species, while a quarter said they had been "directed to inappropriately exclude or alter technical information from a NOAA Fisheries scientific document." Once again, political interference with scientific information appears endemic within the Bush administration, and perhaps nowhere more so than at agencies charged with using science to enforce and implement the Endangered Species Act.

PART THREE

SCIENTIFIC REVELATIONS

"You don't have to wave your Bible to have an effect as a Christian in the public arena. We serve the greatest Scientist. We serve the Creator of all life. We serve the Author of all truth. All we're required to do is proclaim that truth."

DR. W. DAVID HAGER

CHAPTER 11

"CREATION SCIENCE" 2.0

NEARLY FORTY YEARS AGO, in 1966, two talented young political thinkers published an extraordinary book, one that reads, in retrospect, as a profound warning to the Republican Party that went tragically unheeded.

The authors had been roommates at Harvard University, and had participated in the Ripon Society, an upstart group of Republican liberals. They had worked together on *Advance,* dubbed "the unofficial Republican magazine," which slammed the party from within for catering to segregationists, John Birchers, and other extremists. Following their graduation, both young men moved into the world of journalism and got the chance to further advance their "progressive" Republican campaign in a book for the eminent publisher Alfred A. Knopf. In their spirited 1966 polemic *The Party That Lost Its Head,* they held nothing back. The book devastatingly critiqued Barry Goldwater's 1964 presidential candidacy— the modern conservative movement's primal scene—and dismissed the GOP's embrace of rising star Ronald Reagan as the party's hope to "usurp reality with the fading world of the class-B movie."

Read today, some of the most prophetic passages of *The Party That Lost Its Head* are those that denounce Goldwater's conservative backers for their rampant and even paranoid distrust of the nation's intellectuals. The book labels the Goldwater campaign a "brute assault on the entire intellectual world" and blames this development on a woefully wrongheaded political tactic: "In recent years the Republicans as a party have been alienating intellectuals deliberately, as a matter of taste and strategy."

The authors charge that Goldwater's campaign had no intellectual heft behind it whatsoever, save the backing of one think tank, the American Enterprise Institute, which they denounce as "an organization heavily financed by extreme rightists." Continuing in the same vein, they slam William F. Buckley, Jr., for his attacks on leading universities and describe the advent of right-wing anti-intellectualism as "crippling" to the Republican Party. The book further deplores conservatives' paranoid distrust of the "liberal" media and the "Eastern Establishment," and worries that without the backing of intellectuals and scholars, the GOP will prove unable to develop "workable programs, distinct from those of the Democrats and responsive to national problems." If the party wants to win back the "national consensus," the authors argue, it must first win back the nation's intellectuals.

Clearly, *The Party That Lost Its Head* failed in its goal of prompting a broad Republican realignment. The GOP went in precisely the opposite direction from the one these young authors prescribed—which is why the anti-intellectual disposition they so aptly diagnosed in 1966 still persists among many modern conservatives, helping to fuel the current crisis over the politicization of science and expertise. In fact, the chief difference between the Goldwater conservatives and those of today can often seem more cosmetic than real. A massive number of think tanks have now joined the American Enterprise Institute on the right, but in many cases these outlets still provide only a thin veneer of intellectual respectability to ideas that mainstream scholarship rejects.

Certainly, the proliferation of think tanks has not had as a corollary that conservatives now take scientific expertise more seriously. On the contrary, the Right has a strong track record of deliberately attempting to undermine scientific work that might threaten the economic interests of private industry. Perhaps more alarmingly still, similar tactics have also been brought to bear by the Right in the service of a religiously conservative cultural and moral agenda.

The next three chapters demonstrate how cultural conservatives have disregarded, distorted, and abused science on the issues of evolution, embryonic stem cell research, the relation of abortion to health risks for women, and sex education. In the process, we will encounter more ideologically driven think tanks, more questionable science, and more conservative politicians willing to embrace it.

The story begins, however, with a narrative that cuts to the heart of the modern Right's war on science. You see, despite the poignant accuracy of their critique, the authors of *The Party That Lost Its Head*—Bruce K. Chapman and George Gilder—have since bitten their tongues and morphed from liberal Republicans into staunch conservatives. In fact, you could say that they have become everything they once criticized. Once opponents of right-wing anti-intellectualism, they are now prominent supporters of conservative attacks on the theory of evolution, not just a bedrock of modern science but one of the greatest intellectual achievements of human history. With this transformation, the modern Right's war on intellectuals—including scientists and those possessing expertise in other areas—is truly complete.

Three decades ago, no one could have guessed that Bruce Chapman— who did not respond to interview requests for this book—would wind up at the helm of a religiously inspired crusade against evolution. After the publication of *The Party That Lost Its Head,* Chapman carried on his liberal Republican campaign through his involvement in Washington state politics. Elected to the Seattle city council in 1971, he later became secretary of state of Washington and made an unsuccessful stab at the governorship in 1980, running to the left of conservative Democrat (and later ozone depletion denier) Dixy Lee Ray. (Both Chapman and Ray lost in their respective primaries.) Throughout this period, evolution historian and Chapman acquaintance Edward J. Larson has noted, Chapman was a moderate "Rockefeller Republican" to the core.

That changed, however, when Chapman entered the Reagan administration in 1981 as director of the Census Bureau. In a Washington atmosphere in which Reagan himself catered to antievolutionist religious leaders like Jerry Falwell, Chapman moved to the right relatively quickly. Indeed, in Chapman's transformation into a conservative who would absurdly declare evolution a "theory in crisis," which he did in 2003, one can trace key trends in the development of the modern conservative movement, such as the rising influence of the religious Right and the launch of an array of ideological think tanks. Among the latter must certainly be counted Seattle's Discovery Institute, where Chapman currently serves as president and where George Gilder—who underwent a similar ideological transformation, becoming a supply-side economics guru—now serves as a senior fellow.

By June 4, 1983, Chapman could be found publicly condemning liberalism for its "shabby, discredited, sophistical values" and defending "traditional morality." In an article on the "Harvard-trained former liberal," the *New York Times* even singled out Chapman's political shift as emblematic of "a converging of the intellectual left with the religious right within the [Republican Party] under the Reagan banner." Chapman soon left the Census Bureau to work in the White House under Reagan adviser (and later antipornography crusading attorney general) Edwin Meese. "I have become more conservative as I have grown older," he observed at the time.

As the 1980s ended, Chapman initially seemed to veer away from his newfound social conservatism. In the early days of the Discovery Institute—which originated as a Seattle branch of Indianapolis's center-right Hudson Institute—he drew heavily on connections from his moderate Seattle past. The Institute's first slate of directors read "like the guest list for a gathering of liberal Republicans," noted *Seattle Times* columnist Herb Robinson in 1991. Originally, Discovery focused on issues like the economic competitiveness of Seattle and telecommunications policy. The vibe was forward-looking, futuristic, and intellectually contrarian.

Yet much as Chapman himself swung to the right during the Reagan years, Discovery too has turned to religious conservatism. In recent years, it has become home to a reactionary crusade against the theory of evolution that goes under the banner of "intelligent design" (ID). Bringing creationism up to date, ID proponents insist that living organisms show detectable signs of having been designed (that is, specially created) by a rational agent (presumably God), while denouncing "Darwinism" for inculcating atheism and destroying cultural and moral values that had previously been grounded in piety. Such arguments put the ID campaign squarely at the center of a religiously driven culture war, and Chapman has described ID as the Discovery Institute's "No. 1 project." His friend Gilder, meanwhile, has ridiculously pronounced that "the Darwinist materialist paradigm . . . is about to face the same revolution that Newtonian physics faced 100 years ago."

Such declarations appear to have made more moderate Republicans rather uncomfortable. During a November 2004 science journalism conference in Seattle, I had the opportunity to ask former EPA administrator William Ruckelshaus, who once sat on the Discovery Institute's board, what he thought of its antievolutionist activities. Ruckelshaus told me he hadn't been interested in being involved in such a project. (In fact, in a 2000 speech on how to save the Pacific salmon, posted on the Discovery Institute's website, Ruckelshaus called the fish "a marvel of evolutionary adaptation.")

Clearly, Bruce Chapman has presided over an uncomfortable merger between pragmatic, centrist Republicanism and the antievolutionist, culture-warrior wing of the Right. Today, Discovery touts Cascadia, a technology-intensive project to improve transportation in the Pacific Northwest that is funded in part by the Microsoft fortune (through the Bill and Melinda Gates Foundation) even as it seeks to replace one of the cornerstones of biology with what *Wired* magazine has labeled the 2.0 version of creationism. And Chapman—a man who by all accounts cares deeply about ideas and whom the *New York Times* once called "serious and scholarly"—has morphed into a leader of the nation's most prominent religious crusade against modern science.

Intelligent design, as advanced by the Discovery Institute, has many antecedents. An older and more explicitly theological version of the idea holds that the universe itself shows evidence of God's handiwork, a claim that science—which is limited by its methodology to studying natural causes, not allegedly supernatural phenomena—can neither confirm nor refute. Similarly, before the publication of Charles Darwin's *On the Origin of Species by Means of Natural Selection* in 1859, many, indeed most, educated men and women accepted the precepts of "natural theology," an argument by analogy that just as human artifacts like watches show signs of a designer's hand, so do specialized organs like the eye. Perhaps the most famous proponent of this argument was the Reverend William Paley, author of the 1802 work *Natural Theology*.

Darwin read (and was impressed by) Paley as a young student at Cambridge. His *Origin*, however, unfolds as an elaborate rebuttal to Paley's recourse to divine intervention, explaining how complex organs could have evolved through gradual stages from imperfect but still useful antecedents, or from simpler structures that were co-opted for new uses. As Darwin noted in a famous passage from the book's second edition:

> To suppose that the eye, with all its inimitable contrivances for adjusting the focus to different distances, for admitting different amounts of light, and for the correction of spherical and chromatic aberration, could have been formed by natural selection, seems, I freely confess, absurd in the highest possible degree. Yet reason tells me, that if numerous gradations from a perfect and complex eye to one very imperfect and simple, each grade being useful to its possessor, can be shown to exist; if further, the eye does vary ever so slightly, and the variations be inherited, which is certainly the case; and if any variation or modification in the organ be ever useful to an animal under changing conditions of life, then the difficulty of believing that a perfect and complex eye could be formed by natural selection, though insuperable by our imagination, can hardly be considered real.

Providing the linchpin of modern biology, Darwin's work supplanted natural theology's argument from design and left it by the wayside, at least from a scientific standpoint.

Representatives of Bruce Chapman's think tank, however, have plucked the design argument from the annals of intellectual history and pronounced it modern science. Granted, today's technophile ID advocates dress up their arguments "in the idiom of information theory," as leading ID proponent William Dembski has put it, claiming (for instance) that the massive amounts of biological information encoded in DNA could not have arisen through natural selection and must therefore have been designed by an intelligent agent. But judging from ID's poor scientific publication record, it has failed to convince working biologists to join in this quixotic enterprise.

Nevertheless, ID hawkers have crisscrossed the United States arguing that public schools should "teach the controversy" over evolution—a controversy they themselves have manufactured. These advocates have even outlined First Amendment legal strategies to justify their approach. In Ohio, one state where they have enjoyed considerable success, the state board of education adopted a model lesson plan in early 2004 inviting students to "critically analyze five different aspects of evolutionary theory." In fact, the lesson plan contains spurious critiques of evolution that scientific experts have rejected and was explicitly opposed by the National Academy of Sciences. In the Dover Area School District in Pennsylvania, meanwhile, local antievolutionists have actually gone further and explicitly introduced intelligent design into science classes (a tack the Discovery Institute has come to oppose, probably because of its obvious unconstitutionality). In late 2004, the American Civil Liberties Union sued the Dover district, touching off a case that would set a historic legal precedent.

As these activities suggest, ID proponents have adopted many of the same political tactics practiced by the old-school creationists. Granted, ID diverges in some respects from earlier forms of American antievolutionism. It certainly isn't synonymous with "creation science," which provided an allegedly scientific veneer for the biblically based belief that the earth is only between six thousand and ten thousand years old. "Creation scientists" seek to debunk radioisotope dating, which geologists use to determine the age of rocks. As we saw in Chapter 4, they also rely on the feverish nonsense of Flood geology, and wrongly assert that evolution violates the second law of thermodynamics.

Intelligent design officially endorses none of these positions, and its proponents tend to shy away from espousing biblically literalist views in their publications. (With traditional creationists you didn't have to dig as far to find quotations from the Bible.) None of this, however, rescues ID from the broader "creationist" label. Philosopher of science Robert T. Pennock defines creationism as "the rejection of evolution in favor of supernatural design." ID clearly fits this description, even if we must now distinguish between "intelligent design creationism" and the other species that have cropped up in the United States, such as "young-earth creationism" and "creation science."

In fact, the peculiar characteristics of the ID movement are a direct response to the tactical and legal failings of earlier creationists. Its strategies actually represent a natural evolution of the "creation science" movement, proceeding still further in the direction of claiming the mantle of science while denying religious intentions in argument. Although ID has improved and perfected them, one can even detect many rudiments of the current ID approach among earlier advocates of "creation science."

As a legal strategy, "creation science" proved a dramatic failure. In the 1987 case *Edwards v. Aguillard*, seven out of nine justices of the U.S. Supreme Court ruled that a Louisiana law requiring the teaching of "creation science" as a counterpoint to evolution violated the First Amendment by promoting religion. Instrumental in the case was a statement from the *real* scientific community. Seventy-two Nobel laureates signed an amicus brief favoring the overturning of Louisiana's law, arguing that "teaching religious ideas mislabeled as science is detrimental to scientific education."

But though "creation science" failed politically and legally, the ID movement has taken its tactics—recruit Ph.D.s who are also conservative Christians, claim repeatedly to be doing science, and disavow religious motives—to a higher level. As Discovery Institute fellow Francis Beckwith boasts, advocates of ID have "better credentials than their creationist predecessors." "Instead of young-earth creationism, which can be laughed out of the room, intelligent design creationism has a few scientists who are not crackpots defending it," observes Harvard's Steven Pinker, a prominent defender of evolution. "I don't think scientists have woken up to it enough," he adds.

ID officially eclipsed "creation science" as the leading antievolutionist strategy following a classic overreach by traditional creationists in Kansas. In 1999, state board of education members voted to strip evolution and even the Big Bang from state standards, prompting a national outcry. Several of the offending individuals were promptly thrown out of office in the next election. But as Kansas's young-earth creationists beat an embarrassed retreat, ID advocates seized the opportunity. In the wake of the furor, the small Kansas town of Pratt flirted with including ID in its curriculum.

Since then, the ID movement has continued its quest to infiltrate academia, while the Discovery Institute has recruited a wide range of Ph.D.s to serve as fellows. ID claims, repeatedly, to represent a scientific innovation, even though "creation scientists" had always made the Paleyesque argument that living things, in their bodily and organizational complexity, show evidence of the hand of a designer. Even ID's "teach the controversy" program—which advocates instructing public school students in the alleged weaknesses of evolutionary theory, rather than in ID itself, and has actually been adopted in Ohio—had emerged in a rudimentary form among "creation scientists." Following the Supreme Court's *Edwards v. Aguillard* decision, the Institute for Creation Research (ICR) prepared an intriguing evaluation of what the movement should try next. Among other points, ICR noted that "school boards and teachers should be strongly encouraged at least to stress the scientific evidences and arguments *against evolution* in their classes . . . even if they don't wish to recognize these as evidences and arguments *for creationism*." As Glenn Branch, of the anti-ID National Center for Science Education, has observed, this comment shows that the "teach the controversy" strategy was "pioneered in the wake of *Edwards v. Aguillard*."

Clearly, ID proponents follow in the footsteps of their "creation science" forebears, especially when it comes to conveying the impression that they are doing science, instead of trying to advance religious and moral goals. Yet the express strategic objectives of the Discovery Institute; the writings, careers, and affiliations of ID's lead proponents; and the movement's funding sources all betray a clear moral and religious agenda. That might be a mere oddity if ID were actually producing good science, but it isn't. In their book *Creationism's Trojan Horse: The Wedge of Intelligent Design,* philosopher Barbara Forrest and biologist Paul Gross exhaustively demonstrate that the movement is all religion and no science. The remainder of this chapter will take a similarly two-pronged argumentative approach.

First, ID is unmistakably a religious movement. The most eloquent documentation of this comes in the form of a Discovery Institute strategic memo that made its way onto the Web in 1999: the so-called Wedge Document. This seven-page paper represents the antievolutionist equivalent of the tobacco industry documents revealed as a result of litigation, or the American Petroleum Institute's internal memo laying out a strategy to undermine mainstream climate science. The Wedge Document, though, outlines an agenda to undercut science not in the service of corporate goals, but rather to further those based on religion—or as the document states, "to replace materialistic explanations with the theistic understanding that nature and human beings are created by God."

A broad attack on "scientific materialism," Discovery's manifesto asserts that modern science has had "devastating" cultural consequences, such as the denial of objective moral standards and the undermining of religious belief. In contrast, the document states that intelligent design "promises to reverse the stifling dominance of the materialist worldview, and to replace it with a science consonant with Christian and theistic convictions." In order to achieve this objective, the ID movement will "function as a 'wedge'" that will "split the trunk [of scientific materialism] . . . at its weakest points." Much like the strategy implicit in the American Petroleum Institute memo, part of the Wedge strategy involves currying influence with "individuals in print and broadcast media." The document actually expends far more energy outlining media strategies and achievements than in describing a program of scientific research.

The Wedge Document puts ID proponents in an uncomfortable position. Discovery Institute representatives balk at being judged on religious grounds, and accuse those who probe their motivations of engaging in ad hominem attacks. Yet given the express language of the Wedge Document, it is hard to see why we shouldn't take them *at their own word*. Discovery's ultimate agenda—the Wedge—clearly has far more to do with the renewal of religiously based culture by the overthrow of key tenets of modern science than with the disinterested pursuit of knowledge. Discovery's antievolutionist branch, the Center for Science and Culture, was even previously named the Center for the Renewal of Science and Culture.

As it happens, I played a minor role in the history of the Wedge Document. As I worked on a highly critical article about the Discovery Institute for the *American Prospect* in late 2002, the organization still had not officially admitted that the memo was theirs. In an interview for my story, however, Discovery's Stephen C. Meyer, a pro-life religious conservative who directs the Center for Science and Culture, admitted ownership of the Wedge Document for the first time, telling me that it "was stolen from our offices and placed on the Web without permission."

Discovery has since broadly acknowledged the document as its own, describing it as an "early fundraising proposal." The institute mocks the notion that the strategy paper outlines a conspiracy to "replace the scientific method with belief in God," which it attributes to "our somewhat hysterical opponents." Instead, Discovery describes its attack on "scientific materialism" as, of all things, a defense of "sound science."

But though Discovery claims to support science, the Wedge Document makes it clear that the group actually hopes to radically redefine the very nature of scientific inquiry, smuggling assumptions about the supernatural into the very fabric of research and turning science into something much closer to pre-Enlightenment philosophy. The advances of modern science have relied on a "naturalistic" methodology, one that assumes continuous causal processes rather than supernatural interventions. Discovery's radical agenda of reconstituting a religiously imbued science thus represents an assault on modern science itself. In fact, according to creationism historian Ronald Numbers, Discovery's philo-sophical critique of modern science is probably the main thing that sets it apart from older forms of antievolutionism.

It is important to realize how radical this critique actually is. Consider a 2001 article published in *World* magazine and on the Discovery Insti-tute's website by Nancy Pearcey, a senior fellow of the organization and most recently the author of the book *Total Truth: Liberating Christianity from Its Cultural Captivity.* The article imagines that it is the year 2073, and ID's theistic version of science has triumphed. Darwin has been overthrown, and the moral consequences have been dramatic: The threat of a "new era of eugenics" has been averted. *Roe v. Wade* has (apparently) been reversed. Science itself has led to the affirmation of Christian morality: "Human beings have an intrinsic nature and dignity only if the world is an embodiment of the Word, the Logos, the language of a per-sonal Creator. Amazingly, it was the genetic revolution that brought this truth home, transforming the entire American culture." ID theorists, ap-parently, have a very high opinion of themselves, believing that they are fueling a scientific revolution of Copernican proportions. ID proponent Michael Behe has even written that the alleged discovery of design in nature "rivals those of Newton and Einstein, Lavoisier and Schrödinger, Pasteur and Darwin."

But Discovery's critique of "scientific materialism"—Bruce Chapman has denounced the view that "unless you can see it, smell it, taste it, feel it, it doesn't exist"—crumbles on examination. Philosophers of science distin-guish between "methodological naturalism"—science's procedural ap-proach to studying nature by assuming that continuous causal processes occur without supernatural intervention—and "philosophical naturalism," the atheistic conclusion that the supernatural doesn't exist at all. Method-ological naturalism can be justified on purely pragmatic grounds—it works. Indeed, it allows researchers of all religious beliefs to meet on common ground. Philosophical naturalism, in contrast, goes beyond science into the realm of metaphysics. Science, which studies only the natural world, can neither prove nor disprove the existence of the supernatural or God. And wisely, it doesn't try.

In disingenuously pretending that modern science basically amounts to institutionalized atheism, the Discovery Institute wrongly conflates methodological and philosophical naturalism. But many religious scientists—including Brown University cell biologist and *Finding Darwin's God* author Kenneth Miller, a leading critic of ID—accept the former but don't endorse the latter. The truth is that science isn't *necessarily* at war with religion at all, although the ID movement certainly does seem to be at war with modern science.

And just in case the Wedge Document doesn't speak eloquently enough, leading proponents of ID, too, give explicitly religious reasons for their "scientific" advocacy.

The ID movement's central strategist and popularizer, University of California, Berkeley, emeritus law professor and *Darwin on Trial* author Phillip Johnson, turned to Jesus "at the advanced age of 38" and went on to publish several books critical of evolution. Leading ID proponent Jonathan Wells, author of *Icons of Evolution: Science or Myth?*, has written that the words of Unification Church leader Sun Myung-Moon, as well as his own studies and prayers, "convinced me that I should devote my life to destroying Darwinism, just as many of my fellow Unificationists had already devoted their lives to destroying Marxism."

And that's just the beginning. William Dembski, another of ID's leading proponents and armed with Ph.D.s in philosophy and mathematics, recently left Baylor University to head the Southern Baptist Theological Seminary's newly established Center for Science and Theology. Commenting on his appointment to *Baptist Press,* a Southern Baptist national news service, Dembski welcomed the opportunity "to mobilize a new generation of scholars and pastors not just to equip the saints but also to engage the culture and reclaim it for Christ. That's really what is driving me." (Dembski left the position a year later, moving to the Southwestern Baptist Theological Seminary in Fort Worth, Texas.)

And then there's the aforementioned Stephen C. Meyer, a Cambridge history and philosophy of science Ph.D. who seems to have developed ID's philosophical critique of modern science to begin with. A conservative Christian with a background in Republican politics, Meyer has been described as "the person who brought ID to DI" by historian Edward Larson (who was a fellow at the Discovery Institute prior to its antievolutionist awakening). Seeking to institutionalize the ID movement, Meyer turned to the late timber industry magnate C. Davis Weyerhaeuser, a major funder of Christian evangelism in the U.S. through his Stewardship Foundation. Weyerhaeuser provided key "seed money" to establish the Discovery Institute's ID program, according to Larson.

Meyer is also a "university professor" at Palm Beach Atlantic University, in West Palm Beach, Florida, a "Christian liberal arts college" that puts its professors in what can only be described as an intellectual straitjacket. According to the school's "Guiding Principles,"

> All those who become associated with Palm Beach Atlantic as trustees, officers, members of the faculty or of the staff, must believe in the divine inspiration of the Bible, both the Old and New Testaments; that man was directly created by God; that Jesus Christ was born of a virgin; that He is the Son of God, our Lord and Savior; that He died for the sins of all men and thereafter arose from the grave; that by repentance and the acceptance of and belief in Him, by the grace of God, the individual is saved from eternal damnation and receives eternal life in the presence of God; and it is further resolved that the ultimate teachings in this college shall always be consistent with these principles.

ID proponents often denounce evolutionists for being closed-minded and perpetuating groupthink. Yet it appears that Meyer could not accept human evolution and still remain employed by Palm Beach Atlantic University.

The funding of the Discovery Institute, too, betrays its religious agenda. In addition to the Stewardship Foundation (which has generally funded mainstream, moderate evangelical activities), religious right tycoon Howard F. Ahmanson, Jr., has heavily supported the group and sits on Discovery's board of directors. Other ideologically oriented Discovery funders include the Tennessee-based Maclellan Foundation, which describes itself as "committed to the infallibility of Scripture, to Jesus Christ as Lord and Savior, and to the fulfillment of the Great Commission."

As befits this litany of evangelism, ID proponents cannot seem to keep out of churches and the company of old-school creationists. For example, at a 2004 conference held at the Community Bible Church in Highlands, North Carolina, leading ID advocate Michael Behe, a Catholic Lehigh University biochemist and author of *Darwin's Black Box,* spent time with young-earth creationists from the Institute for Creation Research, ministers, and prison evangelist (and former Watergate felon) Chuck Colson. In so doing, Behe closely followed the Wedge Document, which observes that ID proponents must win "a popular base of support among our natural constituency, namely, Christians."

Clearly, intelligent design proponents draw fuel from a radical religious agenda to reform American culture and counteract what they see as the corrosive influence of modern science (and its perceived moral implications). But if ID represents a religiously based strategy, it nevertheless claims to conduct scientific research. However questionable ID's definition of science, then, it is worthwhile asking whether any work that ID's supporters have published actually helps intelligent design qualify as science, or alternatively, whether the ID movement simply aims to dress up its religious agenda in scientific clothing.

The prima facie evidence in this regard does not look very good for ID proponents. In a 2002 resolution, the American Association for the Advancement of Science (AAAS) firmly stated that "to date, the ID movement has failed to offer credible scientific evidence to support their claim that ID undermines the current scientifically accepted theory of evolution." Indeed, literature searches have failed to turn up scientific papers published in peer-reviewed journals that explicitly present research that supports the ID hypothesis. As Brown University's Kenneth Miller, a practicing Catholic, has put it, "The scientific community has not embraced the explanation of design because it is quite clear, on the basis of the evidence, that it is wrong."

The ID movement initially appeared to gain a shred of scientific credibility in late 2004, however, when a review essay explicitly supportive of ID by Stephen Meyer appeared in a little-known taxonomic journal called *Proceedings of the Biological Society of Washington* (that's D.C.). But as the subsequent fallout over Meyer's paper demonstrated, this work's publication represents a startling anomaly and came about through irregular means. The ID movement will have to do much better than this if it wants to be taken seriously on a scientific level (much less have its critiques of evolution taught in public-school classrooms).

The saga of Meyer's paper bears telling resemblances to the story of the Soon and Baliunas paper on climate history, discussed in Chapter 7. In both cases, little-known journals published highly questionable papers, generating massive controversy in the process. Then, apparently thinking better of it, the journals backed away from the work.

Like the Soon and Baliunas article, Meyer's article did not present original scientific research or data; instead, it reviewed and commented on existing literature. Focusing on the well-known "Cambrian explosion"—which occurred roughly 570 to 530 million years ago—Meyer argued that evolutionary theory could not account for the appearance of new organismal forms in a relatively short period of geological time. Instead, Meyer concluded by suggesting that "intelligent," "rational" agents may have been responsible for the "origin of new biological information." Even the Discovery Institute acknowledged that this marked the first time that an article openly advocating ID had been published in a peer-reviewed biology journal (though the group claims previous publications in peer-reviewed books and other outlets).

But soon after the article's publication—which was accompanied by considerable media attention and apparently caused angry journal subscribers to pester the editorial offices demanding an explanation—facts came to light that cast doubt on whether the work should have appeared at all. It turns out that Meyer's paper was accepted for publication by an editor named Richard Sternberg, who has signed a Discovery Institute statement entitled "A Scientific Dissent from Darwinism." Sternberg's credits also include sitting on the editorial board of the

Baraminology Study Group, which studies "creation biology" and whose website is hosted by Bryan College, a fundamentalist Christian school in Tennessee fittingly named after anti-Darwin crusader William Jennings Bryan.

Although Sternberg later created a website on which he protested that (unlike most members of the Baraminology group) he did not subscribe to young-earth creationism, he also did not describe himself as an evolutionist. Sternberg also explained that although he no longer edits the *Proceedings*, his resignation preceded and had no connection to the publication of the Meyer paper.

According to the Biological Society of Washington, which publishes the *Proceedings*, Sternberg "handled the entire review process" for Meyer's paper, a move that is "contrary to typical editorial practices" at the journal, which include review by an associate editor. Sternberg, for his part, counters on his website that "as managing editor it was my prerogative to choose the editor who would work directly on the paper, and as I was best qualified among the editors, I chose myself, something I had done before in other appropriate cases." But the fact of Sternberg's connections to evolution deniers raises questions about how he may have handled Meyer's article.

In any case, the Biological Society of Washington has since backed away from the work, claiming that Meyer's piece represents a "significant departure from the nearly purely systematic content for which this journal has been known throughout its 122-year history" and "does not meet the scientific standards of the *Proceedings*." In so dramatically undermining a paper published by its own journal, the Biological Society of Washington also explicitly endorsed the AAAS resolution on intelligent design. (In response, the Discovery Institute has accused the society of imposing a "gag rule on science.")

Granted, by its own admission, the Biological Society of Washington doesn't specialize in the type of arguments Meyer makes in its journal. So it is conceivable the group was unfair to him. Just to make sure, I sought to learn what paleontologists knowledgeable about the Cambrian explosion thought of Meyer's "science."

After reviewing Meyer's paper at my request, Yale University Cambrian expert Derek Briggs, president of the Paleontological Society, responded by e-mail with what he termed the "obvious" criticism: "Meyer finds explanations for the appearance of evolutionary novelties inadequate . . . so he substitutes one of his own that is totally untestable, so-called intelligent design." Briggs's critique highlights a key reason that ID fails as science. By postulating a supernatural cause involved in the origin and history of life, the ID movement has advanced a mysterious idea that science lacks the tools to evaluate fruitfully.

As an account of the origin and history of life, ID doesn't have any meat to it. It doesn't provide any details that scientists might confirm or refute through future experimentation. And most crucially of all, it doesn't *explain* anything or *predict* anything, a key requirement for successful *scientific* theories. As three of Meyer's scientific critics have noted, "'An unknown intelligent designer did something, somewhere, somehow, for no apparent reason' is not a model."

Another expert who commented on Meyer's paper for me presented a related assessment of both the article and the intelligent design movement generally. "Presumably, this is their best work," wrote Franklin and Marshall College paleontologist Roger Thomas. "Consequently, one must ask, where are the data in support of their position? Where is the fully developed positive case for the necessity of ID, backed by appropriate evidence, that one might expect? It is simply lacking."

A lengthy critique of Meyer's paper in the *Palaeontology Newsletter*, a publication of the UK-based Palaeontological Association, provides yet another nail in the coffin. The author, Ronald Jenner, calls Meyer's paper "slipshod science," writes that it "reads like a student report," and comments that "the only trophy that proponents of ID can really boast about at home is that ID is promoted in a paper that should never have passed the reviewing process." Jenner also remarks on how differently scientists and ID proponents approach problems:

Rather than continuing to trust in the ability of science to make progress, as it always has, Meyer is willing to throw up his hands in bewilderment, and exclaim miraculous intervention of an intelligent designer. That's not the spirit of science. Meyer's paper was neither deep nor comprehensive enough to merit being called an adequate review by any standard, certainly not in view of his profound conclusions.

These words adequately sum up the ID approach, which has often been described as the search to uncover a so-called God of the gaps. Essentially, ID proponents mine the scientific literature, trying to find places where they think they can plausibly charge that evolutionary theory has failed (the Cambrian explosion, for example). Never mind the stunning successes of the theory of evolution (which explains, among myriad other curiosities, why islands feature organisms related to but distinct from nearby mainland populations, or why closely related species have more DNA in common with one another than they do with more distant relatives). And never mind that scientists themselves are currently at work on the outstanding problems and making progress on them. Wherever uncertainty remains in the current evolutionary account—and as we have seen, uncertainty can never be fully dispelled in science—ID theorists swoop in and claim, "God must have done it."

This approach, however, has a devastating drawback. Every time evolutionary theory fills another "gap," critics have to retreat further and admit that they were wrong to invoke supernatural intervention to explain that particular wrinkle of life's history. Oops, God didn't do it after all.

Nevertheless, the "God of the gaps" approach can seem rhetorically convincing to those who, lacking much grasp of the massive number of mysteries that evolutionary theory has already solved, or the proven track record that it therefore enjoys in the scientific community, are greatly impressed to learn of alleged "holes" in the theory.

Despite failed attempts to win scientific backing for ID, this new blossoming of antievolutionism has found dramatic support both on the religious Right and among its political allies. ID critic Barbara Forrest has noted that virtually all of the leading organizations on the Christian Right have embraced or at least shown sympathy for ID, including

James Dobson's Focus on the Family, Phyllis Schlafly's Eagle Forum, the Concerned Women for America, D. James Kennedy's Coral Ridge Ministries, the American Family Association, and the Alliance Defense Fund (a Christian legal group).

ID proponents have also teamed up with conservative Republican legislators to further advance their agenda. ID's most significant supporter has been Pennsylvania senator Rick Santorum, a strong pro-life Catholic who has hopped on the antievolutionist bandwagon even though the head of his church—the late Pope John Paul II—accepted evolution.

In 2001, Santorum teamed up with ID supporters to slip "teach the controversy" language into the so-called No Child Left Behind Act. Singling out evolution in particular, Santorum's amendment to the Senate version of the bill stated that "good science education should prepare students to distinguish the data or testable theories of science from philosophical or religious claims that are made in the name of science." This may sound innocuous enough, but when you learn that the language comes in part from ID movement progenitor Phillip Johnson, who believes that "Darwinism is based on an a priori commitment to materialism, not on a philosophically neutral assessment of the evidence," you see where Santorum is headed.

The Discovery Institute widely heralded the Santorum amendment, claiming that "the Darwinian monopoly on public science education . . . is ending." Santorum himself defended ID in an op-ed article in the conservative *Washington Times,* calling it a "legitimate scientific theory that should be taught in science classes," and interpreted his amendment as stating "how students studying controversial issues in science, such as biological evolution, should be allowed to learn about competing interpretations."

Ultimately, the Santorum amendment did not make its way into the actual No Child Left Behind Act, but similar language reappeared in a nonbinding conference report issued along with the bill (and with congressional endorsement). The conference report stated, in somewhat weaker language, that "where topics are taught that may generate controversy (such as biological evolution), the curriculum should help students to understand the full range of scientific views that exist." Discovery Institute representatives have used this language to claim that the U.S. Congress has endorsed the teaching of ID (which, of course, they insist is a scientific theory). More recently, perhaps seeing it as a political liability, Santorum has backed away from an outright embrace of teaching ID in public schools—even as president Bush, in August 2005, voiced his support for it. (Bush's statement put his science adviser John Marburger, who has stated that ID is not science, in a truly awkward position.) But the president has little real role, other than a symbolic one, in the evolution fight: All of the antievolutionist action today is happening at the state and local level. According to the National Center for Science Education (NCSE), between 2001 and 2004, forty-three U.S. states saw significant antievolution activity within their borders. Much of this activity has been inspired by young-earth creationists, who remain highly motivated and active, but the strategies advanced by the Discovery Institute have increasingly taken precedence. Meanwhile, Republican state political parties have also embraced antievolutionism: A survey by the NCSE found eight state parties with explicitly antievolution platforms or public statements.

Which brings us back to Discovery Institute president Bruce Chapman, a former Republican liberal who veered right and went on to found a think tank that would almost single-handedly lead a war against one of the most robust theories in the history of science: the theory of evolution. On the one hand, Chapman's career suggests a stunning intellectual contradiction. Yet when viewed in a broader historical context, his personal evolution seems quite consistent with trends in the development of the modern Right and its strained relationship with science.

To be sure, the intelligent design movement does not claim an animus against science. Science abusers never do. Rather, the movement seeks to redefine the very nature of science to serve its objectives.

But just like "creation scientists" of yore, ID hawkers have clear and ever-present religious motivations for denying and attacking evolution. And like creationists of yore, they have failed the only test that matters: They simply are not doing credible science. Instead, they are appropriating scientific-sounding arguments to advance a moral and political agenda, one they hope to force into the public-school system.

That is where the true threat emerges. ID theorists and other creationists don't like what the overwhelming body of science has to tell us about where human beings come from. Their recourse? Trying to interfere with the process by which children are supposed to learn about the best scientific (as opposed to religious) answer that we have to this most fundamental of questions. No matter how many conservative Christian scholars Bruce Chapman and the Discovery Institute manage to get on their side, such interference represents the epitome of anti-intellectualism.

Update

This book first appeared just as the Dover, Pennsylvania, evolution trial—technically *Kitzmiller v. Dover Area School District*—built toward a dramatic courtroom confrontation. The case pitted a truly radical school board, which had gone further than any other in the nation in pushing "intelligent design" (ID) on its students, against the American Civil Liberties Union, Americans United for Separation of Church and State, and the Pennsylvania law firm Pepper Hamilton. Everyone called it a replay of the 1925 Scopes trial, but in fact it more closely paralleled 1982's *McLean v. Arkansas*, a case that helped to determine whether ID's predecessor, "creation science," could be constitutionally taught in public school biology classes or whether it violated the First Amendment doctrine of separation of church and state.

The judge in *McLean*, William Overton, wrote a scathing opinion denouncing creation science as religion in disguise; and the judge in the Dover trial, John E. Jones III, more or less did the same with respect to ID. Perhaps the main difference was that Jones's opinion was considerably longer. Released in December 2005 and 139 pages in length, it demonstrated just how rigorously the legal process can, at least sometimes, evaluate "scientific" disputes (a virtue unfortunately not possessed by some journalists who are called upon to cover science in its political context).

Jones's opinion may well represent the death knell of "intelligent design," both as a viable political strategy and as an idea with pretensions to intellectual seriousness. The judge proclaimed that ID does not and cannot constitute science because of its appeal to supernatural explanations and because its advocates do not (with rare exceptions) participate in the scientific process by publishing in peer-reviewed journals. He probed the religious motivations of the "intelligent design" movement and dredged up the "Wedge Document," which he described as "a program of Christian apologetics to promote ID." With all of this, Jones was only getting warmed up.

Drawing on new revelations dug up by the pro-evolution camp in preparation for the trial, Jones also explored the "history and historical pedigree" of the pro-ID textbook introduced into the Dover school district, *Of Pandas and People*. Published by a Christian organization called the Foundation for Thought and Ethics, it was originally authored by several creationists in the late 1980s, with drafts completed before and after the U.S. Supreme Court's 1987 ruling, in the landmark case *Edwards v. Aguillard*, that "creation science" could not be taught in science classes. What Jones found about the early drafts of *Pandas* absolutely devastates ID: In early versions, "creation" had the same definition used in later drafts for "intelligent design"; references to creationism in these earlier versions were then "systematically replaced" with references to ID; and most importantly, these telltale changes came right after the *Edwards* ruling that put an end to "creation science" as a political and legal strategy. From all of this, Judge Jones concluded that ID is simply "creationism re-labeled."

In one of his most powerful passages, Jones underscored just how damaging the ID fight had been to the Dover community:

This case came to us as the result of the activism of an ill-informed faction on a school board, aided by a national public interest law firm eager to find a constitutional test case on ID, who in combination drove the Board to adopt an imprudent and ultimately unconstitutional policy. The breathtaking inanity of the Board's decision is evident when considered against the factual backdrop which has now been fully revealed through this trial. The students, parents, and teachers of the Dover Area School District deserved better than to be dragged into this legal maelstrom, with its resulting utter waste of monetary and personal resources.

This from a judge who is a Republican and a Bush appointee. The passage underscores a point often missed: the evolution battle is not confined to the rarefied world of ideas. It has personal and human consequences. It causes intensive religious strife that can damage and tear apart communities—precisely what happened in Dover, Pennsylvania.

But leading Republicans in the United States, in positions of considerable power, do not apparently possess Judge Jones's powers of discernment. Shortly before the Dover trial went to court, the teaching of "intelligent design" in public school science classes had won support from Senate Majority Leader Bill Frist as well as from President Bush. More recently, Republicans in the House of Representatives elected Congressman John Boehner of Ohio as their new majority leader. In 2002, before winning this role, Boehner coauthored a letter to the Ohio State Board of Education instructing it that students should learn about "differing scientific views on issues such as biological evolution." If Boehner isn't himself an "intelligent design" creationist, he certainly sounds sympathetic to them.

Yet ID's most prominent backer in Congress, Pennsylvania Republican senator Rick Santorum, seemed to respond differently in the wake of the Dover trial. Even as Bush and Frist rushed to support this latest project of the Christian right, Santorum reversed his earlier position in support of teaching ID in our public schools, and retreated to a stance embraced by the Discovery Institute: ID shouldn't officially be taught; rather, schools should teach about the "holes" in evolutionary theory. The catchphrase most frequently used to describe this strategy is "teach the controversy." Santorum also resigned from the advisory board of the Thomas More Law Center, the Michigan-based "public interest law firm eager to find a constitutional test case on ID" (as Judge Jones described the group) that unsuccessfully defended the school board in the Dover case.

"Teach the controversy" now becomes the likely ground on which the seemingly never-ending evolution battle will be fought. In the state of

Kansas in November 2005, the antievolutionist Board of Education voted to adopt a science curriculum that casts doubt on evolution but does not explicitly propose ID as an alternative. But the board also redefined science so that it would not be limited to "natural" explanations, a change that both horrifies scientists and opens the door for ID-type ideas. Meanwhile, other states continue to experiment with various other antievolutionist tactics that may or may not pass constitutional muster. The only thing that seems certain, following the Dover trial, is that more litigation lies ahead—and that creationists will continue, as they have always done, to evolve.

CHAPTER 12

STEMMING RESEARCH

AT THE BEGINNING OF this book, we examined George W. Bush's August 9, 2001, decision on embryonic stem cell research—a case study in how bad scientific information fuels bad policy. Coming as it did in his very first televised address to the nation, Bush's misleading claim about the existence of "more than sixty" embryonic stem cell lines set a tone for how his administration would approach science more generally. Yet in the context of conservative science abuse on the issue of embryonic stem cell research, Bush's "more than sixty" lines assertion represents just one memorable episode in a far broader story.

During Bush's first term in office, the stem cell debate surpassed even global warming among high-profile science policy issues. So it will come as no surprise that religious conservatives deployed a range of "scientific" arguments to justify their views on this issue, rather than merely objecting to embryo research on moral and theological grounds.

First, defenders of the administration (who never seem to acknowledge Bush's pivotal misrepresentation of the number of available stem cell lines) have taken the position that the president's policy permits science to thrive. It clearly doesn't, at least if you bother asking scientists what they think of the matter. And some religious conservatives have gone even further, seeking to explain away the clear scientific need for vigorous research on embryonic stem cells. These advocates claim that so-called adult stem cells*—found in the brain, bone marrow, liver, and other organs—can supplant embryonic ones for research purposes. Yet although this dogma has been broadly embraced by religious conservatives, it has been resoundingly rejected by researchers actually working in the field.

To be sure, the Right's transgressions have been partly counterbalanced by the misleading impression, fostered by some research supporters including the presidential campaign of Democrat John Kerry, that embryonic stem cell research will deliver quick cures. Responsible scientists have shied away from such hype, which itself counts as a misrepresentation of science. Peter Van Etten, president of the Juvenile Diabetes Research Foundation (a leading lobbyist for embryonic stem cell research), has cautioned that it will take at least five years, and probably longer, before we see any medical cures arise from this field.

When we tally up the offenses, however, conservatives once again outdistance any competition. Democrats deserve their share of blame for exaggerating the immediate promise of embryonic stem cell research as a means of sharpening their attacks on Bush's restrictive policy, but they are correct in asserting that this research has broad scientific potential. In contrast, only a political movement that truly disdained science would embrace the stunning fictions of the "adult"-stem-cell-only crusade. At the present moment, those who claim that research on these cells can supplant research on embryonic stem cells appear to be basing their assessment entirely on a leap of faith.

But to grasp the full scope of the Right's distortions, we first need a primer on embryonic stem cell research and why scientists consider it so promising. Biomedical researchers have long known that their favorite lab critters, mice, have both "adult" and embryonic stem cells. But in 1998, several papers extended this line of research to human beings, in the process touching off a massive ethical and political debate.

Most prominently, biologist James Thomson, of the University of Wisconsin, Madison, published a paper in *Science* revealing that he and

---

*The term "adult" stem cell is something of a misnomer, since these cells occur in fetuses, infants, and children as well as adults. The point is that they are not embryonic; they come into existence only as the body begins to develop and its cells become specialized.

coworkers had derived cells from the inner mass of human embryos, which had been donated for research from in vitro fertilization (IVF) clinics. The cells could divide indefinitely in culture (i.e., they were practically immortal) and had the potential to grow into cells from each of the three different "germ layers" of the embryo (what scientists call being *pluripotent*). These attributes suggest that the embryonic stem cells might be able to generate a wide range of replacement tissues for human transplantation—potentially leading to cures for degenerative diseases like Parkinson's and diabetes—while also fueling deep new insights into the processes of human development, including the development of diseases. But there was a catch. To extract the embryonic stem cells, Thomson had to destroy the IVF embryos. With his lab in possession of five embryonic stem cell lines derived from this process, a great controversy had officially begun.

As the debate proceeded, however, it became clear that few laypeople grasped in any detail why scientists want to study embryonic stem cells in the first place. In media coverage, we frequently hear about how these cells could generate specialized tissues for transplantation, along with a long list of diseases that could be helped in this way. But while technically accurate, this presents a woefully incomplete picture of the potential scientific and medical utility of embryonic stem cells.

By far the best explanation that I have heard for scientists' interest in these cells came from Lawrence S. B. Goldstein, a leading researcher in the field from the University of California, San Diego. In a November 2004 speech at Rice University, Goldstein explained that because of their unique attributes, embryonic stem cells could help us bypass four current "bottlenecks" in the development of medical therapies.

First, there aren't enough sources of tissues for transplantation to meet medical needs at present, but we might grow vast amounts of tissues from embryonic stem cells. Second, drugs are extremely expensive to bring to market because of the cost of human trials and because animal trials can often lead scientists down the wrong road, but drug discovery might proceed much more efficiently if we could test drugs in human stem cell preparations. Third, we currently lack a complete understanding of the mechanisms by which many diseases develop, but research on stem cells bearing the genetic signature for various diseases would allow for greater understanding of how these conditions emerge (which, in turn, could suggest new possibilities for treatment). And finally, we see enormous variations among individuals when it comes to their responses to various drugs and other therapies, but certain kinds of stem cells could eventually lead to therapies specially tailored to individual patients.

Remember all of this whenever anyone suggests or implies that embryonic stem cell research has just one purpose. Stem cells are "not a one trick pony," Goldstein explained; rather, the field should be thought of as a "broad enabling technology," a point that has frequently been lost in the debate.

With this background in place, we can begin to understand why George W. Bush's 2001 decision so dramatically constricted the potential of research by limiting federal funding to embryonic stem cell lines already in existence as of August 9, 2001. At least for scientists in need of federal funding, the Bush policy either entirely blocks or substantially impairs all of the different avenues of study outlined by Goldstein, not to mention stifling more basic research aimed at understanding the properties of human embryonic stem cells in the first place.

One might well defend such limitations on moral grounds, but that is not what conservatives and Bush supporters have done. Beyond moral arguments, they also make the thoroughly scientific claim that the Bush policy does not in fact constrict research because the available lines suffice for scientific purposes, painting the president as a morally serious champion of limited but significant scientific research. Bush's policy "provides an effective way to vigorously promote embryonic stem cell research and seek cures for disease without violating respect for nascent human life," wrote President's Council on Bioethics chairman Leon Kass in the *Washington Post* in October 2004. Several months earlier, Bush's secretary of health and human services, Tommy Thompson, had even made the startling assertion that "before anyone can successfully argue that the existing federal stem-cell policy needs to be broadened, we must first exhaust the potential of the stem-cell lines made available within the policy."

In fact, those cells have already been adequately investigated by scientists and found seriously lacking. Even some members of the Bush administration have admitted that the current policy constrains research. On May 14, 2004, National Institutes of Health director Elias Zerhouni informed members of Congress that "from a purely scientific perspective, more cell lines may well speed some areas" of scientific study.

The broadest problem with the Bush policy, as stem cell researcher Evan Snyder explained to me during an interview in San Francisco, is what we might call the "laptop analogy." As I struggled to type down everything that Snyder—a developmental neurobiologist and child neurologist who directs the stem cell research program at the Burnham Institute, in La Jolla, California—had to say about the deficiencies of the available embryonic stem cell lines, he paused and gestured toward my computer. "Can you promise me that when we meet three years from now, you're going to be using the same laptop?" Snyder asked. Of course not, he continued: In three years, computers will have faster RAM, bigger hard drives, and who knows what else. Wings, probably. My trusty Toshiba Satellite will have fallen hopelessly behind the competition.

The same goes for embryonic stem cell research. Bush's policy precluded the use of lines derived in later years through more advanced culturing and derivation techniques, or with different genetic makeups. It froze science in time. "It doesn't take a genius to know that the science I'm doing today is not what I'm going to be doing three years from now," Snyder explained.

What's more, because the Bush-approved lines come from in vitro fertilization clinics and are limited in number, they hardly represent the genetic diversity of America, much less the world. Rather, they contain the genes of affluent, mostly white Americans with fertility problems. Yet without a wide array of genetically distinct embryonic stem cell lines to study, scientists could find themselves wrongly inclined to infer that the quirky behavior of one or a few individual lines reflects the nature of embryonic stem cells in general.

Moreover, since the Bush lines come from embryos left over from IVF treatment—i.e., they were not implanted for some reason—scientists suspect that they may have been flawed or undesirable to begin with. "They're not perfect in a family that already has a medical problem," says Snyder.

And perhaps most devastatingly, if embryonic stem cell research truly aims at curing disease through transplantation, the existing lines probably cannot support that goal. All of the twenty-two lines available under the policy as of this writing grew on a layer of mouse feeder cells, raising concerns about viral contamination and immune system rejection that make the developed cells potentially unsuitable for transplantation into patients (thus knocking out Lawrence Goldstein's first and fourth possibilities for use of these cells). Accordingly, since Bush's 2001 announcement, scientists—constantly switching to new laptops—have begun to develop culturing techniques that don't rely on mouse feeder layers. Lines produced in this more technologically advanced manner, however, will not qualify for federal funding in the United States under Bush's policy.

From the standpoint of understanding and ultimately curing diseases, the existing lines have yet another deficiency, as Stanford University pathologist Irving Weissman explained to me in June 2004 when we met in his office at the Stanford University School of Medicine. Ideally, Weissman explained, scientists would like to have pluripotent stem cell lines containing the genes of a diabetic, an Alzheimer's patient, someone with cystic fibrosis, and individuals suffering from various types of cancer: a stem cell line for virtually every disease. Then, by injecting the human cells into living mice and watching them grow, scientists could observe the step-by-step evolution of the disease over the short lifespan of a mouse rather than the long lifespan of a human (one strategy for pursuing the third scientific possibility as outlined by Goldstein). "You might be able to start to understand, especially with complex genetic diseases like Lou Gehrig's, which gene goes wrong in which order to cause that disease," says Weissman.

The trouble is, many of these disease-specific cell lines could not derive from embryos obtained from fertility clinics, as the federally approved lines do. Instead, scientists would have to obtain them through the process of somatic cell nuclear transfer, sometimes called "therapeutic cloning." It's simple: Find someone suffering from a given disease, extract the nucleus from one of his body cells, implant it in an unfertilized egg, get the egg to start dividing, and then extract pluripotent stem cells from it and start a line. Weissman calls this process a "platform technology" that could dramatically enhance our understanding of diseases through the creation of a disease-specific cell line menagerie. In contrast, Bush and his conservative allies want to criminalize the process of somatic cell nuclear transfer out of the questionable fear that it could lead to reproductive cloning (which the FDA and some states have already banned anyway). "Whoever of you acts to ban this research is responsible for the lives it could save," Weissman warned legislators at a July 2004 Senate hearing.

Such are the limitations of the Bush administration's current "compromise" approach to embryonic stem cell research. The president's misleading promise of far more embryonic stem cell lines than actually existed led to a policy that constricted scientific inquiry even as it purported to support it. Moreover, as the lines available to federally funded scientists slowly approached the twenty-three line maximum, researchers began to realize that many weren't even that useful. "Realistically, it's probably only six or seven that people really use, and feel comfortable with," Evan Snyder told me in 2004.

No wonder that in the 2004 election, Bush's policy suffered a powerful rebuke. In an unmistakable end run around the White House, California passed a ballot initiative licensing a stunning $3 billion in funding for all forms of stem cell research, but with an emphasis on areas the federal government was neglecting (including the controversial therapeutic cloning). From a purely scientific standpoint, the California stem cell initiative—with its unprecedented infusion of state money—may accomplish a Houdini-style escape from the straitjacket of the Bush policy. But it doesn't erase the administration's systematic misrepresentation of the effectiveness of that policy in advancing science.

And even as conservatives tout the Bush policy's alleged contribution to science, they simultaneously make the incongruous argument that embryonic stem cell research itself suffers from hype. Take an article by Eric Cohen, a consultant to the conservative-leaning President's Council on Bioethics, published by *National Review Online* in May 2004. Cohen claims that embryonic stem cell research has been oversold: "The promise of embryonic-stem-cell research is very real but wholly speculative. No human therapies of any kind have yet been developed or tested, and none are on the horizon."

I read this passage to Nobel laureate and stem cell research advocate Paul Berg, of Stanford, who countered, "It's a phony argument because it says, 'Show me, even though I told you you're not allowed to do the experiments.'" Berg added that while human applications do not yet exist, scientists have performed "proof of principle" experiments using mouse embryonic stem cells, which they have been able to differentiate into cells such as dopamine-producing neurons, thereby relieving symptoms of Parkinson's in mice. This is promising stuff, precisely why scientists want to speed forward with embryonic stem cell research.

Another element of the Right's crusade against the "hyping" of embryonic stem cell research concerns Alzheimer's disease. Former president Ronald Reagan's June 2004 death after years of suffering from Alzheimer's dramatically increased attention to the stem cell issue, largely because his wife Nancy had spoken out in favor of lifting the Bush restrictions shortly before her husband's passing. So in the wake of Reagan's death, conservatives promptly set out to debunk the notion that embryonic stem cell research could advance the search for Alzheimer's cures. In the familiar parlance of the Right, conservative talk show host (and Reagan son) Michael Reagan denounced the "junk science argument that stem cell research can lead to a cure of Alzheimer's disease."

In fact, Reagan and other conservatives were attacking a straw man; their argument crumbles once you realize that embryonic stem cell research has more than one purpose. A number of scientists have responsibly stated that because Alzheimer's is a complex brain disease, stem cell *transplantation* therapies—of the sort envisioned to treat Parkinson's, juvenile diabetes, spinal cord injury, and other conditions—probably won't work. Conservatives have taken this to mean that embryonic stem cell research has no application to Alzheimer's, but that hardly follows. Embryonic stem cell research includes not only the search for transplant cures, but also basic scientific study dedicated to understanding in more detail the processes by which diseases develop. Such research could promote cures for Alzheimer's in the long run, even if not through transplantation.

Here is how the science could work. Keeping in mind Irving Weissman's notion of creating a collection of disease-specific embryonic stem cell lines through therapeutic cloning, consider that once scientists have developed such a line from an Alzheimer's patient, they will grow the cells and watch the disease develop. Then, with this new window on a disease in progress, they will hunt for drugs to block one or more of the newly discovered biochemical pathways that contribute to Alzheimer's. This is, of course, precisely what scientists have been contemplating all along, the Right's misrepresentations notwithstanding.

Nevertheless, it must be admitted that pro-research advocates have indeed been guilty of hype in some cases (a form of left-wing science politicization). The problem doesn't arise from scientists' generally responsible statements about the biomedical promise of embryonic stem cell research. That promise certainly exists, despite uncertainties about how long it may take to move research from the laboratory, through clinical trials, and into the doctor's office. Rather, unjustified hype has occurred when politicians and patient advocacy groups have talked incautiously about cures in the near future, downplaying the uncertainties and delivering implicit or even explicit promises that may never actually materialize.

As a canonical example, consider Democratic vice-presidential candidate John Edwards' campaign statement, shortly after the death of Christopher Reeve, that "If we do the work that we can do in this country, the work that we will do when John Kerry is president, people like Christopher Reeve will get up out of that wheelchair and walk again." Edwards raised hopes—possibly false ones—for cures for paralysis in the near future. In the process, he certainly abused science.

Still, the hyping of embryonic stem cell research by Democrats during the 2004 presidential campaign hardly holds a candle to the Right's abuses. So far, we have merely discussed scientific misrepresentations driven by the quest to defend President Bush's policy on embryonic stem cell research (a policy itself built on scientific distortion). But conservative science abuse on this issue doesn't end with mere misrepresentation. The Right has been much more creative than that.

Pro-life conservatives also claim that a scientific *alternative* to embryonic stem cell research exists, research using so-called "adult" stem cells. Like embryonic stem cells, these regenerative body cells certainly hold biomedical promise. But conservatives don't stop at explaining that fact. Rather, they have developed their own corpus of quasi-scientific argumentation to support using these cells to the exclusion or detriment of embryonic ones.

The notion that adult stem cells might substitute for embryonic ones appeals to pro-lifers on a moral level, since no embryos die in this research. It also appeals to them on a rhetorical and strategic level. Religious conservatives would like nothing better than to prevent the destruction of embryos by convincing Americans that *there is simply no need.* Such a claim allows them to debate the stem cell issue within the realm of science, rather than by making morally divisive arguments from pro-life religious premises.

Accordingly, conservatives have repeatedly hyped adult stem cell research. The right's faith-based advocacy in this area has even been likened to "creation science" by Harvard biomedical ethicist Louis Guenin (and sure enough, one prominent adult stem cell cheerleader, conservative bioethics writer Wesley Smith, is a senior fellow at the antievolutionist Discovery Institute). Indeed, the adult stem cell appeal has nestled so deeply in the consciousness of the religious Right that during the second 2004 presidential debate in St. Louis, an audience questioner asked an obviously unprepared John Kerry, "Wouldn't it be wise to use stem cells obtained without the destruction of an embryo?"

Perhaps the most prominent conservative scientist advancing the adult-stem-cell-only argument has been David Prentice, formerly a professor at Indiana State University and now a senior fellow for life sciences at the Family Research Council, a Christian conservative group. Over the past several years, Prentice has made a name for himself by claiming that these cells have "as great, if not greater, potential for biomedical application" than embryonic stem cells. For this reason, Prentice argues that we don't need to bother with morally troubling embryonic stem cells at all. However, Prentice's position stands in stark contrast to the views of both the National Institutes of Health and the star-studded International Society for Stem Cell Research, whose board of directors argued in a June 2004 letter to President Bush that "research on all types of stem cells warrants increased federal funding."

When I interviewed him in the summer of 2004 for an article in the *Washington Monthly,* Prentice described himself as a Christian and "definitely conservative," but added, "that's not how I argue these debates—I'm arguing from the science." Yet his scientific track record on adult stem cells appears rather sparse. Prentice told me that he began collecting scientific references and reviewing the literature in the late 1990s, and had done some research in the area, but hadn't managed to get a publication into print "for the adult stem cell work that we did."

Nevertheless, Prentice has served as an adviser to conservative Christian Republican senator Sam Brownback, of Kansas, a leading adult stem cell booster, as well as Florida Republican David Weldon, a conservative pro-life physician. Prentice has also presented a "commissioned paper" on adult stem cells for the President's Council on Bioethics, arguing that these cells "have demonstrated a surprising ability for transformation into other tissue and cell types."

To see why leading researchers in the field dispute Prentice's claim, it helps first to understand a fundamental principle of human and animal development. Scientists have long known that once our primordial embryonic cells differentiate into specialized body cells, they generally cannot go back. That is why embryonic stem cells represent such a holy grail for researchers—they are "pluripotent," meaning that they still have the potential to become many diverse cell types. Indeed, that is one key reason why embryonic stem cells have so many different potential applications, such as generating a range of different transplant tissues.

Adult stem cells, in contrast, were long considered capable of generating cells only within a single tissue type: their own. Recently, some research has suggested that these body cells, charged with repair of various tissues, might have more plasticity than originally assumed. Prentice and others have seized on this research and publicized it vigorously. Yet leading scientists note that many of the more tantalizing adult stem cell studies either suffer from poor design or have not been replicated by other laboratories, and that the apparent evidence of additional plasticity may be a mere artifact of certain research designs. "Scientifically, there is no independently verified evidence today, none whatsoever, that a pure stem cell of one type—adult tissue, say blood-forming—can turn into another tissue at all," says Stanford's Weissman.

Weissman should know: He was actually the first scientist to isolate the hematopoietic (blood-forming) stem cells found in bone marrow, now recognized as the reason that bone marrow transplants work. As the bone marrow example suggests, human adult stem cells have obviously led to treatments, and will surely lead to more. The scientific study of some types of adult stem cells, particularly hematopoietic ones, also happens to be far more advanced, largely because scientists have known about them longer. Hematopoietic stem cells were isolated in mice in 1988 and in humans four years later. Bone marrow transplants came even earlier than that.

None of this means, however, that adult stem cells can *replace* embryonic stem cells for research purposes, or that they can readily morph into an array of different tissue types: what scientists call transdifferentiation. At present, the jury remains officially out on transdifferentiation, with the most recent science suggesting that at least in some limited cases, it may happen. But even if certain adult stem cells could "transdifferentiate" into *all* different tissue types, thus rivaling the "pluripotency" of embryonic stem cells, they still couldn't foster the same program of research into the processes of early development. After all, embryonic stem cells create the entire body, including its adult stem cells. If you want to study *how* embryonic stem cells differentiate into our various body cells, it obviously makes no sense to study cells that come into existence only at the end of this process.

Moreover, adult stem cells pose further research hurdles. Generally speaking, these cells tend to be evanescent in the body, hard to identify, and available only in small amounts. And they don't grow nearly as robustly as embryonic stem cells. No wonder that when it comes to the relationship between adult and embryonic stem cell research, the University of California-San Diego's Lawrence Goldstein has stated that there is "no strong scientific opinion that says you can do one to the exclusion of the other."

In fact, even the scientist whose work has been most lionized by adult stem cell proponents rejects their agenda. In 2002, Catherine Verfaillie, of the University of Minnesota, published a study in *Nature* suggesting that so-called multipotent adult progenitor cells, a previously unknown type of adult stem cell found in bone marrow, could transdifferentiate into a range of different tissue types. But other scientists have had trouble reproducing Verfaillie's results, even as she herself has decried misrepresentations of her work and its potential. "My research is being misused depending on the point someone wants to get across," she has charged. "They have put words in my mouth."

Nevertheless, adult stem cell boosterism has filtered from scientist-advocates like Prentice to members of Congress. At a July 2004 hearing specifically devoted to adult stem cells, antiabortion senator Sam Brownback, of Kansas, promised the audience that "today's hearing is about miracles," and "today you will see some answers to prayers." I attended the hearing and witnessed a dramatic exchange that, while certainly not miraculous, speaks volumes about the adult stem cell issue.

Testifying that day—July 14, 2004—was a scientist from the University of Alabama, Birmingham, named Jean Peduzzi-Nelson, who had previously downplayed the promise of embryonic stem cells and praised adult stem cells at an October 2003 forum sponsored by the Illinois Catholic Health Association. Peduzzi-Nelson did the same in her Senate testimony, but perhaps because that testimony also declared that "there is no doubt that President Reagan would not favor federal support of research using human embryos," the Democrats in attendance pounced on her. When the question and answer period arrived, senators Ron Wyden, of Oregon, and Frank Lautenberg, of New Jersey, peppered Peduzzi-Nelson with questions aimed at determining the motives behind her adult stem cell advocacy. Lautenberg even asked Peduzzi-Nelson point blank whether she belonged to any pro-life group, while reminding her that she was under oath.

It was, admittedly, a bullying question. It even had something of a McCarthyite ring to it. Still, Peduzzi-Nelson's evasiveness in describing her own beliefs spoke volumes.

First she replied to Wyden by stating, "I am here not on the basis of ethics or politics or anything else," and, "whether I'm pro-life or pro-choice, I wish that all these types of things could be kept out of the discussions." But unfortunately, such matters cannot be ignored, because opposition to embryonic stem cell research—and concomitant advocacy for adult stem cell research as an alternative to it—springs almost entirely from pro-life ethical convictions grounded in religion. Under questioning from Florida senator Bill Nelson on the related topic of therapeutic cloning, Peduzzi-Nelson finally admitted, "I don't believe that you should create human life just to destroy it for some vague scientific purpose," later continuing, "I do consider it needless destruction of human life."

In September of 2004, Brownback held yet another hearing on stem cells. This time, David Prentice, not Jean Peduzzi-Nelson, testified for the religious Right, arguing that "adult stem cells have been shown by the published evidence to be a more promising alternative for patient treatments, with a vast biomedical potential." In a Family Research Council press release heralding the hearing and his appearance, Prentice declared that "adult stem cell research is the best science and thoroughly moral at the same time."

As these words suggest, it sometimes seems as though Christian conservatives like Prentice, believing deeply as they do that embryonic stem cell research is immoral, think that God automatically must have provided a perfectly plausible and ethically acceptable alternative. And that certainly would be convenient. Alas, that doesn't make it true.

Moreover, as Prentice's and Brownback's statements hinted, Christian conservatives have cited a wide array of cases in which patients have allegedly undergone "miracle" cures thanks to adult stem cell treatments. In fact, these episodic claims are highly unreliable. As a rule, new therapies must be tested through clinical trials specifically designed to distinguish truly effective techniques from unreliable ones based solely on anecdotal evidence. When it comes to claims of adult stem cell miracle cures, Elizabeth Blackburn, a renowned researcher at the University of California, San Francisco, and former president of the American Society for Cell Biology, may have put it best. "One person gets up and walks. That doesn't to me constitute a clinical study; that constitutes an anecdote," she told me. "And I'm very happy for the person. But we keep hearing about them, and then they sort of disappear."

Blackburn should know. More than any other scientist, the Australian-born cell biologist—sometimes mentioned as a possible future Nobel Prize winner for her trailblazing discovery of telomerase, the enzyme that fuels replication of cancer cells—has come to symbolize principled opposition to the politicization of embryonic stem cell science.

At the same time that he announced his stem cell policy in August 2001, George W. Bush also announced the creation of the President's Council on Bioethics. The body would be chaired by neoconservative thinker Leon Kass, a frequent critic of new biomedical technologies now at the American Enterprise Institute (he stepped down from his chairmanship in late 2005). When the council's membership was revealed in early 2002, some objected that it seemed to tilt in favor of conservatives and pro-life members. However, Elizabeth Blackburn was named as one member of the council's scientist minority, and her nomination helped to mollify critics.

In perhaps its most prominent report, released in July 2002, the Kass council narrowly recommended a moratorium on cloned embryo research, or therapeutic cloning (a technology crucial to creating disease-specific embryonic stem cell lines but one that the Bush administration, religious conservatives, and Kass himself want to ban). Blackburn sided against Kass and with the dissenting, pro-research minority. As the council turned to other topics, she began to criticize the group's finished reports, including one on the uses of biotechnology for human enhancement and, especially, a January 2004 report entitled *Monitoring Stem Cell Research*.

Then on February 27, 2004, in a huge flare-up, Blackburn and another pro-research member were dismissed from the council (it later became clear that the other member, medical ethicist William May, left at his own request). In their place, three individuals who could be expected to hold more conservative views on embryonic stem cell research and therapeutic cloning were appointed.

Coming shortly after the release of a prominent report by the Union of Concerned Scientists on the Bush administration and science, the dismissal of a distinguished scientist like Blackburn, and the nomination of three apparent conservatives as replacements, touched off a firestorm. While the Family Research Council applauded the move, proclaiming it "good news on the pro-life front," over 170 bioethicists signed a statement protesting Blackburn's ouster. Meanwhile, Leon Kass countered that it was "malicious and false" to suggest that the panel had been stacked, explaining that the group had moved on to different topics requiring different expertise.

In the ensuing days, it became clear that the Blackburn story overlapped, to a significant extent, with the issue of the Right's advocacy for adult stem cell research. One of Blackburn's major beefs with the council concerned the way the relative merits of adult and embryonic stem cell research had been treated in the *Monitoring Stem Cell Research* report. At a January 2004 council session shortly before Blackburn's dismissal, Kass read into the record a detailed statement that she had submitted as commentary on the finished version of the report. Blackburn offered strong cautions about overinterpreting highly preliminary results about the properties of adult stem cells.

If this seemed an implicit critique of the report's approach to the topic, on being dismissed Blackburn grew much more explicit. Writing in the *Washington Post*, she charged that "at council meetings, I consistently sensed resistance to presenting human embryonic stem cell research in a way that would acknowledge the scientific, experimentally verified realities. The capabilities of embryonic versus adult stem cells, and their relative promise for medicine, were obfuscated."

Moreover, shortly after her dismissal, Blackburn published a scientific takedown of two of the council's reports in the journal *PLoS Biology*, coauthored with fellow cell biologist and council member Janet Rowley, of the University of Chicago (who described Blackburn's dismissal from the council as "an important example of the absolutely destructive practices of the Bush administration" to the Associated Press). Among other problems, the two scientists charged that in its hyping of the promise of adult stem cells, *Monitoring Stem Cell Research* "may have ended up distorting the potential of biomedical research." "Many experiments suggesting that adult stem cells have broad plasticity may be incorrectly interpreted owing to an error caused by an experimental artifact of cell fusion present in some unknown proportion of the experiments," Blackburn and Rowley noted, echoing the arguments against transdifferentiation made by adult stem cell guru Irving Weissman.

I sought a response from Leon Kass to these scientific charges, and more specifically, to the question of why Blackburn's criticisms of the *Monitoring Stem Cell Research* report (which have since been publicized) weren't dealt with internally or integrated into the report prior to its official release. Kass requested that I reproduce in full his written answer:

> These charges are utterly baseless, and we have the records to prove it. Dr. Blackburn reviewed and contributed changes to multiple drafts of the scientific chapter of the report and pronounced it generally fit. She signed the report she later criticized, and she did not submit any personal statement dissenting in whole or in part. All but one of her suggestions for change were accepted. At her request, the scientific chapter was reviewed by three outside nationally known stem cell researchers who all pronounced it scientifically accurate, clear, and fair. The only change Dr. Blackburn wanted that we did not accept was a one-sentence assertion: Embryonic stem cells are better than adult stem cells. We thought it was premature and scientifically unwarranted to make such an assertion at this time.

Kass did not provide the records cited above, however. And though points of overlap exist, Blackburn's account differs from his in some key respects. In an interview, she said that she had been "discouraged" from writing a minority report, and added that while the council was "always responsive to my little editorial changes," that wasn't true of more substantive issues. "When I had changes where the report was just barking up the wrong tree scientifically," she said, "there was not a lot of responsiveness." The key issue, she added, was "giving great credence to what I figured was a lot of pretty iffy science we were hearing with respect to adult stem cells."

The *Monitoring Stem Cell Research* report itself contains, as an appendix, a commissioned paper on adult stem cells written by advocate David

Prentice, who also testified before the council. The body of the document, meanwhile, seems quite friendly to the Prentice line. Granted, the report doesn't come out and say that adult stem cells are better than embryonic ones. It does, however, build them up into the position of a rival. Remarking on experiments suggesting the possibility of adult stem cell transdifferentiation, *Monitoring Stem Cell Research* says that the research has "ignited debate about the relative merits of embryonic stem cells and adult stem cells: which is more valuable, both for research and (especially) for clinical treatment?"

This "debate," of course, truly exists only in the minds of embryonic stem cell research opponents (who already think they've won it). For most scientists working in the area, both types of cells have biomedical promise, but they are not in competition with each other. Indeed, in some cases they may ultimately prove complementary. "When they say, 'a debate among scientists,' that's a wonderful device that conservatives can use," says Blackburn.

More generally, *Monitoring Stem Cell Research* reads as a political document that defends the Bush stem cell policy. Consider its treatment of the president's clear misrepresentation of the number of available lines. "In August 2001, President Bush announced that 'more than sixty genetically diverse stem cell lines' (or stem cell preparations) already existed, and so would be eligible for funding under his policy," the report notes. The key parenthetical remark here has been added by the President's Council on Bioethics with the apparent goal of getting the president out of hot water. Bush himself had drawn no such distinction.

In fact, in defending the Bush policy shortly after its announcement, Health and Human Services secretary Tommy Thompson had pronounced all of the Bush lines "diverse, robust, and viable for research." He issued no warning at the time about a difference between "lines" and "preparations." Only after considerable stonewalling did Thompson finally admit to Congress that far fewer lines than claimed were ready to go.

In short, while Elizabeth Blackburn and Leon Kass contradict each other regarding precisely what happened in the editing of the contested *Monitoring Stem Cell Research* report, some things cannot be disputed. First, the report itself takes an apologetic approach to the controversial

Bush stem cell policy. Second, it suggests an equivalence between adult and embryonic stem cells in terms of their medical promise. And for the latter point, the report was publicly accused of possibly "distorting the potential of biomedical research" by two members of the council itself, who happen not only to be distinguished scientists, but specialists in cell biology.

And that's not all. We also know that whatever the reasons may have been for her dismissal from the council, Elizabeth Blackburn and another pro-research member, William May, were replaced by a slate of individuals with far more reliably conservative views on embryonic stem cell research and the related therapeutic cloning. Johns Hopkins surgeon Benjamin Carson, for instance, has signed a statement calling therapeutic cloning "morally offensive since it involves creating, killing, and harvesting one human being in the service of others." (Fellow signatories included such religious Right leaders as James Dobson and Gary Bauer.) Another new council member, political scientist Diana Schaub, of Loyola College, in Maryland, has similarly written (in a review of the President's Council on Bioethics' first report, no less) that "cloning is an evil, and cloning for the purpose of research actually exacerbates the evil by countenancing the willful destruction of nascent human life." And then there's Berry College political scientist Peter Lawler, who has made his opposition to abortion clear in print.

Leon Kass did not even disagree when I asked him—citing the Family Research Council's description of the three new appointees as "Good News on the Pro-Life Front"—whether these new council members generally view embryonic stem cell research and therapeutic cloning more skeptically than the members they replaced (Blackburn, May, and former council member Stephen Carter, who left the group in 2002). Kass replied in writing:

Perhaps so regarding Blackburn and May (but not Carter), though the issue has not come up so I don't know for sure. Do not be misled by the spin that single-issue advocacy groups—scientific or other—want to place on such matters. As the transcripts of our meetings show, the three new members are highly thoughtful, independent-minded and nonideological people, whose views cannot be predicted, not by the Family Research Council, not by journalists.

No one is saying, of course, that the new appointees are robots. Still, their past writings give a reasonable indication of their conservative-leaning moral convictions on embryo-related issues. By bringing them on while dropping a talented scientist like Elizabeth Blackburn, the President's Council on Bioethics clearly shifted further to the political right.

As the pro–adult stem cell campaign shows, it would be highly simplistic to call today's religious conservatives "antiscience." Rather, on issues of moral consequence to them, these conservatives have shown an impressive capacity to draw on sympathetic experts capable of providing "scientific" counterarguments. Alas, as in the case of David Prentice and adult stem cells, these arguments generally remain well outside of the mainstream, and often distort the true state of scientific understanding.

The next chapter demonstrates that neither Prentice nor the Discovery Institute's antievolutionists are unique in their quest to generate scientific arguments sympathetic to a religiously conservative moral agenda. When it comes to claims that abortion poses health risks to women, as well as arguments against various forms of contraception, we see a remarkably similar pattern. Not content to oppose abortion or premarital sex on moral grounds alone, religious conservatives have tried to adduce data and scientific arguments to advance their case. They draw on Christian conservative scientists to make sure that these arguments sound convincing, and on think tanks and political connections to give them currency.

Update

Without a doubt, the most important development on the stem cell front has been the House of Representatives' bipartisan 2005 vote to loosen the restrictive Bush policy. The vote, by a 238–194 margin, saw much aisle-crossing on the Republican side and represented a stinging rebuke to the president. It clearly demonstrates how the embryonic stem cell issue has become a serious liability for conservative Republicans who continue to oppose funding for research that could potentially lead to significant medical advances.

In the context of the House debate, conservatives once again attempted to argue that adult stem cell research presents a scientific alterative to embryonic work, but this rhetoric failed to prevent a strong vote against Bush's policy. Yet despite a pledge by Senate Majority Leader Bill Frist, as of this writing the stem cell issue still had not come up for a vote in the U.S. Senate, leaving Bush's policy in place and sparing the president the embarrassment of having to use a veto.

It would be foolhardy to try to predict the bill's fate in the Senate, except to say that if Frist does not allow a vote, Democrats will surely use stem cells as a 2006 campaign issue. Meanwhile, states have continued to innovate, passing legislation that often bails out the federal government for its inaction. And research opponents have come up with new ways of misusing science to undermine stronger federal funding of embryonic stem cell research. No longer do they simply talk about adult stem cells; instead, a wide range of other "alternatives" have since been floated, such as removing stem cells from "dead" embryos or "de-differentiating" body cells back to the point at which they are less specialized and more like blank-state embryonic cells. These new ideas share some commonalities: They are all speculative and relatively untested, and scientists do not agree, at least at this writing, that they can justifiably supplant more typical embryonic stem cell research. To dub them "alternatives" is premature at best.

We have also recently learned that the United States is not as far behind in the stem cell field as previously surmised: Cutting-edge work out of South Korea, where scientists had claimed to have derived stem cell lines from cloned embryos, now appears to have been faked. As the Korean scandal mounted, some conservatives tried to use it to more broadly discredit stem cell science or the scientific peer review process. Yet if anything, the Korean scandal showed that the scientific process, in a broad sense, *works*. True, questionable work got published, but it was eventually exposed by peers and colleagues. In the face of what looks like intentional deception, that's the best that we can hope for.

Finally, in late 2005 Leon Kass stepped down from the President's Council on Bioethics. His replacement, Georgetown University bioethicist Edmund Pellegrino, is a conservative Roman Catholic and opponent of embryonic stem cell research but has received praise from both sides of the aisle for being evenhanded. The council in general, however, maintains its unmistakably conservative bent.

CHAPTER 13

SEXED-UP SCIENCE

EVER SINCE HE WAS A KID, Joel Brind found himself drawn to science. At age ten, in 1961, he got his hands on an issue of *Life* magazine that described how the electron microscope had provided a new window on the cell and its curiously shaped organelles and structures. "Then and there I decided to become a biochemist," Brind recalled in a 2000 essay in the magazine *Physician*, which happens to be a publication of Dr. James Dobson's Focus on the Family, a powerful organization of the religious Right. You see, Brind may have received his biochemistry Ph.D. from New York University in 1981, but in 1985 came a bigger milestone: He found Jesus. Soon, Brind recognized the "noble task" God had chosen for him: He would demonstrate the biological connection between a woman's having an abortion and then contracting breast cancer later in life, thereby dissuading countless women from killing their unborn children. "With a new belief in a meaningful universe, I felt compelled to use science for its noblest, life-saving purpose," Brind wrote.

A professor of human biology and endocrinology at Baruch College of the City University of New York, Brind—who did not respond to interview requests—has pursued this quest with seemingly singular determination. In 1999, he even cofounded a think tank to support his theory, the innocuously named Breast Cancer Prevention Institute, which pushes the so-called ABC (abortion–breast cancer) link. Brind also advises the

Coalition on Abortion/Breast Cancer, which likens the government's "cover-up" of a link between abortion and breast cancer to the notorious Tuskegee syphilis scandal. A search of the National Library of Medicine's PubMed database, however, shows that Brind has published little in scientific journals on the alleged abortion–breast cancer link other than a number of letters and one review and meta-analysis of existing studies. More frequently, he seems to air his views in pro-life newsletters and other publications, and he has watched them gain traction among religious conservatives such as the aforementioned Florida Republican physician, Rep. David Weldon. Pushed by pro-lifers, several states have even passed laws requiring that women learn of breast cancer risks before having an abortion.

Initially, the notion of an added risk of breast cancer due to abortion may have seemed scientifically plausible. After all, full-term pregnancy has a protective effect against breast cancer in younger women. Moreover, studies dating back to 1957 had found varying results on whether abortion raises the risk of breast cancer.

However, scientists had long cited methodological flaws in the positive studies, which relied on the questionable assumption that women would report the truth about their abortion histories to researchers despite the considerable social stigma. Then in 1997, a massive *New England Journal of Medicine* study of 1.5 million Danish women discounted the ABC link. Because Denmark maintains detailed medical registries, including abortion records, the study avoided the shortcomings of previous work. "You want to have the information on induced abortion from an objective source—from the registry, not from the woman," explains Harvard epidemiologist Karin Michels. "The scientific community felt this was by far the best study that had been done to date, and really settled the issue," adds Lynn Rosenberg, an epidemiologist at Boston University who has participated in the ABC debate.

But Brind refused to accept the Danish study, alleging methodological flaws, and hasn't changed his tune since. Meanwhile, the evidence has grown stronger and stronger. When over one hundred experts gathered by the National Cancer Institute (NCI) reaffirmed in 2003 that abortion "is not associated with an increase in breast cancer risk," Brind alone dissented. In fact, the very convening of these experts in the first place seems to attest to Brind's influence. Arguably his biggest coup came in 2002, when following a letter from Rep. Chris Smith (R-N.J.) and twenty-seven other pro-life members of Congress, the National Cancer Institute removed an online fact sheet that discounted any association between abortion and breast cancer, later updating it with one suggesting that studies on the issue were inconclusive. The change brought howls of outrage from abortion rights and breast cancer advocates, and later, NCI assembled its workshop to reinvestigate what most scientists already considered a closed issue.

So while Brind may carry on his campaign, he has clearly lost the scientific debate. Moreover, the NCI incident has contributed to an unprecedented mobilization of scientists against the Bush administration. Cited by the Union of Concerned Scientists, Rep. Henry Waxman, and others, it has served as Exhibit A in the story of White House manipulation of science.

By now, Joel Brind's should seem a familiar story. In Chapter 11, we examined the longest-running and hardest-fought political science battle in American history: over the teaching of evolution. No simplistic tale of science versus religion, the struggle over the content of biology education in this country demonstrates a gradual change in creationist tactics over time, as deeply religious groups and individuals have increasingly appropriated scientific credentials and arguments to make their case, a strategy that has culminated in the "intelligent design" movement. Similarly, in Chapter 12, we surveyed the campaign to elevate adult stem cells over embryonic ones for the purposes of research. Once again, religiously conservative adult stem cell advocates have employed "science" in service of what can only be described as a moral agenda.

Brind also seems to fit this pattern, and so do other conservative scientists of the religious Right, some of whom have enjoyed even greater sway with the Bush administration and GOP politicians. While often scientifically questionable, their claims—that abortion causes mental illness and other negative health outcomes in women, that social science supports the push for "abstinence only" education programs, that condoms are not very effective in preventing HIV and other STDs, that emergency contraception may make younger teens more promiscuous—now frequently furnish the Right's leading arguments on these issues.

In short, where religious conservatives may once have advanced their pro-life and socially traditionalist views through moral arguments, they now increasingly adopt the veneer of scientific and technical expertise. This development represents a "maturation" of conservative Christian thinking, notes Michael Cromartie, an expert on the religious Right at the Ethics and Public Policy Center. "They are saying that their faith is not just a pietistic private exercise, but that it has implications in the world of education, or politics, or the world of science."

With the development described by Cromartie, today's scientists of the Christian Right may have gained a PR edge. But they have largely abandoned a distinguished model for how devout religious believers should act in the scientific arena: Ronald Reagan's Surgeon General C. Everett Koop. Because of Koop's antiabortionist views, liberal Democrats like Henry Waxman loudly protested his appointment. But by the time Koop left his position in 1989, Waxman had recanted. That is because on issues of concern to the religious Right—AIDS and the purported health risks of abortion—Koop approached the evidence dispassionately, even if it meant angering his ideological compatriots. "I am the nation's surgeon general, not the nation's chaplain," Koop explained in 1989.

The Joel Brinds of the world, though, have diverged from the Koop model. And even as Brind tries to link abortion to breast cancer, the prolific (and pro-life) David Reardon, of the Illinois-based Elliot Institute, has launched a parallel campaign to show, contra Koop and reams of science, that abortion *does* cause mental illness, chemical dependency, and a range of other poor health outcomes in women. It is true that women sometimes feel temporarily depressed or guilty after an abortion. But the notion that abortion regularly causes severe or clinical mental problems has been rejected by, in addition to Koop, a group of experts convened by the American Psychological Association (APA).

Unlike some of the Christian Right's other scientists, Reardon and his coauthors have successfully published a number of peer-reviewed scientific papers in their area of advocacy (though these have largely failed to sway other scientists). Reardon also granted me a long interview as I worked on an article for the *Washington Monthly,* defending his scientific work while also describing his theological worldview. Later, he followed up with a lengthy e-mail message explaining how he differs from others in the pro-life movement. "I can honestly say that I care about both women and their children and am constantly lecturing pro-lifers about the need to make their concern for women at least equal to their concern for the baby," he wrote.

Reardon emerged on the scene in 1987 with his book *Aborted Women, Silent No More,* billed as a "comprehensive review" of abortion's aftereffects that included testimonials from "aborted women." Conservatives widely applauded the book. The next year, Reardon founded yet another of the quasi-academic think tanks and advocacy institutes that seem omnipresent when it comes to the Christian Right and science: the Elliot Institute, whose full name is—fittingly enough—the "Elliot Institute for Social Sciences Research." At the time, Reardon had a background in electronic engineering.

Yet he soon began to conduct actual scientific research, publishing his first peer-reviewed study in 1992. In 1995 he acquired a Ph.D. in biomedical ethics from Pacific Western University and now calls himself "Dr. David C. Reardon" in print. The U.S. Government Accountability Office recently investigated Pacific Western University in a report on "diploma mills and other unaccredited schools," and indeed, the institution does not appear in the Council for Higher Education Accreditation's database of institutions accredited by recognized U.S. accrediting organizations. To my questions about his degree, Reardon replied that he put in "a substantial amount of work" for it. By e-mail, he added that "my coauthors have degrees from accredited universities and many teach at major universities."

Whatever his credentials, Reardon has published extensively in scientific journals, and his arguments have had significant influence. In 2004, conservative Republican representative Joe Pitts, of Pennsylvania, sponsored a bill in the House of Representatives to provide $15 million in federal funding for research on postabortion depression, even though the APA calls abortion "a safe medical procedure that carries relatively few physical or psychological risks." In fact, numerous experts convened by the APA in 1990 to study this issue—including Nancy Adler, of the University of California at San Francisco; Henry David, of the Transnational Family Research Institute; Brenda Major, of the University of California at Santa Barbara; and Nancy Felipe Russo, of Arizona State University—say that Reardon's work fails to overturn the APA consensus position. Reardon has, in fact, published studies showing statistical correlation between abortion and negative mental health and other outcomes. But Reardon regularly implies or asserts that abortion generally *causes* such results, a far stronger and less defensible claim.

Consider a 2003 Reardon study published in the *Canadian Medical Association Journal.* Reardon and coauthors report that women who undergo abortions are admitted for psychiatric care more frequently than those who do not. But as numerous critics have pointed out, such statistics do not prove that abortion causes mental problems. In a rebuttal of the study, Brenda Major noted that Reardon's group failed to control for the different life circumstances of women who choose abortion and those who bring their pregnancies to term. (Women who choose abortion, for example, are less likely to be married or in an intimate relationship, factors themselves linked to poorer mental health.) As Major noted, "It is inappropriate to imply from these data that abortion leads to subsequent psychiatric problems"—especially given a range of other well-conducted studies that find few postabortion emotional problems for most women.

Reardon countered that in our interview. "Proving causation is always very difficult," he says. "You can't do a blind random study where you impregnate a certain number of women and forcibly abort half." What you can do, however, is use statistical analysis to control for confounding factors. And sure enough, "in well-designed studies that control for variables Reardon fails to take into account, legal abortion is not found to be associated with degradation in mental health," notes Arizona State University psychologist Nancy Felipe Russo. Indeed, as far as confounding factors go, consider the fact that thanks to antiabortion advocacy, women seeking to terminate unwanted pregnancies often have to face virulent protesters outside of clinics, a potentially traumatic experience that only intensifies the difficulty of the abortion experience.

Perhaps Reardon's different reading of the data on abortion and mental health relates to his pro-life advocacy. At the website Afterabortion.org, Reardon's Elliot Institute devotes a whole section to proposed legislation that would restrict abortion or require various types of warnings and disclosures before a woman could obtain one. In fact, Reardon appears to believe, for religious reasons, that abortion *must* cause harm. In a 2002 essay in the conservative journal *Ethics & Medicine,* Reardon defended what he called the "neglected rhetorical strategy" of opposing abortion on the grounds that it hurts women, instead of simply because it is morally wrong. He also noted that "because abortion is evil, we can expect, and can even know, that it will harm those who participate in it. Nothing good comes from evil."

That is theological, not scientific, thinking. Moreover, it suggests an inclination on Reardon's part to assume the answer to the question about abortion's risks, and then conduct research designed to support that conclusion. Asked about that passage, Reardon stated that he is able to "put on different hats" and act as a scientist and as a religious believer at different times. If that were so, one would expect his conclusions to resemble more those of C. Everett Koop.

The closely connected issues of sex education and condom effectiveness pose a similar challenge for religious conservatives, who for moral reasons do not want juveniles or the unmarried having sex, but for political reasons wish to bolster their claim with appeals to science. Chief among these is the argument that abstinence is the only absolutely effective means of preventing pregnancy and sexually transmitted diseases (STDs), and that "comprehensive" sex-ed programs—which by definition teach the benefits of both abstinence and contraceptive use, rather than exclusively telling students to avoid sex until marriage—are ineffective, or even dangerous.

That is where Dr. Joe S. McIlhaney comes in. McIlhaney is president of the Austin-based Medical Institute for Sexual Health, which he founded in 1992. The institute doesn't do much original research, instead interpreting data from sources like the CDC and NIH. The group strongly promotes teaching teenagers to abstain from sex, while criticizing educational programs that promote "safe sex" and questioning the ability of condoms to prevent most STDs (at least at an effectiveness level that the Medical Institute deems acceptable). Under the Bush administration, McIlhaney has also served on the advisory committee to the director of the Centers for Disease Control and Prevention and the President's Advisory Council on HIV and AIDS. You might call him well connected.

In an interview, McIlhaney, a Christian, strenuously protested being characterized as a scientist motivated by religion. "We talk about sexual abstinence, and immediately we're classified as just what you told me about, Christian conservative," he countered. "We're a medical, scientific organization." Technically, of course, McIlhaney is right to single out abstinence as the only foolproof way to protect against sexually transmitted diseases and unintended pregnancy. But McIlhaney goes far beyond this indisputable assertion to cast doubt on less perfect, but nevertheless effective, means of prevention: condoms and comprehensive sex education programs, which teach teens about abstinence but also about contraception (since some will inevitably fail to abstain).

McIlhaney cites the "failure" of comprehensive sex-ed programs. Yet if these programs have failed, it is hard to know what constitutes success. Douglas Kirby, a teen sexuality expert at the California-based research organization ETR Associates, explains that strong data show that some comprehensive sex-ed programs work to change behavior by increasing contraceptive use or delaying or reducing the frequency of sex among teens. McIlhaney knows that, but he counters that "just because you've increased condom use by maybe five or ten or twenty percent for kids in a program, does not mean you're actually going to impact the STD rate for them." But definitively showing that these programs—already proven effective in the short term—affect long-term STD or pregnancy rates would require expensive longitudinal studies with huge sample sizes that funders are rarely willing to bankroll. In short, it is an unreasonable demand.

Yet ironically, the evidence is even scantier when it comes to the effectiveness of abstinence education. "There are no studies meeting reasonable criteria that show that any program has delayed the initiation of sex,"

Kirby told me in 2004. He is hardly alone in this assessment. In September 2004, the sexual health group Advocates for Youth released a study of federally funded abstinence education programs, finding that U.S. states' own reviews of these programs showed "few short-term benefits and no lasting, positive impact. A few programs showed mild success at improving attitudes and intentions to abstain. No program was able to demonstrate a positive impact on sexual behavior over time."

Nevertheless, the Bush administration has humored McIlhaney and other critics of comprehensive sex education. The Centers for Disease Control and Prevention even went so far as to do away with an initiative called Programs That Work, which listed sexual education programs that had been shown to be effective in the peer-reviewed scientific literature. According to an analysis by Congressman Henry Waxman's office, as of 2002, Programs That Work had identified five "comprehensive" sex education programs and no "abstinence only" ones.

And it is not just that these abstinence programs do not work. They seriously and inexcusably misinform students about sex, perpetuating dangerous myths and stereotypes. In late 2004, Waxman's office released a report on federally supported abstinence education programs that have been heavily embraced by the Bush administration and will be funded to the tune of $170 million in 2005. In evaluating the curricula of these programs, the report found that the vast majority exaggerated the failure rates of condoms, spread false claims about abortion's health risks (including mental health problems), and perpetuated sexual stereotypes ("girls are weak and need protection"; "men are sexually aggressive and lack deep emotions"; and so forth). Perhaps most outrageously, one curriculum even claimed that "sweat" and "tears" could transfer the HIV virus. You might think that this would be a fringe claim even on the Right, but Senate majority leader Bill Frist, himself a physician, repeatedly refused to repudiate the notion of such transmissions in an interview with ABC's George Stephanopoulos in December 2004. Only when pressed did Frist finally admit that it would be "very hard" to get AIDS from sweat and tears.

As the misleading content of many abstinence programs suggests, a closely related prong of the Right's attempt to use "science" to undermine

responsible sex education involves a crusade to highlight condom ineffectiveness. For instance, McIlhaney's Medical Institute publishes a brochure
questioning whether condoms make sex "safe enough." In an interview,
McIlhaney chuckled softly when repeatedly asked about how well condoms work. "It's just this simple sort of little latex device, and we're talking about the futures of young people," he said.

Yet according to a 2001 review by the NIH, condoms have only a min-
imal slippage and breakage rate (roughly 2 to 4 percent), and have been
positively proven effective in the prevention of HIV, gonorrhea, and unwanted pregnancy. For numerous other STDs, data was deemed inadequate as of the time of the NIH's review. However, the study noted that
condoms have been shown to block "particles of similar size to those of
the smallest STD viruses," suggesting that consistent and proper condom
use should protect against these other diseases as well.

Moreover, emphasizing the inadequacies of condoms can paint a skewed
picture when their more widespread use among teens and adults who do
not abstain from sex would certainly help prevent the spread of STDs. "To
say condoms don't work is really a misleading statement. To say they don't
work perfectly is accurate. But in sexual situations where there's either a
known or perhaps unknown high risk of exposure, do condoms lower that
risk? The answer is yes," says Ward Cates, a physician who heads the Institute for Family Health at the nonprofit Family Health International.

Once again, McIlhaney's criticisms, and those of his ideological compatriots, seem to have had considerable influence with the Bush administration. Both the CDC and the State Department's Agency for
International Development have altered informational materials on condoms to downplay their effectiveness. In the case of the CDC, the edited
fact sheet removed a discussion of research showing that education about
condoms *does not* increase sexual activity, as well as removing information
about how to use condoms properly. This manipulation of data has
shown up regularly in the litany of complaints about the Bush administration's abuses of science.

The sad fact is that Joe McIlhaney actually represents a relatively moderate position among conservatives who support a pro-abstinence, anti-
condom agenda. He sounds like a flaming liberal compared to Tom

Coburn, a conservative physician elected in 2004 to the U.S. Senate from Oklahoma and formerly cochair of Bush's President's Advisory Council on HIV and AIDS (on which McIlhaney also sits). Following the NIH's 2001 review on condom effectiveness, Coburn declared in a press release that "Condoms Do Not Prevent Most STDs," and claimed that the report had finally unmasked "the 'safe' sex myth for the lie that it is." In fact, Coburn had taken the lack of data about condom effectiveness in preventing certain conditions as proof of *ineffectiveness*, despite strong grounds for inferring the contrary.

We have already seen how James Inhofe politicizes and abuses the science of climate change. Inhofe's Oklahoma counterpart, Coburn, carries the tradition into the public health arena. In fact, Coburn has been a true innovator in the "scientific" crusade against the condom based on claims that it doesn't work well. "I do think that he has taken a leadership role in concocting many of these things," says Heather Boonstra, a senior public policy associate at the Alan Guttmacher Institute, a nonprofit reproductive health research group.

As a former member of Congress, Coburn advocated warning labels on condom packages and even called for the resignation of CDC director Jeffrey Koplan, claiming that he had withheld "the truth of condom ineffectiveness." No wonder that when Bush named Coburn to cochair his HIV/AIDS advisory group, the *Atlanta Journal Constitution* scathingly noted, "The message that condoms work has been proved to save lives; likewise, any false message to the contrary has real potential to kill people. Those who would spread and defend that false message become conspirators in unnecessary deaths."

The Right's appeal to "science" to advance its agenda on sexual behavior extends far beyond opposition to condoms. It also lies in the background of the Food and Drug Administration's wildly controversial May 2004 decision to overrule a committee of scientific advisers and block the sale of "Plan B"—an emergency contraceptive, or "morning after" pill—over the counter. But to understand what may have been the FDA's single most controversial move during George W. Bush's first term, you first have to make the acquaintance of yet another of the religious Right's scientists and a man who played a key role in the decision: the pro-life obstetrician and gynecologist W. David Hager, a part-time professor at the University of Kentucky College of Medicine who also happens to serve on the advisory board of McIlhaney's Medical Institute for Sexual Health and has served on the Physicians Resource Council for James Dobson's Focus on the Family.

Of all the characters introduced in this chapter, chances are greatest you have heard of Hager, so widely was he attacked by women's groups and even chain e-mails following his 2002 appointment to a seat on the FDA's Reproductive Health Drugs Advisory Committee (a tenure that was extended in 2004). Given his track record, Hager's nomination had "controversy" written all over it. Prior to his appointment, Hager had lobbied the FDA to reverse its approval for the so-called abortion pill RU-486, criticized the birth control pill for promoting promiscuity, and appeared to endorse a modest form of faith healing. A book he coauthored with his ex-wife Linda Carruth Hager entitled *Stress and the Woman's Body*—published by Fleming H. Revell, a division of Baker Publishing Group, whose books "furnish resources to all . . . who seek to live for the Lord and worship him"—actually suggests Bible readings for treating premenstrual syndrome.

Granted, Hager talks about science, too. The book's chapter on PMS begins with three case studies from Hager's medical practice and a medical discussion of the condition. At the end, however, it offers a "spiritual perspective" as well, warning women that they must also "address the spiritual ramifications of this problem." This dovetails with the book's preface, where Hager explains, "Because Linda and I are Christians, we will direct you toward a personal relationship with Jesus Christ, along with other things you can do to increase the peace and joy in your life." Hager also authored the book *As Jesus Cared for Women,* which goes even further by endorsing concepts like "demonic possession," "miracle" births, and "miraculous" healing and extols the alleged healing power of prayer. Of course, Hager warns that "Even with divine intervention, the healing process may take a long time."

These prior statements and activities—each suggestive of a strong tendency to blend religion and medicine—made Hager's nomination a flashpoint. And in fact, Hager's role in the FDA's high-profile decision to block over-the-counter approval of Plan B suggests that women's groups such as the National Organization for Women and Planned Parenthood that protested his appointment were justified in doing so. The episode also shows just how much influence a single politicized member of a federal advisory committee can potentially have, thereby legitimating widespread complaints about the Bush administration's packing of these committees (discussed in more detail in the next chapter).

Consisting of two tablets of the synthetic hormone levonorgestrel, the drug known as Plan B dramatically reduces the chance of getting pregnant when used as soon as possible after sexual intercourse. It has virtually no serious side effects. At a December 16, 2003, FDA advisory meeting of Hager's committee and over-the-counter drug experts—who had convened to decide whether the drug should be made more broadly available—one doctor even called Plan B "the safest product that we have seen brought before us."

Nevertheless, Hager personally refuses to prescribe emergency contraception out of the pro-life concern that it might interfere with the implantation of a fertilized egg. Plan B's manufacturer, Barr Pharmaceuticals, acknowledges such a possibility. However, the drug works chiefly by blocking (or delaying) ovulation following unprotected sex, not by interfering with implantation.

No one disputes, however, that in a country where nearly half of pregnancies are unintentional, a number amounting to about three million each year, broader access to Plan B could have a dramatic effect. The Alan Guttmacher Institute estimates that emergency contraception prevented more than one hundred thousand unintended pregnancies and fifty-one thousand abortions in the year 2000, when Plan B could be acquired only by prescription.

Yet in May 2004, the FDA turned down Barr Pharmaceuticals' application to make Plan B available over the counter. The agency's ruling—which ignored a 23-4 recommendation from its scientific advisers (Hager was in the minority) as well as recommendations by expert FDA staff—was quickly denounced as a political sop to the religious Right. "Normally, I don't criticize my successors at FDA, but I think that was not a good call," says Donald Kennedy, executive editor in chief of *Science* magazine and formerly commissioner of the FDA under President Carter.

The FDA's stated reason for ignoring its scientific advisers—that Barr had not provided adequate data on how younger adolescents would use the drug—was largely argued by Hager at the December advisory committee meeting. Early on, Hager raised questions about "age restriction on the availability" of Plan B. He continued to do so on multiple occasions, finding encouragement from Dr. Louis Cantilena, chairman of the FDA's Nonprescription Drugs Advisory Committee, who directed the meeting. This culminated in a long digression to discuss the possibility of age restrictions during a vote on an unrelated topic.

In its "Not Approvable" letter—signed not by a lower-level staffer, as is customary, but by Dr. Steven Galson, acting director of the FDA's Center for Drug Evaluation and Research—the FDA appeared to seize on this interlude. Galson's letter observed that despite the vote to approve the drug, "some members" of the committee (i.e., Hager and Cantilena) had raised questions about "inadequate sampling of younger age groups." Hager thus appears to have played a key role in providing the FDA higher-ups with a scientific-sounding rationale for their controversial decision.

Indeed, Hager has himself admitted to this central role, though he also grants some of the credit to God. In an October 29, 2004, speech delivered at Kentucky's Asbury College, an evangelical Christian school that he attended and where he now sits on the board of trustees, Hager stated that after the committee's vote on Plan B, "I was asked to write a minority opinion that was sent to the commissioner of the FDA." Hager then went on to describe its contents:

The opinion I wrote was not from an evangelical Christian perspective. I pointed out that we don't have long term data in adolescent women, we don't have information to say what will happen with these young girls if they use emergency contraception and then don't come to their doctors for screening for other diseases and to be counseled against being sexually active. . . . *I argued it from a scientific perspective, and God took that information, and he used it through this minority report to influence the decision.* You don't have to wave your Bible to have an effect as a Christian in the public arena. We serve the greatest Scientist. We serve the Creator of all life. We serve the Author of all truth. All we're required to do is to proclaim that truth. [Italics added]

Divinely inspired or not, the rationale adopted by the FDA for blocking broader access to Plan B has since come under scathing criticism. It is true that Barr's study, conducted at family planning clinics and pharmacies, managed to include only 29 out of 585 subjects between the ages of fourteen and sixteen, in part because younger adolescents are not as sexually active as those in older age groups. Yet citing this concern is a "political fig leaf," says James Trussell, director of the Office of Population Research at Princeton, who cast one of the twenty-three votes in Plan B's favor. "There were no data on that age group for the sponge when it was approved. There were no data for that age group for the female condom," says Trussell. "That's just talking contraceptives. If you talk about all the other drugs, there's no data either."

The scientific subtext of the Hager-FDA concern—also cited by religious Right groups such as the Concerned Women for America and conservative members of Congress—is that young girls might become more promiscuous if they had access to Plan B or cease using ordinary contraception, thus exposing themselves to more sexually transmitted diseases. Yet that wasn't the case with older women in the studies presented, or in a recent randomized study of emergency contraception use by over two thousand women published in the *Journal of the American Medical Association*. Moreover, "there's no evidence that this drug has different side effects in younger girls," says Dr. Alastair Wood, of Vanderbilt University, who sits on the FDA's Nonprescription Drugs Advisory Committee and voted for Plan B's approval. In an interview, Hager himself agreed that there is no particular reason to think that young teens would behave differently with the drug, but points to the lack of evidence. "I'm saying we don't have that information," he says.

Yet it would be virtually impossible, and probably unethical, for Barr to explicitly target young adolescents in a future study. "Obviously, you can't go advertising for women who have had unprotected sex to please come, you know, to a counter," noted Dr. Carole Ben-Maimon, of Barr Pharmaceuticals, at the FDA advisory committee hearing. "Visualize the consent form for a study in which you're going to study fourteen- to sixteen-year-olds with emergency contraception, specifically," quips Wood.

Discussing his role in the FDA's decision, Hager commented, "You're making it out as though I'm some kind of powerful individual, and I'm not." If George W. Bush's FDA was truly set on blocking broader access to Plan B, it certainly could have found another scientific-sounding excuse in Hager's absence. But in this case, it was Hager who delivered that excuse, precisely what everyone who criticized his nomination had feared in the first place. Of course, Hager insists that he can keep his religious beliefs and science separate. "I make diagnoses based on evidence-based medicine," he told me. Perhaps, but on Plan B he also demands far more evidence than his scientific peers consider reasonable.

On many of the issues discussed above, ranging from alleged links between abortion and breast cancer to condom effectiveness, critics such as Henry Waxman and the Union of Concerned Scientists have lamented the Bush administration's willingness to abuse science and distort data. However, these critics have not always recognized the deeper issue. Had the administration not been able to rely on the expertise and input of Christian conservative scientists—whose views often contrast starkly with those of mainstream researchers—such actions would probably not have been possible. Certainly they would not have seemed so easily justifiable.

In his famous 1896 work *A History of the Warfare of Science with Theology in Christendom,* Cornell University president Andrew Dickson White cast the past as a long struggle between scientific truths and religious dogmas. Needless to say, his thesis today seems rather simplistic. White does not appear to have anticipated the emergence of "creation science," much less individuals like Joel Brind, David Reardon, and W. David Hager—religious men who believe that they serve "the greatest Scientist" and seek to advance their views *through* science, not by defeating or suppressing it.

Ironically, the dramatic triumph of science itself has inadvertently created strong incentives for politicization and abuse. Americans have great faith in the scientific enterprise, but throw up their hands in despair and confusion when scientists appear to disagree on key issues. And the Christian scientists cited above do disagree, making forceful arguments for their own positions that only other experts can competently refute. By cherry-picking the expertise they provide, and ignoring the weight of scientific opinion, conservative politicians have indeed found the ideal "fig leaf" for their decisions. Instead of the best knowledge and expertise available, they can now go, whenever they choose, with the most convenient.

Update

Of all the stories related in this chapter, the issue of Plan B "emergency contraception" has been the most incapable of staying out of the news. Since this book first appeared, two whistleblowers have left the Food and Drug Administration to protest the agency's continuing failure to make this drug more widely available: Susan Wood, former director of the FDA's Office of Women's Health, and Frank Davidoff, an advisory committee member and physician who had voted for the drug's approval. Both accuse the FDA of corrupting science and inexcusably delaying a decision on Plan B. And their allegations gained considerable support from a 2005 Government Accountability Office report on the Plan B case, which exposed either (depending upon how used to this stuff you are by now) alarming malfeasance or just more typical Bush administration misdeeds with respect to science.

The GAO's audit labeled the FDA's behavior in the Plan B case "unusual"—government code for suspicious—because the agency overruled both its scientific advisers and its own expert staff in order to deny the drug's approval for over-the-counter availability. From 1994 to 2004, GAO noted, this was the only one of 23 cases in which the FDA had overturned its advisory committee on an over-the-counter drug approval application.

And the GAO report went further. It noted that the Plan B decision's alleged foundation—that insufficient data existed on the question of whether the drug might pose risks to younger adolescents—was "novel and did not follow FDA's traditional practices." In the past, FDA had no problem extrapolating data from older to younger age groups (as indeed its own scientific advisory committee had overwhelmingly recommended in the Plan B case). But most startlingly, GAO found that the "not approvable" decision may have been preordained: High-level management personnel were "more involved" than usual, and the decision itself may even have been made before the agency's internal reviews had been completed. In short, politics may have come first, followed by a post hoc scientific-sounding rationale.

As this chapter demonstrates, that scientific rationale, at least in the context of the FDA's deliberations, appears to have originated from W. David Hager. And sure enough, we can now see from Hager's much-discussed "minority opinion" letter to then-FDA commissioner Mark McClellan that in it he had made a similar argument. In an August 2005 note to reproductive health advocates, former FDA commissioner Lester Crawford provided the Hager letter (which had been the subject of much media speculation) and stated that it had come in unsolicited and had in fact always been part of a publicly available docket at the agency. Crawford downplayed the letter's significance; but recall that Hager had previously stated his belief that with God's help, his missive had influenced the FDA's decision. Whether coincidentally or otherwise, the letter makes precisely the argument that ultimately prevailed, and that the GAO had described as a novel rationale for FDA decision-making: It raises concerns about a lack of data on how younger adolescents will use the drug.

Meanwhile, Hager's term on the FDA's Reproductive Health Drugs advisory committee came to a close in 2005. But the FDA has since further delayed any approval of Plan B by sending the decision into a regulatory rulemaking process with no fixed endpoint in sight. As of this writing, we definitely have not heard the last of it.

The Plan B story dominates all others, but when it comes to science abuses on matters of reproductive health, the Bush administration remains busy. In mid–2005, the U.S. Department of Health and Human Services launched a website, entitled 4Parents.gov, ostensibly designed to tell parents how to talk to their teens about sex and (of course) to promote abstinence. In designing the site, the government partnered with a nonprofit group, called the National Physicians Center for Family Resources, whose website promotes the questionable notion of a link between abortion and breast cancer. Sure enough, scientists examining the new 4Parents.gov site on behalf of Rep. Henry Waxman quickly pointed out that it was rife with misinformation. For instance, the site repeatedly exaggerated the ineffectiveness of condoms and made unsupported claims like the following: "Parents must make it clear to their children that oral sex is as dangerous in terms of disease as is intercourse." According to Waxman's team of experts, that's simply false. Without promoting adolescent oral sex, our government ought to be able to state the facts about it accurately. (No indication exists that the 4Parents.gov website was subsequently revised in light of these expert criticisms.)

Meanwhile, a potential new front has opened in the Christian right's war on sexual health. Two pharmaceutical companies stand on the verge of a great breakthrough: marketing vaccines to protect against certain strains of the human papilloma virus (HPV), a sexually transmitted virus that can lead to cervical cancer. Already, Christian conservatives—who in the past have used the threat of HPV to attack the effectiveness of condoms—are queuing up to oppose mandatory vaccinations, fearful that they will contradict an abstinence-only message. Family Research Council head Tony Perkins has stated that he would not have his own daughter inoculated, and Focus on the Family has declared that "The seriousness of HPV and other [sexually transmitted infections] underscores the significance of God's design for sexuality to human well-being." In other words, rather than a public health scourge to be fought, at least some conservative Christians appear to view sexually transmitted diseases as God's way of punishment for moral transgression.

Meanwhile, the Bush administration has placed another of the Christian right's scientists on the Centers for Disease Control and Prevention committee that, as of this writing, had not yet released its recommendations on how the HPV vaccine should be administered. His name is Reginald Finger, and he was formerly "medical issues analyst" with Focus on the Family.

PART FOUR

THE ANTISCIENCE PRESIDENT

Although scientific input to the government is rarely the only factor in public policy decisions, this input should always be weighed from an objective and impartial perspective to avoid perilous consequences. Indeed, this principle has long been adhered to by presidents and administrations of both parties in forming and implementing policies. The administration of George W. Bush has, however, disregarded this principle.

UNION OF CONCERNED SCIENTISTS, "RESTORING SCIENTIFIC INTEGRITY IN POLICYMAKING"

CHAPTER 14

BUSH LEAGUE SCIENCE

O N FEBRUARY 18, 2004, the conservative war on science, which had been gathering momentum for decades, finally jolted the media and American public to attention. All it took was a little star power.

That morning, the Cambridge, Massachusetts–based Union of Concerned Scientists (UCS) held a phone-in press conference to announce a dramatic development. Over sixty leading scientists and former government officials, among them twenty Nobel laureates, had signed a statement denouncing the administration of George W. Bush for misrepresenting and suppressing scientific information and tampering with the process by which scientific advice makes its way to government officials. Examples included distorting the science of climate change, quashing government scientific reports, and stacking scientific advisory panels. "Other administrations have, on occasion, engaged in such practices, but not so systematically nor on so wide a front," the statement read.

The UCS document, which solely targeted George W. Bush, did not purport to address a modern conservative or Republican war on science. But in fact, that is what it had uncovered. From acid rain to global climate change, from "creation science" to "intelligent design," conservatives had been meddling with science for years, largely on behalf of their pro-industry and religiously conservative supporters. Now an impressive roster of scientists and former policymakers had laid the problem at the doorstep of a conservative Republican president.

We have already encountered a number of the UCS statement signatories in these pages, and not by accident. Many are leading scientists who have had disconcerting past run-ins with politics. Nobel laureates Mario Molina and F. Sherwood Rowland, who watched the Gingrich Congress challenge their work on ozone depletion, signed the UCS statement. So did many climate experts, including Rosina Bierbaum, Michael Oppenheimer, Michael MacCracken, and James McCarthy, who work in the single most politicized scientific field today. So did Gordon Orians and Stuart Pimm, familiar with "sound science" attempts to undermine the Endangered Species Act. And so did Jack Gibbons, who had seen the Office of Technology Assessment, which he had shepherded for more than a decade, cast aside by conservatives who never understood its value. Science politicization, Gibbons told me, has "never been this blatant or this bad. We almost wistfully think back to the Reagan years."

Other distinguished UCS statement signatories included celebrity scientists such as Paul Ehrlich and E. O. Wilson, and luminaries whose contributions to science and policy alike span nearly half a century, such as physicists W. K. H. Panofsky and Richard Garwin. The latter had earned their stripes in historic political-science battles, over issues like "Star Wars" and the supersonic transport. No doubt these long experiences had only heightened their concerns about George W. Bush.

Finally, during the UCS press conference, another big name joined the cause: Russell Train, the lifelong Republican and environmentalist who could no longer fathom the direction his political party had taken. Forty-nine Nobel laureates, 63 National Medal of Science recipients, and 171 members of the National Academy of Sciences, along with thousands of others, would eventually sign on to the UCS document.

The mobilization of these scientists was no freak occurrence. In the words of UCS senior scientist Peter Frumhoff, a "palpable unease" in the scientific community had been mounting long before the UCS statement appeared. Over the first three years of Bush's presidency, a wide array of cases had piled up in which the administration had been accused of abusing science. Many have already been discussed—and the charges substantiated—in the preceding pages. All in all, the totality of accusations suggested not just a pattern, but a sweeping and unprecedented threat to the role of science in policymaking, and even to the legitimacy of science itself.

From the very beginning of the Bush administration, scientists had their concerns. Prior to the attacks of September 11, 2001, the White House had alienated them with its approach to embryonic stem cell research, global climate change, and missile defense. Moreover, the president's team took over a year to fill key scientific posts: Food and Drug Administration commissioner, National Institutes of Health director, and surgeon general. Finding an FDA commissioner took almost twenty months. Clearly, this was not a priority.

Furthermore, much like the Reagan administration years earlier, the Bush team showed little concern about bringing science into the White House, delaying for months the naming of a presidential science adviser. The Senate did not confirm John Marburger, a physicist and former Brookhaven National Laboratory director, as well as former president of the State University of New York at Stony Brook, until late October 2001, long after Bush had already staked out a position on the leading science-based issues of the day. (In contrast, the Senate confirmed Clinton's first science adviser, Jack Gibbons, on January 28, 1993, just eight days after the president's inauguration.)

Marburger's difficulties, and scientists' concerns, only mounted when it became clear that his position had been diminished. For over a decade, the national science adviser, who heads the White House Office of Science and Technology Policy (OSTP), had held the rank of "assistant to the president," like the national security adviser and other top White House aides. But as Marburger conceded to me in a 2001 interview, "that title was never offered to me." At the time, George H. W. Bush's science adviser, D. Allan Bromley, and Clinton's second science adviser, Neal Lane, voiced serious concerns about the downgrading of Marburger's role and how it might impede his access to the president.

Blows to the office structure of OSTP—science's beachhead in the executive office of the president—raised further hackles. Though previous

OSTP directors had four Senate-confirmed deputies, Marburger chose just two: one for science and one for technology. This ruled out associate directors for the environment as well as for national security and international affairs, and seemed a sure sign of declining influence. Marburger insisted that the restructuring came at his own initiative, but science policy insiders had trouble believing it. "Everybody in Washington knows that the number of Senate-confirmed appointments you control is a direct measure of your capacity to participate," says Harvard's John Holdren, who served on President Clinton's Committee of Advisors on Science and Technology and signed the UCS statement.

Along with worries about science advice to the president, concerns also arose about the quality of advice to the government more generally. In September 2002, a front-page *Washington Post* exposé revealed a "broad restructuring" of scientific advisory committees within the Bush Department of Health and Human Services so as to align them ideologically with the White House. It is hard to say what triggered this activity, but perhaps the administration had chosen to heed the advice of former Advancement of Sound Science Coalition executive director Steven Milloy, who in May 2001 had warned in the *Washington Times*, a leading conservative newspaper, about a "lingering infestation" of Clinton-friendly "science moles" within the federal government.

Subsequent news reports fingered additional cases of administration politicization of federal advisory committees—a little-noticed alphabet soup of boards, panels, and study groups sometimes dubbed the "fifth branch" of American government. It wasn't just appointees like W. David Hager. In one oft-cited case, William R. Miller, a University of New Mexico psychologist tapped to join a National Institute on Drug Abuse advisory panel, had to answer questions about his stance on abortion and whether he had voted for Bush. He hadn't, and he didn't end up on the panel.

The assault on these little-known committees—hundreds of which advise various branches of the federal government on science and technology issues—greatly piqued scientists, who view them as a chief entry point of science into the political process. In early 2003, *Science* executive editor-in-chief Donald Kennedy denounced the Bush administration's willingness to stack the scientific deck in an editorial entitled "An Epidemic of Politics." "What's unusual about the current epidemic is not that the Bush administration examines candidates for compatibility with its 'values,'" Kennedy argued. "It's how deep the practice cuts; in particular, the way it now invades areas once immune to this kind of manipulation."

Advisory committees weren't the half of it. In late 2002, the *New York Times* reported on several cases in which reproductive health information on government websites had shifted, apparently to appease religious conservatives. We have already discussed these examples, which include information about condom effectiveness and the relationship (or lack thereof) between abortion and breast cancer. Meanwhile, a short report by Democratic representative Nick Rahall, ranking member of the House Committee on Resources, drew attention to a series of alleged science abuses by the Bush Department of the Interior, several of which related to the Klamath River Basin water fight. In particular, Rahall singled out the case of Mike Kelly, the whistleblower scientist who charged, after the 2002 Klamath fish kill, that the science underlying the disastrous low flow regime had been inadequate but that his concerns had been suppressed for political reasons.

Then in August 2003, Democratic representative Henry Waxman followed up with a far more sweeping report covering advisory committees, the Departments of the Interior and Health and Human Services, global warming, stem cell research, and much else. "The subjects involved span a broad range, but they share a common attribute: the beneficiaries of the scientific distortions are important supporters of the President, including social conservatives and powerful industry groups," noted the Waxman report in its introduction. Waxman had definitely gotten the political side of the story right.

All of this information was publicly available, but scientists were also talking to colleagues in federal agencies and hearing troubling stories. In particular, such tales had come to the ears of Kurt Gottfried, a physicist at Cornell University and chair of the Union of Concerned Scientists. Gottfried pushed the group to get involved, and with the support of other UCS leaders, this initiative led to a day-long meeting on November 11, 2003, in Washington, D.C. There, a number of scientists, most of whom would

later sign the UCS statement, assembled to compare notes, trying to articulate precisely how the Bush administration had crossed a new line.

Later, critics would allege that the UCS statement represented yet another manifestation of a peculiar (and rather embarrassing) phenomenon:
the tendency of some academics to sign petitions without really considering what they are endorsing. "I have too much experience with all sorts of
distinguished people who, when some document is thrust before them,
simply sign it, usually without reading it," the first President Bush's late
science adviser, D. Allan Bromley, a UCS critic, told me. Yet this argument about academic groupthink does not apply very well to the UCS
statement. Hardly off the cuff, the document instead sprang from a day-
long conference in which scientists assembled to analyze and articulate
their concerns, and to draft a document that accurately expressed them.
After conferring with their colleagues, the participants recognized the
prevalence of science abuses across a wide variety of fields and disciplines.
"The pattern was shockingly similar," says Harvard climate expert James
McCarthy, a UCS statement signatory.

Given the distinguished roster of scientists assembled by the Union of
Concerned Scientists, the group could safely expect to draw considerable
attention. Yet not even the UCS suspected that the science politicization
issue would have so much traction—that it would, in fact, figure in the
upcoming presidential campaign.

After the *New York Times* posted an online story about the UCS declaration, it apparently became the most e-mailed article on the paper's website.
Soon editorials appeared in major newspapers about the scientists' critique,
even as John Marburger, leaping to the president's defense, dismissed a
lengthy UCS accompanying document prepared to substantiate the charges
as a "conspiracy report" (a description he defended in response to written
questions submitted for this book). "I think there are incidents where people have got their feathers ruffled," Marburger told the *New York Times.*
"But I don't think they add up to a big pattern of disrespect."

But those convinced of an antiscience "pattern" would soon have another key piece of evidence. Just days after the UCS statement's release,
on February 27, came the "Blackburn affair," in which the White House

replaced two President's Council on Bioethics members supportive of research on human embryos with three individuals expected to view such research far more skeptically or even to oppose it outright. Though set in motion prior to the release of UCS's report, the bioethics council shakeup was interpreted in the media as an arrogant and dismissive response to it. Soon over 170 bioethicists signed a statement of their own protesting Blackburn's ouster.

Before long, another round of all-out "science wars" began, reminiscent of the 1984 fights over "Star Wars." On March 4, Marburger found himself debating his predecessor, Neal Lane, Clinton's second science adviser and a UCS statement signer, on the Diane Rehm Show. "We think the allegations are almost entirely wrong and wrong in detail," Marburger flatly stated.

But having listened to the segment, Rehm's next guest, Howard Gardner, a Harvard psychologist, wasn't impressed. "It's kind of pathetic," said Gardner, "to hear Dr. Marburger, who is the president's science adviser, trying to refute eighty-five different accusations, in each case saying, 'Well, you really have to know the details.'" Then he continued:

The question is, what's the big picture? What is your role, Dr. Marburger, and at which point would you say, "I'm out of here, because science has nothing to do with the way we're doing things any more"? Now, I don't know enough to be able to argue each of those cases, but I think what every listener has to ask, whatever your work is, "What's the line that you wouldn't cross because then you couldn't look at yourself in the mirror any more?" And I actually feel very sorry for Marburger, because I think he's probably enough of a scientist to realize that he's basically become a prostitute.

The debate got ugly quickly, then. But Gardner's harsh language notwithstanding, he had made a powerful point. In trying to rebut examples one by one, Marburger had put himself in the position of trying to hold back an ocean. In effect, he had taken a "corpuscular" approach to the case against the Bush administration, rather than stepping back to look at the weight of the evidence.

A registered Democrat, and by all accounts eminently qualified for the role of presidential science adviser, Marburger cut a rather tragic figure in the whole dispute. He had some prominent allies—the late D. Allan Bromley, who denounced the UCS's "politically motivated statement" to the *New York Times,* and lobbyist and former GOP Science Committee chair Robert Walker—but not very many. The George C. Marshall Institute, too, came to the science adviser's aid with a press conference at the National Press Club, largely dedicated to debunking the criticisms made by UCS and accusing the group of having its own (liberal) political agenda. Robert Walker headlined.

The debate peaked on April 2, when Marburger's office released a detailed rebuttal to the charges expressed in a report the Union of Concerned Scientists had published to back up its scientist sign-on statement. Calling the allegations "wrong and misleading," Marburger stated that they "certainly do not justify the sweeping [conclusion]" that science abuses under Bush had been systematic and pervasive. "It is my hope that the detailed response I submit today will allay the concerns of the scientists who signed the UCS statement," Marburger wrote.

That hope turned out to be a stunningly false one. Instead, with his rebuttal Marburger may have condemned himself to be remembered, much like "Star Wars" defender George Keyworth, as a science adviser who failed to speak truth to power or to stick up for the nation's scientists.

In a written response to questions for this book, Marburger—a staid and matter-of-fact character whom I once watched deliver an absolutely bloodless speech at Rice University on science budget trends—took a philosophical outlook on being called a "prostitute" and his general controversial standing in the scientific community. "You have to remember that I was a university president for fourteen years," he replied. "During that time I dealt with outstanding scientists and scholars who were essential to the mission of the university on a wide range of controversial issues and I felt no resentment." He went on to call the scientists who had criticized the administration "colleagues whom I respect." Following the spat over the UCS statement, Marburger even held a private meeting with some of his distinguished critics, a typical university administrator's strategy for dealing with "ruffled feathers."

But unfortunately, Marburger's for-the-record response to charges of political interference with science—officially prepared by his office for members of Congress—only inflamed the situation. Gathering together responses from federal agency officials to the UCS's charges, and having his office conduct a "thorough investigation into all the allegations," Marburger summarily dismissed virtually every UCS accusation as factually incorrect. But against a backdrop of mounting concern among scientists over the administration's practices, this simply failed to pass the smell test. Indeed, the "concerned" scientists assembled by UCS did not back down in the least on reading what Marburger had to say. "One supposes he was ordered to produce a rebuttal, but they could have produced a more nuanced rebuttal than that crass, heavy-handed, and grossly wrong one that they issued," says Harvard's Holdren.

First, if Marburger truly aimed to deal with the perceived problem of political science abuse, then his response took far too narrow a tack. He challenged the UCS report that had accompanied the scientists' statement of concern, but that document provided just one of many windows on Bush administration science abuses, and not necessarily a full picture. For example, while Henry Waxman had criticized Bush's misrepresentation of the number of available embryonic stem cell lines in his report, the UCS did not. And neither report had brought up the Bush administration's attempts to undermine the World Health Organization's work on the link between diet (and particularly sugar-sweetened drinks) and health.

And even on issues where Marburger did defend the administration, most of his answers fell far short of the mark. For example, the UCS report addressed, at great length, the subject of climate change. It charged that the Bush administration had "consistently sought to undermine the public's understanding of the view held by the vast majority of climate scientists" that human greenhouse gas emissions are fueling global warming. In response, Marburger pointed out that Bush had acknowledged the scientific consensus view in a June 11, 2001, Rose Garden speech. But even then, Bush artfully avoided a direct acknowledgement that climate change resulting from human activity is happening *now*. And usually, the administration did not even admit this much, preferring to magnify remaining scientific uncertainties without accurately describing what scientists know with confidence.

Worse, the White House continued to misrepresent the state of scientific understanding on the issue of climate change even *after* Marburger's assurance that the president had acknowledged the consensus view. Consider Bush's responses to *Science* during the 2004 presidential race, after the magazine asked both campaigns whether "human activity" is "increasing global temperatures." "In 2001, I asked the National Academy of Sciences to do a top-to-bottom review of the most current scientific thinking on climate change," Bush replied. "The nation's most respected scientific body found that key uncertainties remain concerning the underlying causes and nature of climate change." Bush went on to quote passages about scientific uncertainty from that report without mentioning its strong conclusions about global warming resulting from human activity.

Such flagrant misrepresentation goes far beyond mere dishonesty. It demonstrates a gross disregard for the welfare of the American public, whom Bush represents, and for the population of the entire globe, whose fate depends in large measure on the behavior of the American behemoth. Yet in response to questions, Marburger defended Bush's reply to *Science* and even commented, "I think it is perfectly consistent to be concerned about the emission of greenhouse gases and to express uncertainty about the relation of those emissions to 'global temperatures.'" Consistent, perhaps, but the selective emphasis on uncertainty clearly distorts the conclusions stated by the National Academy of Sciences and other expert bodies.

Subsequently, the administration proceeded along a similar tack with respect to the alarming warming of the Arctic region. On November 4, 2004, the *Washington Post* reported that the administration had worked actively to prevent the Arctic Climate Impact Assessment—a report commissioned by a council of eight nations, including the United States, and drawing on the work of some three hundred scientists—from serving as a basis for policy recommendations to address the melting of the Arctic region that is now underway. According to the *Post*, the administration

questioned whether enough "evidence" existed to justify such recommendations and "repeatedly resisted even mild language that would endorse
the report's scientific findings." Those findings, of course, blamed recent
Arctic warming largely on human greenhouse gas emissions. (The State
Department, much like Marburger, denied the accuracy of the *Post* report
while failing to engage on the details.)

Marburger's response also foundered on a key science abuse case study
related to global warming. Relying on an internal EPA memo leaked to
the National Wildlife Federation and first reported on by the *New York
Times,* the UCS charged that the White House had attempted to edit the
climate change section of a draft version of the EPA's *Report on the Environment,* an ambitious public education document, to distort the science.
(Ultimately, the EPA decided not to include a section on climate change
in its final report.) The UCS report discussed the leaked EPA memo at
length and published the document itself as an appendix, thus making it
a centerpiece of the group's argument. That wasn't a bad choice: The
memo makes for some disturbing reading.

Because of White House edits, the memo complains, the draft *Report
on the Environment* "no longer accurately represents scientific consensus
on climate change." The White House had "discarded" references to conclusions by the National Academy of Sciences, and used "natural variability . . . to mask the scientific consensus that most of the recent
temperature increase is likely due to human activities." Moreover, the
memo noted, "uncertainty is inserted (with 'potentially' or 'may') where
there is essentially none." Finally, and as discussed in Chapter 7, the
memo complained that the White House had excised a "hockey stick" diagram of the thousand-year temperature record and instead inserted a reference to the dubious paper by Willie Soon and Sallie Baliunas.

Moreover, these weren't just suggestions, but commands. According to
the EPA memo, the White House had insisted that "no further changes
may be made" to the draft report, forcefully asserting its right to doctor
the conclusions expressed by EPA scientific experts.

In his response, Marburger did not question the authenticity of the EPA
memo, readily downloadable from the website of the National Wildlife Federation. Rather, he simply ignored it, claiming, incredibly, that an "ordinary

review process indicated that the complexity of climate change science was not adequately addressed in EPA's draft document." But whoever wrote the EPA memo hardly considered the process "ordinary." Neither did Republican former EPA administrator Russell Train, who commented of the affair, "I can state categorically that there never was such White House intrusion into the business of the E.P.A. during my tenure."

Marburger's failure even to address this revealing memo hardly suggests an honest and genuine engagement with the UCS's charges. In response to questions for this book, he did not remedy this defect. Asked why EPA scientists would have leaked the memo if they had considered the White House's review process "ordinary," Marburger replied, "I can't speak to the motivations of individuals, or to their understanding of the process they were involved in." Imagine a lawyer taking such a strategy in court and ignoring the other side's key piece of evidence. He or she would lose the case hands down.

To be sure, Marburger did deflect some of the UCS's charges. For example, the group claimed that in late 2003, the Department of the Interior had improperly replaced a team of experienced scientists working on the Missouri River ecosystem with a scientific "SWAT team" expected to reach a different conclusion concerning actions needed to protect several endangered species. But as Marburger noted, the new team included many of the earlier scientists, and its recommendations did not differ much from previous ones. Fair enough.

Yet on other familiar issues—"sound science" reforms to the Endangered Species Act, claims related to abortion and breast cancer, advisory committee politicization—Marburger's defense once again fails. Consider a few more detailed examples:

Undermining the Endangered Species Act. In its report, the UCS noted that the Bush administration had supported a bill to inject "sound science" into the Endangered Species Act, legislation that "would make it harder to list threatened and endangered species, in particular by greatly limiting the use of population modeling." (This wasn't Greg Walden's bill, discussed in Chapter 10, but an earlier, similar version.) In responding to UCS, Marburger either totally missed or totally evaded the point.

His rebuttal document contained a three-paragraph defense of the administration's endangered species *policies,* but never even addressed the central concern about the abuse of *science*: that the administration had endorsed changes to the ESA's science standards that wildlife biologists have overwhelmingly opposed. And in response to questions for this book, Marburger simply promised that "the administration supports using the best available science in making environmental decisions." That's not much consolation given its endorsement of the "sound science" bill, which would alarmingly modify the ESA's current "best available science" standard.

Abortion and Breast Cancer. In what has become one of the best-known case studies in the scientific argument against the Bush administration, the UCS charged that an online National Cancer Institute fact sheet had, for several months, suggested the possibility of a link between abortion and breast cancer—a link that almost certainly does not exist. The offending document stated:

> The possible relationship between abortion and breast cancer has been examined in over thirty published studies since 1957. Some studies have reported statistically significant evidence of an increased risk of breast cancer in women who have had abortions, while others have merely suggested an increased risk. Other studies have found no increase in risk among women who had an interrupted pregnancy.

This clearly suggests the possibility of a risk to women, depicting the science as unsettled. An outraged letter from Henry Waxman and other members of Congress described the statement as "nothing more than the political creation of scientific uncertainty."

The objectionable document first appeared in late 2002, after a previous and much more thorough fact sheet—one that accurately discounted an "ABC" connection by actually bothering to evaluate the merits of conflicting studies—had been taken down in June of that year, apparently in response to a complaint by pro-life members of Congress. The story doesn't end there, though: The offending fact sheet didn't last long either. After a February 2003 NCI expert workshop that weighed the "ABC" question and once again found that "induced abortion is not associated with an increase in breast cancer risk," a new fact sheet appeared in March 2003, once again discounting any risk. The question then becomes, why did the controversial document ever appear in the first place?

In his response, Marburger never even acknowledged the posting of the temporary and misleading document that had spurred so much controversy. Instead, he simply noted that the NCI had pulled its original fact sheet—the accurate one—"when it became clear that there was conflicting information in the published literature." (When asked, he declined to elaborate on this response by citing actual "conflicting" studies.)

This assertion about the "published literature" puts Marburger in a ridiculous position. Between June 2002 and February 2003, the NCI had taken three different stances on abortion and breast cancer: no risk, conflicting studies/potential risk, no risk. But the complexion of the scientific literature had remained essentially unchanged during this brief time period. The only thing that had changed was the administration's willingness to distort science.

Political Vetting of Advisory Committee Candidates. Relying on previous accounts, the UCS told the story of William Miller, who had been asked his views on abortion and whether he had voted for Bush in an interview for a position on a National Institute on Drug Abuse advisory panel. Miller had complained about the interview to numerous journalists, and by e-mail provided his "best recollection" of what happened for this book:

> I received a telephone call from a White House staffer who said that I was being considered for appointment to the National Drug Abuse Advisory Council, and that he needed "to vet you to determine whether you hold any views that might be embarrassing to the President." The sequence of questions was:
>
> 1. Do you support faith-based initiatives? (Yes. This is apparently why I was being considered. I believe my name was raised by the Faith-Based Initiatives office, who had called me earlier and asked if I would be willing to be nominated.)
> 2. Do you favor legalization of drugs? (No. At this point the staffer said, "Good, you're two for two.")

3. Do you support needle exchange? (Yes, there is clear evidence that it reduces the spread of infection. "Now you're two for three. The President opposes needle exchange on moral grounds regardless of the consequences.")

4. Do you support capital punishment for drug kingpins? (No. I am opposed to capital punishment for any reason. "Hmm. I suppose it might be all right if you're opposed to capital punishment in general on moral grounds, and not just for drug kingpins.")

5. Are you pro-life? (No. I am not pro-abortion, but I do support the right to choose.)

6. Did you vote for President Bush? (No)

7. Why didn't you support the President? (I discussed domestic and international policy issues on which we differ.)

"I'll need to check to see if your views are acceptable, and I'll get back to you."

I never heard back.

In his "rebuttal," Marburger did not contradict Miller's firsthand account. Rather, the White House science adviser simply denied that a "litmus test" had been the reason Miller didn't receive a callback. Fine, maybe he didn't get the position for other reasons. But clearly the real issue, which Marburger did not acknowledge or address, was that someone asked Miller these inappropriate and offensive questions in the first place. And in fact, following the release of the UCS report, other scientists emerged from the woodwork to report undergoing similar politicized vetting.

As these examples suggest, a close reading of Marburger's response suggests that the president's science adviser has opted to defend the castle, rather than to engage seriously with critics. This may say something about how Marburger's rebuttal was prepared. After investigating UCS's charges—a process that Marburger says included "briefings with senior agency officials in which I participated"—the science adviser did little more than present the government's official line on each separate incident, in the process minimizing or simply ignoring the scientists' complaints.

But by proceeding in this way, Marburger has already lost the argument. He accuses the UCS of failing to "seek and reflect responses or explanations from responsible government officials," but he never provides any good reason for trusting these officials, rather than all of the scientists who went public with their charges (in some cases risking retribution) and the independent journalists who told their stories (and who *had* generally sought responses from the government, in accordance with standard journalistic practices).

At best, Marburger simply asks us to believe one account of events, for which there is little evidence, instead of another, for which the evidence is considerable. This leaves readers of the UCS report and Marburger's rebuttal with a stark choice. Either a large number of government and university scientists are wrong, as are the journalists who deemed them credible, or else Marburger is. Unfortunately for Marburger, he happens to be a White House official obviously engaged in damage control.

Moreover, he represents an administration that did not even bother to respond substantively to claims of science politicization (by Henry Waxman and numerous others) until they had been blared all over the media and endorsed by twenty Nobel laureates. When asked by a *New York Times* reporter about the Waxman report in August 2003, Bush spokesman Scott McClellan simply dismissed the document as "riddled with distortion, inaccuracies, and omissions." "I'd be particularly interested in what the omissions were," quips Waxman.

In short, judging by the volume of complaints, the selectivity of Marburger's response, his inability to soundly rebut most of the charges that he did tackle, and the fact that even under the Reagan administration alleged science abuses had not been nearly as widespread or numerous, we can infer that the Bush administration *almost certainly* had politicized science to an unprecedented degree. While no precise means of quantifying science abuse exists, the vast number of credible allegations strongly point in that direction, as does the historical analysis provided earlier in this book. Even the Gingrich Republicans, after all, controlled only Congress, not the executive branch. They had little means of interfering with science in federal agencies or with their advisory committees.

But what about the presidency of Bill Clinton? Conservatives have often brought charges of political interference with science against the last Democratic administration. A 2003 book published by the George C. Marshall Institute entitled *Politicizing Science: The Alchemy of Policymaking* contains a number of such allegations, particularly related to the issue of global warming. Thus, for example, Patrick Michaels charges that the U.S. National Assessment on climate change impacts represented "junk science" because "the models that serve as its basis are inconsistent with observations." We have already addressed this criticism in Chapter 7. Similarly, Joseph P. Martino discusses the Klamath River Basin dispute and concludes that the Fish and Wildlife Service made a decision in that case that was scientifically unjustified (and this was, of course, a Fish and Wildlife Service filled with Clinton holdovers acting very early in the Bush administration). Once again, we have dealt with this criticism previously.

Politics certainly brushed up against science during the Clinton administration, just as it does in every presidency. But as the foregoing instances show, in many cases where conservatives have cried "science abuse" they have actually been the guilty party. Weed out these examples and you're not left with much of a rap sheet for the Clinton administration. And of course, scientists hardly mobilized against Clinton the way they did against George W. Bush.

Moreover, it just so happens that we have a perfect case study that can be used to contrast the Clinton and second Bush administrations' respective approaches to science: the question of whether the government should support controversial needle-exchange programs to reduce the spread of HIV among (and by) intravenous drug users. Evaluations by the National Institutes of Health, the World Health Organization, the Centers for Disease Control, and other respected health bodies have found that these programs not only work, they *don't* encourage more drug abuse. As a report from the National Academy of Sciences put it in 1995, "well-implemented needle exchange programs can be effective in preventing the spread of HIV and do not increase the use of illegal drugs."

Despite powerful evidence of their effectiveness, neither the Bush II nor Clinton administrations had the guts to support these programs with federal funding. But only one administration felt compelled to abuse science to justify its stance. In announcing the decision to leave the funding of needle exchange programs up to "local communities" in 1998, Health and Human Services Secretary Donna Shalala fully acknowledged the science up front. "We have concluded that needle-exchange programs, as part of a comprehensive HIV-prevention program, will decrease the transmission of HIV and will not encourage the use of illegal drugs," she stated—even as she went on to explain, awkwardly, that the programs would not be supported: "We had to make a choice. It was a decision. It was a decision to leave it to local communities."

In contrast, the Bush administration simply twisted the science. In an extraordinary February 2005 editorial, the *Washington Post* revealed that to justify the decision to oppose needle-exchange programs (which are especially disliked by religious conservatives), a Bush official directed the paper "to a number of researchers who have allegedly cast doubt on the pro-exchange consensus." So the *Post* actually called up these scientists and found that, lo and behold, they think no such thing. The Bush administration peddled other questionable evidence to the paper, too, which the *Post* also skewered. It was just the latest case in which the Bush administration had sought to distort scientific information in order to justify a policy stance, precisely what the Clinton administration managed to avoid on the very same issue.

And if these considerations all suggest political interference with science has reached a nadir with the Bush administration, others do as well. Over the course of this book, we have seen a convergence of multiple trends that have all triggered increasing attacks on science: the rise of conservative think tanks; dogged attempts by industry groups to find new means of battling over the scientific basis of regulations, rather than regulations themselves; and a growing tendency among religious conservatives to find a "scientific" argument on each moral question of interest to them. All of these developments have merged under the Bush administration, which has shown strong sympathies with both industry groups and social conservatives even as both groups (and the think tanks that provide their intellectual firepower) have increasingly turned to using, or rather abusing, science to their own advantage. Considered this way,

Bush administration abuses and distortions of science come to look more and more like a large-scale political strategy.

In this context, Marburger's denial only made the problem worse. He should have stood up as science's defender or else resigned. Instead, he helped provide a cover for the administration, which pretended that no problem existed even as science abuses became more and more corrosive and damaging.

Because of their sweeping, systemic nature, and because they have pervaded much of the federal government, the Bush administration's abuses push the issue of political interference with science to the point of crisis.

First, these abuses spread vast amounts of misinformation to the American public. If women cannot trust the National Cancer Institute for accurate information on breast cancer risks, then where can they turn? Similarly, the Bush administration has misinformed the public about the reality of human-induced global warming, about the number of stem cell lines that would be available for research under Bush's policy, and much else. In short, the administration has rendered itself untrustworthy when it comes to scientific claims, thus seriously eroding its credibility.

But perhaps even more serious than misinforming the public, we must rate the disturbing implications of the Bush administration's actions for the relationship between science and policy, and (relatedly) for the role of scientists serving in government. Politicized questioning of advisory committee nominees, and politicized editing of government reports, suggest a willingness to torque analyses to make them seem supportive of preexisting policy positions. This approach flagrantly undermines the proper role of science in government: as a valuable resource to inform decision-making. When politicians use bad science to justify themselves, rather than good science to make up their minds, we can safely assume that wrongheaded and even disastrous decisions lie ahead.

The actions of the Bush administration, and especially its meddling with the activities of scientists at federal agencies, even threaten to impair government itself. If agencies like the EPA become viewed as irretrievably corrupted by politics—simply places that receive orders barked by White House officials from across town—they could find themselves unable to recruit the talent they need to address the nation's technical problems in areas ranging from pollution control to drug safety to homeland security.

Finally, all of these assaults culminate in a severe blow to science itself. By failing to respect the integrity of science, and instead repeatedly undercutting it and employing it opportunistically, the Bush administration erodes public confidence in the scientific endeavor and leaves it crippled and undermined. This fosters outright relativism about the value of science as opposed to other ways of knowing—outright "faith," for example. "What's intriguing about the Bush administration, given their views on most issues," explains Thomas Murray, president of the bioethics think tank the Hastings Center, "is that they have a postmodern take on science. It's the first postmodern science administration we've ever known. They don't seem to understand science, quite frankly—or if they do, they really seem not to care. They just want to use it for political purposes."

And lest this concern about a postmodern approach to science seem overly alarmist, consider that we have actually seen this type of thinking take root. In a famous October 2004 *New York Times* article on the Bush administration, journalist Ron Suskind described his encounter with a "senior adviser" to the president:

> The aide said that guys like me were "in what we call the reality-based community," which he defined as people who "believe that solutions emerge from your judicious study of discernible reality." I nodded and murmured something about enlightenment principles and empiricism. He cut me off. "That's not the way the world really works anymore," he continued. "We're an empire now, and when we act, we create our own reality."

This adviser seems to have been referring more to foreign policy, but the lesson carries over to science as well. Who needs careful, painstaking inquiry into the nature and causes of problems, or how the world works, when power and ideology can mold the nature of truth itself? Whoever is speaking here has become the most dangerous sort of relativist. To such a person, the realities accessed (however imperfectly) by scientific inquiry can mean nothing if those in power so declare. And these are our leaders.

Given this disdain for the "reality-based community," it will come as no surprise that after Marburger's rebuttal to the Union of Concerned Scientists, the Bush administration promptly proceeded to misuse, abuse and interfere with science some more. Scientists at the Department of Health and Human Services were barred from consulting with the World Health Organization without prior political approval. The Food and Drug Administration ignored its scientists and blocked broader access to Plan B emergency contraception. The National Marine Fisheries Service embraced the dubious notion of counting hatchery fish to determine the viability of wild salmon populations. And so on.

Perhaps most alarming of all, partisans rushing to aid and defend the Bush administration against charges of science abuse took stances *in favor* of the politicization of scientific advisory committees. In responding to the *Washington Post*'s inquiries back in September 2002, Department of Health and Human Services spokesman William Pierce had already stood up for the proposition that advisory committees should be staffed in a manner consistent with the president's political views. Soon this became, in a sense, the party line. Testifying before a National Academy of Sciences panel in July 2004, Michigan Republican congressman Vernon Ehlers, himself a physicist and generally regarded as a champion of science, defended the practice of asking advisory committee appointees about their voting records and party affiliation. "I think it's an appropriate question. I don't think scientists should consider themselves a privileged class—that politics is for everyone else and not for them," Ehlers stated. In effect, he blessed the notion of dividing science into "Republican" and "Democratic" camps. (To some extent, scientists may well divide in this way, but there is no reason to make matters worse.)

When the National Academy released its final report, it unequivocally rejected Ehlers's position. "It is no more appropriate to ask S&T [science and technology] experts to provide nonrelevant information—such as voting record, political-party affiliation, or position on particular policies—than to ask them other personal immaterial information, such as hair color or height," wrote the committee. But the damage had been done. A prominent Republican *and scientist* had taken a stance in favor of classifying scientists by political affiliation, welcoming this view into the party mainstream and further eroding the necessary distinction between the findings of science and the political decisions made because or in spite of those findings.

Ehlers wasn't the only Bush defender to countenance making science still more politically vulnerable in the future. In late September 2004, Robert Walker, former Republican chairman of the House Committee on Science and acting at the time as a Bush campaign representative, noted ominously that the science community could face a "push back at some point in the future" for its criticisms of the administration. In an interview, Walker clarified that he was not issuing a "threat"; rather, he said, he was warning scientists of the consequences of their actions. "Where you are willing to utilize half-truths as a part of your agenda," Walker said, "the fact is that you reduce your credibility." Walker's statement about half-truths is unobjectionable, but as we have seen, the scientists' charges against the Bush administration were largely accurate. Furthermore, one wonders whether Walker ever considered applying his observation to the multitude of half-truths promulgated by the Bush administration and its Republican allies in Congress.

With the Bush administration showing no inclination to mend its ways, soon science became an issue in the 2004 presidential campaign. Forty-eight Nobel laureates, many of them signers of the original UCS statement, officially endorsed the candidacy of Democrat John Kerry, and a group called Scientists and Engineers for Kerry–Edwards emerged, cochaired by Harold Varmus (president and CEO of Memorial Sloan-Kettering Cancer Center, former director of the National Institutes of Health, and corecipient of a Nobel Prize for studies of the genetic basis of cancer) and several other distinguished scientists. A related organization called Scientists and Engineers for Change even took to the campaign trail, deploying leading scientists to deliver speeches in crucial swing states such as Ohio and Florida. At least one courageous scientist who prominently criticized Bush in the days before the election—NASA expert James Hansen, who spoke out on the subject of global warming—was actually on the government payroll. It was a repeat, albeit imperfect, of 1964, when leading scientists had challenged Barry Goldwater and his alarmingly anti-intellectual followers.

Meanwhile, Kerry repeatedly lambasted Bush for limiting the federal funding of embryonic stem cell research, an issue that served as a proxy for broader discontent among America's scientists. In his Democratic national convention acceptance speech, Kerry posed the question, "What if we have a president who believes in science, so we can unleash the wonders of discovery like stem cell research to treat illness and save millions of lives?" The stem cell issue would not suffice to lift Kerry to victory, and it hardly helped that the Democratic candidate often unrealistically spoke as though cures for debilitating diseases might be right around the corner. Still, the emergence of stem cell research as a central issue in the Democratic campaign signaled a new national attention to the Bush administration's distortions and abuses of science.

Denials have since persisted about the scope of the problem, of course. When large numbers of the nation's most distinguished scientists leave their laboratories to protest political interference with science, one would think that such an action would cause those accused to reflect on the harm they might be doing. But instead, critics have thrown the charge right back in the scientists' faces by citing their association with the Union of Concerned Scientists. "It protests that Bush is being political with science, but the union is itself political with science," wrote Gregg Easterbrook, of the *New Republic*, a frequent contrarian commentator on science issues.

While the Union of Concerned Scientists has traditionally been associated with "liberal" causes like environmentalism and arms control, the protest against the Bush administration's abuse and politicization of science was of a different order altogether. Indeed, the scientists who teamed up with the group were not all liberals—witness Republicans Russell Train and Richard Garwin—and the UCS's activities merely represented the most prominent attempt by scientists to draw attention to the problem of political science abuse. In truth, most of the group's charges against the Bush administration had been originally exposed by journalists, with advocacy organizations like the UCS, and Democrats like Henry Waxman, later compiling the examples of abuse into reports.

In the final analysis, those defending the administration would have us blindly trust its word over that of independent scientists and reporters, even when the administration's arguments have been exposed as largely lacking in substance and tailored to a political agenda. Yet only the most dedicated Bush loyalist would weigh the evidence in this way. Anyone else would consider the vast array of charges against the Bush administration, the multiple sources from which they have arisen, their detailed substantiation by journalists and other investigators, and the credibility of the scientists and journalists who have endorsed them as different strands of evidence that, woven together, suggest a very damning "big picture."

At least on an analytical level, then, the debate over whether the Bush administration had inappropriately interfered with and undermined science—and more importantly, whether it had done so to an unprecedented extent—ended with the critics victorious. On a political level, however, the debate merely created more denials and polarization, even triggering disturbing defenses of an even deeper entanglement between science and politics on the part of Vernon Ehlers and others.

Having exhaustively discussed both the modern Right's abuses of science and their consequences, then, we now turn in a more constructive direction: toward suggesting what can be done before the problem of political science abuse becomes thoroughly intractable—if it has not become so already.

WHAT WE CAN DO

THE POLITICAL MISUSE AND ABUSE of science presents a severe challenge to modern democratic governments, which depend on a creative tension between elected representatives on the one hand, and unelected technocratic elites on the other. While we cannot allow scientific experts to rule us directly, we nevertheless need them desperately. Our leaders simply cannot do their jobs competently without considerable reliance on expertise that they themselves do not possess. But political interference with science corrupts the channels of communication between credible experts and policymakers, weakening and ultimately destroying this necessary relationship.

The advent of the modern conservative movement, its takeover of the Republican Party, and its ultimate triumph under the administration of George W. Bush have brought us to a point where a true divorce between democratic government and technocratic expertise seems conceivable. Indeed, it appears to be actually underway. To those who understand that such a split will lead to economic, ecological, and social calamity, it presents a terrifying prospect— and leaves us with only two options.

First, we could cry out warnings to conservatives, begging that they step back from the abyss before it is too late. Apparently, the forty-nine Nobel laureates who endorsed the Union of Concerned Scientists' statement were unable to cry out with a loud enough voice to achieve this objective. Scientists must continue to issue warnings and to decry abuses of science, but we should not delude ourselves that their jeremiads, however convincing, will solve the problem.

Instead, we must push for safeguards that strengthen the role of legitimate expertise in informing government decision-making, protect that expertise from manipulation and abuse, and more generally seek to restore a spirit of candor and collaboration between the scientific community and our elected officials. A number of groups have called for such steps, demanding both new institutions and new laws to safeguard the role of science in policymaking. In particular, a report by the Federation of American Scientists, *Flying Blind*, has exhaustively analyzed the chaos that ensues when expert input and political decision-making become separated—a case in point being George W. Bush's 2001 stem cell decision—and proposed helpful solutions touching both Congress and the executive branch of government.

First and most obviously, we must revive the Congressional Office of Technology Assessment, or a close equivalent. Members of Congress clearly lack an impartial and credible source of scientific analysis and expertise, and this deficiency has created a vacuum that has often been filled by politicized sources. For those who claimed that OTA did not always work according to the congressional schedule, a revival of the office can include any necessary restructuring. As Arizona State University science policy scholar David Guston notes of OTA studies, "They don't always have to be two hundred pages long and eight months in coming, but you can't just pick up the phone and call your bud."

In fact, some scientists and politicians have begun to clamor for OTA's return. The authors of a 2003 collection, *Science and Technology Advice for Congress*, outline a range of options for improving the science savvy of elected representatives, from simply resurrecting OTA to creating a similar organ in the Government Accountability Office or Congressional Research Service. They also suggest increasing the role of the well-respected but undeniably slow-paced National Academy of Sciences.

Meanwhile, Democratic representative Rush Holt, of New Jersey, a physicist, has introduced several bills outlining different approaches for restoring an OTA-like capacity to Congress. "One of the reasons for defunding OTA was that people like Gingrich accused it of being partisan," says Holt. "And I would argue that because they did away with it, it made it possible for science on Capitol Hill to become partisan."

As we have seen, in the wake of OTA's demise, Congressional Republicans held "science court" hearings pitting industry-friendly scientists against the mainstream—a tradition that continues in the hands of James Inhofe and others. As another needed reform, Congress should implement mechanisms to ensure full disclosure of any potentially relevant conflicts of interests by witnesses invited to testify at hearings *at the time of their testimony*. Such a step would at least partially deter the worst excesses of the "science court" tradition.

And just as science advice to Congress needs strengthening, so does the role of science in the executive branch of government. As far as advice to the president goes, the science adviser must regain the rank of assistant to the president, while the White House Office of Science and Technology Policy (OSTP) must regain its previous strength. As the Federation of American Scientists notes, Congress should also consider acting to raise the stature of OSTP further, increasing both its prominence within the White House and its public role. The ever increasing importance of scientific information to political decision-making justifies such a promotion.

Similarly, we must safeguard scientific advisory committees, which have proven particularly vulnerable to political manipulation. Proposed legislation by Democratic representatives Henry Waxman and Bart Gordon would move in this direction by barring political litmus tests for committee membership, and requiring tough disclosure and conflict-of-interest policies. This "scientific integrity" bill also has a number of other commendable features, such as extending whistleblower protection to federal employees who allege abuses of science, so that they will not face the threat of retaliation.

With respect to all of these proposed reforms to strengthen and safeguard the nation's scientific advisory apparatus, we should bear in mind the eloquent warning of Lewis Branscomb, of Harvard's Kennedy School of Government, a UCS statement signer and physicist who has served in scientific advisory capacities in both Democratic and Republican administrations:

The integrity of the science advisory process cannot withstand overt actions to censor or suppress unwanted advice, to mischaracterize it, or to construct it by use of political litmus tests in the selection of individuals to serve on committees. Nor can it survive threats to the job security of scientists in government when they attempt to call such political interventions to the attention of Congress or the press. Science advice must not be allowed to become politically or ideologically constructed. If we fail in the attempt to preserve the integrity of science in democratic governance, a strong source of unity in the electorate, based on common interest in the actual performance of government, will be eroded. Policymaking by ideology requires that reality be set aside; it can be maintained only by moving towards ever more authoritarian forms of governance.

Is Branscomb's worry about creeping authoritarianism overblown? Simply recall the words of the anonymous Bush administration official who, disdaining the "reality-based community," averred that "we're an empire now, and when we act, we create our own reality," and you will see that we have plenty to worry about.

Steps to safeguard science advice, however, are just the beginning. Still in the arena of legislative reforms, we must roll back the incursions of the "sound science" regulatory reform movement. Measures like the Data Quality Act, which enable attacks on scientific information that are designed to create paralysis by analysis, must be repealed. The "peer review" superstructure recently erected under the Data Quality Act must be dismantled pending a governmentwide study, by the National Academy of Sciences or another competent body, of the proper role for peer review in regulatory decision-making. If such an inquiry were to find a need for significant reforms in spite of the regulatory delay that they would cause, then such reforms should be considered, with an emphasis on sensitivity to the differing needs of individual federal agencies.

We must also remain vigilant in opposition to proposed laws, such as reforms to the Endangered Species Act, that use 'science' as a way to tie the hands of government regulation agencies. As a general principle, elected representatives have no business specifying, in minute detail, how federal agencies should evaluate scientific information. We staff these agencies with scientific experts for a reason. Let's let them do their jobs.

These proposals alone, however, cannot solve the problem. We must also work to reduce the current incentives for science politicization, and even consider steps to deter political science abuses in the future.

Political attacks on science succeed, at least in part, because they confuse the public and policymakers, leading them to believe that a scientific "controversy" exists where one actually does not, or that widely discredited claims are still given serious consideration in the world of science. This would not happen so frequently, however, if journalists—the chief purveyors of scientific information to the American public in controversial and politicized areas—performed their job better.

Throughout this book we have seen repeated examples of strategic attempts to spin reporters. The 1998 American Petroleum Institute memo, discussed in Chapter 7, discussed a plan to "maximize the impact of scientific views consistent with ours with Congress, the media and other key audiences." Similarly, the Discovery Institute's Wedge Document explicitly discussed media strategies. And no wonder: The evidence suggests that many journalists reporting on science issues fall easy prey to sophisticated public relations campaigns. For instance, in a 2004 paper published in the journal *Global Environmental Change,* the scholars Maxwell T. Boykoff and Jules M. Boykoff analyzed coverage of global warming in the *New York Times,* the *Washington Post,* the *Wall Street Journal,* and the *Los Angeles Times* between 1988 and 2002. During this fourteen-year period, climate scientists successfully forged a powerful consensus on human-caused climate change. But reporting in these four major papers failed to reflect this consensus.

The Boykoffs analyzed a random sample of 636 articles. They found that a majority, 52.7 percent, gave "roughly equal attention" to the scientific consensus view that human activity contributes to climate change and to the opposed (and often industry-supported) view that natural fluctuations suffice to explain the observed warming. By comparison, just 35.3 percent of articles emphasized the scientific consensus view while still presenting the other side in a subordinate fashion (a far more appropriate story structure). Finally, 6.2 percent emphasized the industry-friendly view (simply absurd), and a mere 5.9 percent focused on the consensus view without providing the industry/"skeptic" counterpoint (justifiable, perhaps, but probably not ideal in all circumstances).

Most intriguing, the Boykoffs' study found a shift in coverage between 1988, when climate change first garnered wide media coverage, and 1990. During that period, journalists broadly moved from focusing on scientists' views of climate change to providing "balanced" accounts. During this same period, the Boykoffs noted, climate change became highly politicized, and a "small group of influential spokespeople and scientists emerged in the news" to question the mainstream view that industrial emissions are warming the planet. The authors conclude that the U.S. "prestige press" has produced "informationally biased coverage of global warming . . . hidden behind the veil of journalistic balance."

Reporters need to understand better how science abusers exploit the journalistic norm of "balance"—demanding equal treatment for fringe or widely discredited views—and adjust their writing accordingly. Let's face it: Journalistic "balance" has no corollary in the world of science. On the contrary, scientific theories and interpretations survive or perish based on the process of peer review, by which scientific claims are carefully scrutinized before being published in reputable journals; on whether the results of scientific experiments can be replicated by other scientists; and ultimately, on whether they win over scientific peers. When consensus builds, it is based on repeated testing and retesting of an idea.

For this reason, journalists should treat fringe scientific claims with considerable skepticism and find out what major peer-reviewed papers or assessments have to say about them. Moreover, they should adhere to the principle that the more outlandish or dramatic the claim, the more skepticism it warrants. The fact is, nonscientist journalists can all too easily fall for scientific-sounding claims that they are unable to evaluate adequately on their own.

That doesn't mean that scientific consensus is right in every instance. There are famous examples of cases in which it was proved wrong: for instance, for many decades the consenus opinion in geology rejected Alfred Wegner's now accepted theory of continental drift. In the vast majority of modern cases, however, scientific consensus can be expected to hold up under scrutiny precisely because it has emerged from a lengthy and rigorous process of professional skepticism and criticism. At the very least, journalists covering science-based policy debates should familiarize themselves with this professional proving ground, learn what it says about the relative merits of competing claims, and "balance" their reports accordingly. In doing so, they will thwart and expose many of the most severe forms of science abuse.

When it comes to deterrence of future abuse of science, we must consider other measures as well. In particular, the repeated abuse of science that we have seen on the part of self-interested corporations, and their assorted minions, strongly suggests a systemic problem. Industry groups play the "science" card because it works (if only for a time, as Big Tobacco learned) and because they can generally get away with it. As Brown University clinical associate professor David Egilman observed at a July conference sponsored by the Center for Science in the Public Interest, when it comes to industry manipulation and suppression of science, "The penalties for getting caught never approach the cost advantages of increased profits. There are rarely criminal penalties for deaths and injuries." As Egilman suggests, as a society we must use the legal system more vigilantly to deter corporate abuse of science, recognizing their serious social costs.

And just as science-abusing corporations must be fought in the courts, science-abusing religious conservatives—who would misinform our children about the origin of the human species and about virtually everything having to do with sex—must be fought in the schools, the educational system, and the public arena more generally. Here the challenge truly becomes staggering. Short of massive educational reform, we can begin by supporting the few beleaguered groups, like the National Center for Science Education, that combat the religious Right in its attempt to commandeer science to serve a religious agenda.

We must also mobilize the natural defenders of Enlightenment values: scientists themselves, who all too often fail to engage antievolutionists and other know-nothings in defense of what they hold dear. True, groups like the National Academy of Sciences and American Association for the Advancement of Science have shown an historic willingness to step up when it counts, especially with powerful friend-of-the-court briefs in creationism lawsuits. But scientists have too often failed to counter creationist efforts at the local level, preferring to remain in their ivory towers. Moreover, while scientific societies have battled antievolutionists for decades, they must bring their activist senses up to date, and also battle the spread of misinformation in sex education courses and other areas.

Legal reforms, new levels of activism, raising journalistic standards—all of these measures will help beat back science abuse. In the end, however, we cannot escape the reality that we face a political problem, one that requires explicitly political solutions.

Ideally, Republican moderates like John McCain and Arnold Schwarzenegger would serve as emissaries to the right wing of their party, warning of the dangers of science abuse. Yet if these moderates have attempted such a step, we can detect no evidence of its effectiveness. Rather, we see the opposite. The Bush administration has alienated and spurned moderate Republicans such as former EPA administrator Christine Todd Whitman and former treasury secretary Paul O'Neill, who wanted to take global warming seriously rather than hide behind distortions and evasions of reliable scientific consensus.

In this context, and considering its track record, we have no choice but to politically oppose the antiscience right wing of the Republican Party. This does not necessarily entail an outright partisan agenda. Encouraging the electoral success of Republican moderates with good credentials on science could potentially have just as constructive an effect as backing Democrats.

But if we care about science and believe that it should play a crucial role in decisions about our future, we must steadfastly oppose further political gains by the modern Right. This political movement has patently demonstrated that it will not defend the integrity of science in any case in which science runs afoul of its core political constituencies. In so doing, it has ceded any right to govern a technologically advanced and sophisticated nation. Our future relies on our intelligence, but today's Right— failing to grasp this fact in virtually every political situation in which it really matters, and nourishing disturbing anti-intellectual tendencies— cannot deliver us there successfully or safely. If it will not come to its senses, we must cast it aside.

INTERVIEWS

Chapter One: James Battey, director, National Institute on Deafness and Other Communication Disorders (National Institutes of Health), and chair, Stem Cell Task Force (National Institutes of Health), July 26, 2004 (telephone; interview conducted originally for *American Prospect*); Paul Berg, Cahill professor of biochemistry, emeritus, Stanford University School of Medicine, June 9, 2004 (in person); Rosina Bierbaum, dean, School of Natural Resources and Environment, University of Michigan, December 6, 2004 (telephone); Donald Kennedy, executive editor-in-chief, *Science,* June 22, 2004 (in person); Steven Pinker, Johnstone Family professor, Department of Psychology, Harvard University, May 11, 2004 (in person); Thomas Roskelly, spokesman for the Annapolis Center for Science-Based Public Policy, November 1, 2004 (telephone).

Chapter Two: Adam Finkel, visiting research scholar, Woodrow Wilson School of Public and International Affairs (Princeton University), October 6, 2004 (telephone); Robert Frosch, senior research associate, Belfer Center for Science and International Affairs, John F. Kennedy School of Government (Harvard University), May 10, 2004 (in person); John Holdren, director, Science, Technology and Public Policy Program, Belfer Center for Science and International Affairs, John F. Kennedy School of Government (Harvard University), May 11, 2004 (in person); Brian Leiter, Joseph D. Jamail centennial chair in law, professor of philosophy, and director of the Law & Philosophy Program, University of Texas School of Law, November 16, 2004 (telephone).

Chapter Three: Richard L. Garwin, IBM fellow emeritus, Thomas J. Watson Research Center, November 29, 2004 (e-mail); Sheila Jasanoff, Pforzheimer professor of science and technology studies, Harvard University, John F. Kennedy School of Government, May 11, 2004 (in person); Peter J. Kuznick, associate professor of history and director, Nuclear Studies Institute, American University, May 3, 2004 (telephone); W. K. H. Panofsky, professor and director emeritus, Stanford Linear Accelerator Center, June 10, 2004 (in person); Russell Train, chairman emeritus, World Wildlife Fund, May 6, 2004 (in person); Henry Waxman (D-CA), ranking minority member, House Committee on Government Reform, May 17, 2004 (in person).

Chapter Four: Stephen G. Brush, distinguished university professor of the history of science, University of Maryland, April 30, 2004 (telephone); Sidney Drell, professor and deputy director, emeritus, Stanford Linear Accelerator Center, June 10, 2004 (in person); James Gilbert, distinguished university professor, Department of History, University of Maryland, April 26, 2004 (telephone); John C. Green, professor of political science; director, Ray C. Bliss Institute of Applied Politics, University of Akron, April 26, 2004 (telephone); Michael Oppenheimer, Albert G. Milbank professor of geosciences and international affairs, Woodrow Wilson School and Department of Geosciences, Princeton University, December 3, 2004 (telephone); W. K. H. Panofsky, professor and director emeritus, Stanford Linear Accelerator Center, June 10, 2004 (in person).

Chapter Five: Rosina Bierbaum, dean, School of Natural Resources and Environment, University of Michigan, September 25, 2003 (telephone; interview conducted originally for *Boston Globe*); John Gibbons, former director, Office of Technology Assessment, June 3, 2004 (in person); Roger Herdman, director, National Cancer Policy Board, Institute of Medicine, September 25, 2003 (telephone; interview conducted originally for *Boston Globe*); Amo Houghton, former member of Congress, September 25, 2003 (telephone; interview conducted originally for *Boston Globe*); Henry Kelly, president, Federation of American Scientists, September 24, 2003 (telephone; interview conducted originally for *Boston Globe*); James McCarthy, Alexander Agassiz professor of biological oceanography and director, Museum of Comparative Zoology, Harvard University, May 11, 2004 (in person); Naomi Oreskes, associate professor, Department of History and Program in Science Studies, University of California, San Diego, January 25, 2005 (telephone); Bob Palmer, former minority staff director, House of Representatives Committee on Science, April 29, 2004 (in person); William Schlesinger, dean, the Nicholas School of the Environment and Earth Sciences, Duke University and James B. Duke professor of biogeochemistry, April 23, 2004 (telephone); Rick Tyler, spokesman, Speaker Newt Gingrich, September 29, 2003 (telephone; interview conducted originally for *Boston Globe*); Robert Walker, chairman, Wexler & Walker, and former chairman of the House of Representatives Committee on Science, December 2, 2004 (telephone).

Interview Requests Declined or Unanswered: Newt Gingrich.

Chapter Six: Gary Bass, executive director, and Sean Moulton, senior information policy analyst, OMB Watch, October 20, 2004 (in person); Lisa Bero, professor, Department of Clinical Pharmacy, School of Pharmacy and Institute for Health Policy Studies, School of Medicine, University of California, San Francisco, June 8, 2004 (in person); Adam Finkel, visiting research scholar, Woodrow Wilson School of Public and International Affairs (Princeton University), April 6, 2004 (telephone); Lisa Heinzerling, professor, Georgetown University Law Center, February 12, 2004 (telephone); William Kovacs, vice president, environment, technology and regulatory affairs, U.S. Chamber of Commerce, March 18, 2004 (telephone; interview conducted originally for *Washington Monthly*); William O'Keefe, CEO, George C. Marshall Institute, April 13, 2004 (telephone); Thomas Roskelly, spokesman for the Annapolis Center for Science-Based Public Policy, November 1, 2004 (telephone); David Vladeck, associate professor, Georgetown University Law Center, December 1, 2004 (telephone); Robert Walker, chairman, Wexler & Walker, and former chairman of the House of Representatives Committee on Science, December 2, 2004 (telephone).

Chapter Seven: Eric J. Barron, dean, College of Earth and Mineral Sciences and distinguished professor of geosciences, Pennsylvania State University, April 20, 2004 (telephone); Jeffrey Connor, communications director, Congresswoman Jo Ann Emerson, November 29, 2004 (telephone); Christopher Horner, senior fellow, Competitive Enterprise Institute, March 22, 2004 (telephone; interview originally conducted for *American Prospect*); Senator James Inhofe, chairman, Committee on Environment and Public Works (as represented by committee staff), written response received March 24, 2004 (via e-mail; interview originally conducted for *American Prospect*); Donald Kennedy, executive editor-in-chief, *Science*, June 22, 2004 (in person); Richard Lindzen, Alfred P. Sloan professor of meteorology, Department of Earth, Atmospheric and Planetary Sciences, Massachusetts Institute of Technology, April 15, 2004 (telephone); Michael Mac-Cracken, former executive director, U.S. Global Change Research Program National Assessment Coordination Office, March 9, 2004 (in person); Jerry D. Mahlman, senior research fellow, National Center for Atmospheric Research, March 5, 2004 (telephone; interview originally conducted for *American Prospect*); Frank Maisano, former spokesman, Global Climate Coalition, January 18, 2005 (telephone; interview originally conducted for *Mother Jones*); Michael Mann, assistant professor, Department of Environmental Science, University of Virginia, March 2, 2004 (telephone; interview originally conducted for *American Prospect*), December 15, 2004 (telephone); James McCarthy, Alexander Agassiz professor of biological oceanography and director, Museum of Comparative Zoology, Harvard University, May 11, 2004 (in person); William O'Keefe, CEO, George C. Marshall Institute, March 22, 2004 (telephone; interview originally conducted for *American Prospect*); Naomi Oreskes, associate professor, Department of History and Program in Science Studies, University of California, San Diego, January 25, 2005 (telephone); Michael Oppenheimer, Albert G. Milbank professor of geosciences and international affairs, Woodrow Wilson School and Department of Geosciences, Princeton University, March 2, 2004 (telephone; interview originally for *American Prospect*); Thomas Roskelly, spokesman for the Annapolis Center for Science-Based Public Policy, November 1, 2004 (telephone); Edward S. Sarachik, Department of Atmospheric Sciences, University of Washington, May 3, 2004 (telephone); William Schlesinger, dean, the Nicholas School of the Environment and Earth Sciences, Duke University and James B. Duke professor of biogeochemistry, April 23, 2004 (telephone); Stephen Schneider, professor, Department of Environmental Sciences, and codirector, Center for Environmental Science and Policy, Stanford University, March 30, 2004, (telephone; interview originally for *American Prospect*); John M. Wallace, professor, Department of Atmospheric Sciences, University of Washington, April 27, 2004 (telephone); Tom Wigley, senior scientist, National Center for Atmospheric Research, March 29, 2004 (telephone; interview originally for *American Prospect*).

Interview Requests Declined or Unanswered: Sallie Baliunas, Willie Soon.

Chapter 8: Gary Bass, executive director, and Sean Moulton, senior information policy analyst, OMB Watch, October 20, 2004 (in person); Jeffrey Connor, communications director, Congresswoman Jo Ann Emerson, November 29, 2004 (telephone); Sherry Ford, senior communications manager, crop protection, Syngenta America Inc., March 23, 2005 (responses received by e-mail); Stanton Glantz, professor of medicine and director, Center for Tobacco Control Research and Education, University of California, San Francisco, March 31, 2004 (telephone; interview originally conducted for *Washington Monthly*); Tyrone Hayes, professor, Department of Integrative Biology, University of California, Berkeley, December 13, 2004 (telephone); Lisa Heinzerling, professor, Georgetown University Law Center, February 12, 2004 (telephone); William Kovacs, vice president of environment, technology and regulatory affairs, U.S. Chamber of Commerce, March 18, 2004 (telephone; interview originally conducted for *Washington Monthly*); Thomas McGarity, W. James Kronzer Chair in Trial and Appellate Advocacy, University of Texas School of Law, March 1, 2004 (telephone; interview originally conducted for *Washington Monthly*); David Michaels, professor, Departments of Occupational and Environmental Health and Epidemiology, George Washington University School of Public Health and Health Services, September 22, 2003 (telephone); James O'Reilly, adjunct professor of law, University of Cincinnati College of Law, March 16, 2004 (telephone; interview originally conducted for *Washington Monthly*); Sidney Shapiro, University Distinguished Chair in Law, Wake Forest University, March 10, 2004 (telephone); Rena Steinzor, Jacob A. France research professor of law and director, Environmental Law Clinic, University of Maryland School of Law, March 4, 2004 (telephone); Jim Tozzi, advisory board member, Center for Regulatory Effectiveness, March 17, 2004, March 24, 2004, April 2, 2004, February 17, 2005 (in person); David Vladeck, associate professor, Georgetown University Law Center, December 1, 2004 (telephone); Wendy Wagner, Joe A. Worsham Centennial Professor, University of Texas School of Law, March 5, 2004 (telephone; interview originally conducted for *Washington Monthly*).

Interview Requests Declined or Unanswered: Steven Milloy.

Chapter Nine: Kelly Brownell, director, Yale Center for Eating and Weight Disorders, February 20, 2004 (telephone); Carlos Camargo, associate professor, Department of Epidemiology, Harvard School of Public Health, February 2, 2005 (telephone); Philippe Grandjean, professor and chair of environmental medicine, University of Southern Denmark, March 23, 2004 (telephone); Shiriki Kumanyika, professor of epidemiology, Center for Clinical Epidemiology and Biostatistics, University of Pennsylvania School of Medicine, February 13, 2004 (telephone); David Ludwig, director, Obesity Program, Children's Hospital Boston, and associate professor of pediatrics, Harvard Medical School, February 24, 2004 (telephone); Gary Myers, professor of neurology and pediatrics, Department of Neurology, University of Rochester Medical Center, March 11, 2004 (telephone; interview originally conducted for *American Prospect*); Marion Nestle, Paulette Goddard Professor of Nutrition, Food Studies, and Public Health at New York University, February 11, 2004 (telephone; interview originally conducted for *Mother Jones*); Kaare Norum, Institute for Nutrition Research, University of Oslo, February 10, 2004 (telephone; interview originally conducted for *Mother Jones*); William Pierce, spokesman, Department of Health and Human Services, February 23, 2004 (telephone;

interview originally conducted for *Mother Jones*); Nicola Pirrone, head of Rende division, Italian National Research Council, Institute for Atmospheric Pollution, March 4, 2005 (telephone); Deborah Rice, Environmental Health Unit, Maine Bureau of Health, March 12, 2004 (telephone; interview originally conducted for *American Prospect*); Bruce Silverglade, director of legal affairs, Center for Science in the Public Interest, February 9, 2004 (in person; interview originally conducted for *Mother Jones*); William Steiger, director of the Office of Global Health Affairs and special assistant to the secretary for international affairs, U.S. Department of Health and Human Services (with William Pierce), March 8, 2005 (telephone); Rena Steinzor, Jacob A. France Research Professor of Law and director, Environmental Law Clinic, University of Maryland School of Law, March 4, 2004 (telephone); Jim Tozzi, advisory board member, Center for Regulatory Effectiveness, March 17, 2004 (in person); Margo Wootan, director of nutrition policy, Center for Science in the Public Interest, January 24, 2005 (telephone); Derek Yach, professor and head, Division of Global Health, Yale School of Public Health, September 22, 2004 (telephone).

Interview Requests Declined or Unanswered: Andrew Briscoe/The Sugar Association, Steven Milloy, Willie Soon.

Chapter Ten: Michael Bean, attorney and chair, wildlife program, Environmental Defense, February 25, 2004 (telephone; interview originally conducted for *Legal Affairs*); Holly Doremus, professor, University of California, Davis School of Law, February 19, 2004 (telephone; interview originally conducted for *Legal Affairs*); Mike Kelly, former fisheries biologist, National Marine Fisheries Service, August 26, 2004 (telephone); Douglas Markle, professor of fisheries, Department of Fisheries and Wildlife, Oregon State University, August 2, 2004 (telephone); Peter Moyle, professor of fish biology, Department of Wildlife, Fish, and Conservation Biology, University of California, Davis, August 26, 2004 (telephone); Ransom Myers, Killam Chair of Ocean Studies, Dalhousie University, August 3, 2004 (telephone); Dennis Murphy, director, Graduate Program in Ecology, Evolution, and Conservation Biology, University of Nevada, Reno, February 26, 2004 (telephone; interview originally conducted for *Legal Affairs*); Gordon Orians, professor emeritus, Department of Biology, University of Washington, February 18, 2004 (telephone; interview originally conducted for *Legal Affairs*); Robert Paine, professor emeritus, Department of Biology, University of Washington, August 9, 2004 (telephone); Stuart Pimm, Doris Duke Professor of Conservation Ecology, Nicholas School, Duke University, April 26, 2004 (telephone); J. B. Ruhl, Matthews & Hawkins Professor of Property and associate dean for academic affairs, Florida State University College of Law, February 16, 2004 (telephone; interview originally conducted for *Legal Affairs*).

Chapter Eleven: Stephen G. Brush, distinguished university professor of the history of science, University of Maryland, April 30, 2004 (telephone); Glenn Branch, National Center for Science Education, February 23, 2005 (e-mail); Derek Briggs, Department of Geology and Geophysics, Yale University, October 13, 2004 (e-mail); James Gilbert, distinguished university professor, Department of History, University of Maryland, April 26, 2004 (telephone); Edward Larson, Talmadge Chair of Law and Russell Professor of American History, University of Georgia, November 1, 2004 (telephone); Jeffrey

McKee, professor, Department of Anthropology, Ohio State University, October 13, 2004 (in person); Ronald Numbers, Hilldale and William Coleman Professor of the History of Science and Medicine, University of Wisconsin, Madison, October 18, 2004 (telephone); Steven Pinker, Johnstone Family Professor, Department of Psychology, Harvard University, May 11, 2004 (in person); Roger D. K. Thomas, John Williamson Nevin Professor of Geosciences, Franklin and Marshall College, January 12, 2005 (e-mail). In addition to interviews, much of this chapter was inspired by the conference "Evolution and God: 150 years of love and war between science and religion," Case Western Reserve University, October 15–17, 2004.

Interview Requests Declined or Unanswered: Bruce Chapman.

Chapter Twelve: James Battey, director, National Institute on Deafness and Other Communication Disorders (National Institutes of Health), and chair, Stem Cell Task Force (National Institutes of Health), July 26, 2004 (telephone; interview conducted originally for *American Prospect*); Paul Berg, Cahill Professor of Biochemistry, Emeritus, Stanford University School of Medicine, June 9, 2004 (in person); Elizabeth Blackburn, Morris Herzstein professor of biology and physiology, Department of Biochemistry and Biophysics, University of California, San Francisco, December 13, 2004 (telephone); Robert Goldstein, chief scientific officer, Juvenile Diabetes Research Foundation International, May 24, 2004 (telephone; interview originally conducted for *American Prospect*); Lawrence S. B. Goldstein, professor of cellular and molecular medicine at the University of California, San Diego, School of Medicine, June 16, 2004 (telephone; interview originally conducted for *American Prospect*); Louis Guenin, lecturer on ethics in science, Department of Microbiology and Molecular Genetics, Harvard Medical School, May 10, 2004 (in person); Leon Kass, chairman, President's Council on Bioethics, March 20, 2005 (written responses to questions received by e-mail); Thomas Murray, president, the Hastings Center, October 6, 2003 (telephone); David Prentice, senior fellow for life sciences, Family Research Council, July 13, 2004 (telephone; interview originally conducted for *Washington Monthly*); Evan Snyder, program director, stem cells and regeneration, the Burnham Institute, June 8, 2004 (in person; interview originally conducted for *American Prospect*); Irving Weissman, Professor of pathology and director, Stem Cell Institute, Stanford University School of Medicine, June 9, 2004 (in person). In addition to interviews, parts of this chapter were influenced by the conference "Stem Cells: Saving Lives or Crossing Lines, Human Embryonic Stem Cell Policy," James A. Baker III Institute for Public Policy, Rice University, November 20–21, 2004.

Chapter Thirteen: Nancy Adler, professor of medical psychology, University of California, San Francisco, June 24, 2004 (telephone; interview originally conducted for *Washington Monthly*); Heather Boonstra, senior public policy associate, the Alan Guttmacher Institute, December 2, 2004 (telephone); Willard Cates, Jr., president and chief executive officer, Institute for Family Health, Family Health International, July 27, 2004 (telephone; interview originally conducted for *Washington Monthly*); Michael Cromartie, vice president, Ethics and Public Policy Center, June 30, 2004 (telephone; interview originally conducted for *Washington Monthly*); Frank Davidoff, editor emeritus, *Annals of Internal Medicine*, June 17, 2004 (telephone; interview originally conducted for *Mother Jones*); W. David Hager, professor, Department of Obstetrics & Gynecology, University of Kentucky, June 22, 2004 (telephone; interview originally conducted for *Mother Jones*); Donald Kennedy, executive editor-in-chief, *Science*, June 22, 2004 (in person); Michael F. Greene, professor of obstetrics, Gynecology and Reproductive Biology, Harvard Medical School, May 21, 2004 (telephone; interview originally conducted for *Mother Jones*); Stanley Henshaw, senior fellow, Alan Guttmacher Institute, June 28, 2004 (telephone; interview originally conducted for *Washington Monthly*); Douglas Kirby, senior research scientist, ETR Associates, June 25, 2004 (telephone; interview originally conducted for *Washington Monthly*); Brenda Major, professor, department of psychology, University of California, Santa Barbara, June 24, 2004 (telephone; interview originally conducted for *Washington Monthly*); Joe S. McIlhaney, Jr., founder and president, Medical Institute for Sexual Health, July 14, 2004 (telephone; interview originally conducted for *Washington Monthly*); Karin Michels, associate professor, Department of Epidemiology, Harvard School of Public Health, June 25, 2004 (telephone; interview originally conducted for *Washington Monthly*); David Reardon, director, Elliot Institute, July 12, 2004 (telephone; interview originally conducted for *Washington Monthly*); Lynn Rosenberg, professor of epidemiology, Boston University School of Public Health, June 28, 2004 (telephone; interview originally conducted for *Washington Monthly*); Nancy Felipe Russo, regents' professor, Department of Psychology, Arizona State University, June 28, 2004 (telephone; interview originally conducted for *Washington Monthly*); Felicia Stewart, co-director, Center for Reproductive Health Research & Policy, University of California, San Francisco, June 2, 2004 (telephone); Nada Stotland, professor of psychiatry and professor of obstetrics and gynecology, Rush Medical College, July 1, 2004 (telephone; interview originally conducted for *Washington Monthly*); James Trussell, professor of economics and public affairs and director, Office of Population Research, Princeton University, June 19, 2004 (telephone; interview originally conducted for *Mother Jones*), Alastair Wood, associate dean and professor of medicine, Vanderbilt University School of Medicine, May 26, 2004 (telephone; interview originally conducted for *Mother Jones*).

Interview Requests Declined or Unanswered: Joel Brind.

Chapter Fourteen: D. Allan Bromley, Sterling Professor of the Sciences, Yale University, April 26, 2004 (telephone); John Gibbons, former director, Office of Technology Assessment, former science adviser to the President (Clinton) and director, Office of Science and Technology Policy, June 3, 2004 (in person); John Holdren, director, science, technology and public policy program, Belfer Center for Science and International Affairs, John F. Kennedy School of Government (Harvard University), May 11, 2004 (in person); Kevin Knobloch (president), Peter Frumhoff (senior scientist), and Suzanne Shaw (director of communications), Union of Concerned Scientists, May 10, 2004 (in person); Neal Lane, Edward A. and Hermena Hancock Kelly University Professor, Rice University, and former assistant to the President for science and technology, September 23, 2003 (telephone); James McCarthy, Alexander Agassiz Professor of Biological Oceanography and director, Museum of Comparative Zoology, Harvard University, May 11, 2004 (in person); John H. Marburger III, science adviser to the President and director, White House Office of Science and Technology Policy, December 22, 2004 (e-mail response to questions); William R. Miller, distinguished professor, Departments of Psychology & Psychiatry, University of New Mexico, February 15, 2005 (e-mail); Thomas Murray, president, the Hastings Center, October 6, 2003 (telephone); Robert Walker, chairman, Wexler & Walker, and former chairman of the House of Representatives Committee on Science, December 2, 2004 (telephone); Congressman Henry Waxman (D-CA), ranking minority member, House of Representatives Committee on Government Reform, May 17, 2004 (in person).

Epilogue: David Guston, professor of political science and associate director, Consortium for Science, Policy, and Outcomes, Arizona State University, September 16, 2003 (telephone); Congressman Rush Holt (D-NJ), April 29, 2004 (in person).

CREDITS

This book grew in part out of a series of similarly themed articles by the author, including the following:

"The Science Gap," *Boston Globe* (Ideas), October 5, 2003.

"Beware 'Sound Science': It's doublespeak for trouble," *Washington Post* (Outlook), February 29, 2004.

"Earth Last: Environment and Public Works Committee Chairman James Inhofe is waging a war on science so extreme that even the Bush administration seems moderate by comparison," *American Prospect*, May 2004.

"Paralysis by Analysis: Jim Tozzi's regulation to end all regulation," *Washington Monthly*, May 2004.

"Sucker Punch: How conservatives are trying to use a conflict over obscure fish to gut the science behind the Endangered Species Act," *Legal Affairs*, May/June 2004.

"Research and Destroy: How the religious right promotes its own 'experts' to counter mainstream science," *Washington Monthly*, October 2004.

"Christian Science? Siding with an antiabortion doctor, the FDA rejects easy access to a morning-after pill," *Mother Jones*, September 2004.

"Some Like It Hot," *Mother Jones*, May/June 2005.

NOTES

1: THE THREAT

1   *The success of science depends:* Interview with Steven Pinker, May 11, 2004.

2   *Bush's nationally televised claim:* George W. Bush, "Remarks by the President on Stem Cell Research," August 9, 2001. Available online at: http://www.whitehouse.gov/news/releases/2001/08/20010809-2.html (accessed November 22, 2004).

   should not underwrite research: Quoted in Richard Lacayo, "How Bush Got There; Months of debate—and one lucky break—led to the President's compromise. The inside story," *Time*, August 20, 2001.

   in vitro fertilization clinics: Much confusion exists over the origins and status of the excess in vitro fertilization (IVF) embryos from which embryonic stem cell lines derive. It hardly helps that much variety exists in the IVF industry when it comes to the treatment of excess embryos. Luckily, a systematic study of the treatment of excess embryos at fertility clinics has been undertaken (see Gurmankin et al., "Embryo disposal practices in IVF clinics in the United States," *Politics and the Life Sciences*, 2004, vol. 22, no. 2). This analysis shows that most U.S. clinics willingly preserve excess embryos, and more than half of those give couples the option of donating them for medical research. More than half also present the option of simply disposing of the excess embryos. At least in many cases, then, IVF couples find themselves in the position of choosing whether their excess embryos will go to science, or in the trash.

3   *made this decision possible:* Quoted in Richard Lacayo, "How Bush got there; months of debate—and one lucky break—led to the President's compromise. The inside story," *Time*, August 20, 2001.

   global telephone survey: Interview with James Battey, July 26, 2004.

   Paul Berg vividly explained: Interview with Paul Berg, June 9, 2004.

   The Bush White House either didn't know: "It is clear, in retrospect, that the White House sent Bush out on national television without having vetted (or even understood) the biological status of the cell lines he had embraced as the foundation of his compromise policy," journalist Stephen H. Hall notes in his definitive book on the stem cell debate, *Merchants of Immortality: Chasing the Dream of Human Life Extension* (New York: Houghton Mifflin, 2003), p. 304.

   and not bothering to consult: Interview with Rosina Bierbaum, December 6, 2004.

4   *roughly a third of whom:* In August 9 and August 10–11 polls, the Gallup Organization asked Americans whether they had watched any of Bush's speech. The first time, 32 percent replied yes, the second time, 45 percent did. See Matthew C. Nisbet, "Public opinion about stem cell research and human cloning," *Public Opinion Quarterly,* 2004, vol. 68, no. 1, pp. 131–154. See Table 4, page 141.

6   *science moles:* Steven Milloy, "Lingering infestation of science moles," *Washington Times,* May 30, 2001.

somebody with a doctorate: Nicholas Lemann, "The Controller: Karl Rove is working to get George Bush reelected, but he has bigger plans," *New Yorker,* May 12, 2003.

In political spats with environmentalists: For a thorough treatment of some of the abuses of science that have gone hand in hand with conservative antienvironmentalism, see Paul R. Ehrlich and Anne H. Ehrlich, *Betrayal of Science and Reason: How Anti-Environment Rhetoric Threatens Our Future* (Washington, D.C.: Island Press, 1996).

7   *Chris Cannon of Utah:* Representative Cannon is an executive committee member of the Congressional Western Caucus, which on February 10, 2004, held an issues conference that included on the agenda the question of "Sound Science and Public Policy" (see http://chriscannon.house.gov/wc/meetings/Conference/soundscience.htm, accessed October 18, 2004). Action items included "promote a Sound Science Caucus to focus on this issue" and "Support Rep. Walden's H.R. 1662 as legislation that exemplifies sound science and public policy." The proposed law, a reform of the science provisions of the Endangered Species Act, is discussed at length in Chapter 10.

a conservative-leaning group: The Annapolis Center does not reveal its funding sources on its website. However, a *Wall Street Journal* report from 1997 noted that the group was largely funded by members of the National Association of Manufacturers (NAM); see John J. Fialka, "Panel Judging EPA's Proposed Air Regulations Receives Most of Its Funding From the Regulated," *Wall Street Journal,* January 16, 1997. In an interview on November 1, 2004, Annapolis Center spokesman Thomas Roskelly said that the group still receives funding from NAM but not nearly as much. He also emphasized that the funds the center receives from industry are "unrestricted," meaning that "industry cannot buy a particular report or a particular outcome." One of the group's funders has been ExxonMobil, which gave a total of $120,000 in 2002 (corporate giving report on file with author) and $182,500 in 2003 (giving report also on file with author).

an early proponent: Wallop's online bio on the website of Frontiers of Freedom can be read here: http://ff.org/about/mwbio.html (accessed October 18, 2004). For an account of Wallop's lobbying in favor of research into space-based lasers, see Gregg Herken, *Cardinal Choices: Presidential Science Advising from the Atomic Bomb to SDI* (Stanford: Stanford University Press, 2000), p. 209, and Frances Fitzgerald, *Way Out There in the Blue: Reagan, Star Wars and the End of the Cold War* (New York: Simon & Schuster, 2000), pp. 121–124.

Politicizing Science: Michael Gough, ed., *Politicizing Science: The Alchemy of Policymaking,* (Washington, D.C.: George C. Marshall Institute; and Stanford, CA: Hoover Institution Press, 2003).

Silencing Science: Steven Milloy and Michael Gough, *Silencing Science* (Washington, D.C.: Cato Institute, 1998).

Greenpeace has suggested: Greenpeace asserts that "Genetic engineering of the food we eat is an inherently risky process. Current understanding of genetics is extremely limited and scientists do not know the long-term effects of releasing these unpredictable organisms into the environment and people's diets." See http://www.greenpeace.org/international_en/campaigns/intro?campaign_id=3992 (accessed December 4, 2004).

8   *no adverse health effects:* "Safety of Genetically Engineered Foods: Approaches to Assessing Unintended Health Effects" (2004), Food and Nutrition Board, Institute of Medicine, Board on Agriculture and Natural Resources, Board on Life Sciences. Available online at: http://www.nap.edu/books/0309092094/html (accessed August 5, 2004).

Marion Nestle has written: Marion Nestle, *Safe Food: Bacteria, Biotechnology, and Bioterrorism* (Berkeley: University of California Press, 2003).

some animal rights activists: For instance, in an online fact sheet entitled "Alternatives: Testing Without Torture," People for the Ethical Treatment of Animals (PETA) cites computer modeling as one alternative to animal testing. See http://www.peta.org/factsheet/files/Factsheet-Display.asp?ID=87 (accessed October 6, 2004).

8   *a remarkable piece of science fiction:* Interview with Donald Kennedy, June 22, 2004.
When it's academics who wield the power: Interview with Steven Pinker, May 11, 2004.

9   *according to polls:* In a November 2004 poll, the Gallup Organization found that 45 percent of Americans agree with the statement, "God created human beings pretty much in their present form at one time within the last 10,000 years or so." This is consistent with previous Gallup polls using similar question wording. See National Center for Science Education, "Public view of creationism and evolution unchanged, says Gallup," November 19, 2004. Available online at http://www.ncseweb.org/resources/news/2004/US/724_public_view_of_creationism_and_11_19_2004.asp (accessed December 4, 2004).

George W. Bush supported: "Opinions of candidates on teaching creationism alarm some scientists," *St. Louis Post-Dispatch,* August 27, 1999. Interestingly, President Bush's science adviser John Marburger described evolution as a "cornerstone of modern biology" in an online colloquy sponsored by the *Chronicle of Higher Education* in 2003. See: http://chronicle.com/colloquylive/2004/03/science (accessed April 11, 2004).

Pennsylvania senator Rick Santorum: Rick Santorum, "Illiberal Education in Ohio Schools," *Washington Times,* March 14, 2002, describing intelligent design as a "legitimate scientific theory that should be taught in science classes."

routinely publishes articles: See, for example, Hunter Baker, "The professor's paroxysm," *National Review Online,* March 15, 2004, available online at http://www.nationalreview.com/comment/baker200403150909.asp (accessed October 18, 2004), and John G. West, "Evolving double standards," *National Review Online,* April 1, 2004, available online at http://www.nationalreview.com/comment/west200404010900.asp (accessed October 18, 2004).

10   *studies suggest that ties to industry:* For a very helpful discussion of this problem, see Richard Horton, "The Dawn of McScience," *New York Review of Books,* vol. 51, no. 4, March 11, 2004. Horton cites Justin E. Bekelman et al., "Scope and impact of financial conflicts of interest in biomedical research," *JAMA,* January 22, 2003, pp. 454–465, a review of previous studies of the relationship between scientists and industry that found that "financial relationships among industry, scientific investigators, and academic institutions are widespread. Conflicts of interest arising from these ties can influence biomedical research in important ways."

A 1998 analysis: Deborah Barnes and Lisa Bero, "Why review articles on the health effects of passive smoking reach different conclusions," *Journal of the American Medical Association,* May 20, 1998, vol. 279, no. 19, noting, "These findings suggest that the tobacco industry may be attempting to influence scientific opinion by flooding the scientific literature with large numbers of review articles supporting its position that passive smoking is not harmful to health," and "Of the 106 articles in our study, 77% failed to disclose the sources of funding for the research."

12   *the quack "scientist":* See Valery N. Soyfer, *Lysenko and the Tragedy of Soviet Science* (New Brunswick, 1994: Rutgers University Press).

2: POLITICAL SCIENCE 101

14   *outright myths they tell:* For a thorough debunking of some of science's favorite myths, consult Daniel Greenberg, *Science, Money, and Politics: Political Triumph and Ethical Erosion* (Chicago: University of Chicago Press, 2001).

15   *science and technology studies:* Science and technology studies blend historical, sociological, philosophical, and other approaches to studying the scientific endeavor. For a good primer on this topic, see Sheila Jasanoff, Gerald E. Markle, James C. Peterson, and Trevor Pinch, eds., *Handbook of Science and Technology Studies* (Thousand Oaks, California: Sage Publications, 1995).

a question scholars have debated: It is worth noting that the sides in this "debate" have often been ill-defined. Those who study science in a sociological or "social constructivist" vein don't necessarily deny that science, at least on some level, accesses and describes the world as it actually exists. See, e.g., Sheila Jasanoff and Brian Wynne, "Science and decisionmaking," in *Human Choice and Climate Change: The Societal Framework,* S. Rayner and E. Malone, eds. (Columbus, Ohio: Batelle Press, 1998), 1–87 (noting that "social and cultural commitments are built into every phase of knowledge production and consequent social action, even though enormously effective steps are often taken to eliminate the traces of the social from the scientific world. The forms in which environmental knowledge and policy issues are publicly expressed have to be seen against this backdrop as historically and socially contingent, even though they are equally constrained by nature"); George Levine, "What is science studies for?" in *Science Wars,* Andrew Ross, ed. (Durham: Duke University Press, 1996) (noting that "we would be fools to behave as though there is no knowledge of the natural world to be had and that science has no better shot at it than any other professionals, or nonprofessionals"); and Stanley Aronowitz, "The politics of the science wars," also in *Science Wars* (noting that "The critical theories of science do not refute the results of scientific discoveries since, say, the Copernican revolution or since Galileo's development of the telescope. What it does challenge is the notion that science and its discoveries are exempt from ideology critique, deconstruction, or historical investigation that might be trained on any other discourse: literature and art, politics, social scientific theory, and so forth").

16 *so-called "academic pork":* See Chris Mooney, "'Academic pork': educators fault, praise payments," *Boston Globe,* August 19, 2001.

examinations of the lives: For example, in 2003 an appropriations amendment authored by Rep. Patrick Toomey, a Pennsylvania Republican, sought to block funding for several grants that had been peer reviewed and approved by the National Institutes of Health. Four of the studies involved sexual behaviors: "a study on San Francisco's Asian prostitutes and masseuses, a study of sexual habits of older men, a study on mood arousal and sexual risk taking, and a study on American Indian and Alaskan transgendered individuals." Quoted in Ted Agres, "Politicizing research or responsible oversight?" *Scientist,* July 14, 2003. Available online at http://www.biomedcentral.com/news/20030714/05 (accessed October 27, 2004).

scientists continually grasp: See Greenberg, *Science, Money, and Politics,* noting, "More money for more science is the commanding passion of the politics of science. More is deemed better, including the production of more scientists from a university system that is well supported by, but ingeniously decoupled from, the general economy" (p. 3).

the science says: Interview with Robert Frosch, May 10, 2004.

17 *Here is my definition:* I am indebted to freelance journalist Dylan Otto Krider for this definition, which draws on a suggestion he made to me. The breakdown of different types of science abuse also relies on input from David Meyer.

some tried to catalogue: See a "Report on the misuse of science in the administrations of George H.W. Bush (1989–1993) and William J. Clinton (1993–2001)," by the students of ENVS 4800, Maymester 2004, at the University of Colorado. It can be read online here: http://sciencepolicy.colorado.edu/admin/publication_files/resourse–1429-ENVS%204800%20Report.pdf (accessed October 4, 2004).

18 *Numerous examples of this phenomenon:* For examples see the Union of Concerned Scientists, "Scientific integrity in policymaking: an investigation into the Bush administration's misuse of science," February 2004, especially the sections "Censoring information on air quality" and "Addressing multiple air pollutants"; available at http://www.ucsusa.org/global_environment/rsi/page.cfm?pageID=1322 (accessed October 4, 2004).

stalled the release: See Cass Peterson, "White House stalls acid rain report; study stresses stronger action," *Washington Post,* August 18, 1984.

altering the scientific testimony: See Philip Shabecoff, "Scientist says budget office altered his testimony," *New York Times,* May 8, 1989.

19 *a 2001 report:* "Climate Change Science: An Analysis of Some Key Questions," Committee on the Science of Climate Change, Division on Earth and Life Studies, National Research Council (Washington, D.C.: National Academy Press, 2001).

19 *confirms the robust conclusion:* As the report puts it, "The IPCC's conclusion that most of the observed warming of the last 50 years is likely to have been due to the increase in greenhouse gas concentrations accurately reflects the current thinking of the scientific community on this issue. The stated degree of confidence in the IPCC assessment is higher today than it was 10, or even 5 years ago, but uncertainty remains because of (1) the level of natural variability inherent in the climate system on time scales of decades to centuries, (2) the questionable ability of models to accurately simulate natural variability on those long time scales, and (3) the degree of confidence that can be placed on reconstructions of global mean temperature over the past millennium based on proxy evidence. Despite the uncertainties, there is general agreement that the observed warming is real and particularly strong within the past 20 years. Whether it is consistent with the change that would be expected in response to human activities is dependent upon what assumptions one makes about the time history of atmospheric concentrations of the various forcing agents, particularly aerosols."

conservatives have cited the uncertainties: Furthermore, some have argued that the NAS report itself overstated scientific uncertainties with respect to global warming; see Kevin Trenberth, "Climate variability and global warming; National Research Council report," *Science,* July 6, 2001.

20 *thus increasing controversy:* Some have questioned the very notion that scientific research can "reduce uncertainty" in science-based policy disputes, instead asserting that uncertainty tends to increase as policy-relevant scientific results come under greater scrutiny. See Daniel Sarewitz, "How science makes environmental controversies worse," *Environmental Science & Policy* 7 (2004), 385–403.

Science may be able to guide policymakers: George Brown, "Environmental science under siege: fringe science and the 104th Congress," a report to the Democratic Caucus of the Committee on Science, October 23, 1996.

21 *policymakers may have to make a decision:* For another defense of decision-making even in the face of "considerable" uncertainty, see Roger A. Pielke, Jr., "Room for doubt," *Nature,* vol. 410, March 8, 2001 ("As a general principle, science and technology will contribute more effectively to society's needs when decision-makers base their expectations on a full distribution of outcomes, and then make choices in the face of the resulting—perhaps considerable—uncertainty").

In a 2001 letter: George W. Bush, "Letter to Senators Hagel, Helms, Craig, and Roberts," March 13, 2001. Available at http://www.whitehouse.gov/news/releases/2001/03/20010314.html (accessed October 18, 2004).

In a caustic 2001 editorial: "Faith-Based Reasoning," *Scientific American,* June 2001.

due to technical shortcomings: In a July 2003 study, the American Physical Society found that "intercepting missiles while their rockets are still burning would not be an effective approach for defending the U.S. against attacks by an important type of enemy missile." American Physical Society, "Report of the APS study group on boost-phase intercept systems for national missile defense." Available online at http://www.aps.org/public_affairs/popa/reports/nmd03.cfm (accessed April 11, 2004). Granted, intercepting ballistic missiles in their "boost phase" represents only one currently discussed approach to national missile defense, the others including "terminal" (intercepting missiles in the final phase before they hit their target) and "midcourse" (intercepting missiles as they travel above the atmosphere through space). The last of these approaches characterizes the Bush administration's current "missile defense" program, but it has multiple flaws, as the eminent physicist Richard Garwin has noted in *Scientific American* (see Garwin, "Holes in the missile shield," *Scientific American,* October 25, 2004). The "midcourse" approach runs aground, notes Garwin, because simple enemy countermeasures—such as releasing decoys along with the actual nuclear warhead—would overwhelm the missile defense system. See also Lisbeth Gronlund, George N. Lewis, and David C. Wright, "The continuing debate on national missile defenses," *Physics Today,* December 2000.

Philosophers of science have struggled: Perhaps the most definitive take on the so-called demarcation problem comes from philosopher Larry Laudan, who does indeed dub it a "pseudo-problem" (see Laudan, "The demise of the demarcation problem," 1983, reprinted in Michael Ruse ed., *But Is It Science? The Philosophical Question in the Creation/Evolution Controversy* (Amherst, New York: Prometheus Books, 1996). Laudan thinks we can still distinguish between good and bad science, or at any rate better and worse science, by evaluating whether claims are "well confirmed" and theories are "well tested." But he proposes that "If we would stand up and be counted on the side of reason, we ought to drop terms like 'psuedo-science' and 'unscientific' from our vocabulary; they are just hollow phrases which do only emotive work for us."

While Laudan's take on the demarcation problem has won widespread acceptance among philosophers of science, that doesn't mean everyone has given up using the term "pseudo-science" entirely. Philip Kitcher, another respected philosopher of science, recognizes the difficulties of demarcation but nevertheless considers the term "pseudoscience" meaningful, writing that "if a doctrine fails sufficiently abjectly as a science, then it fails to be a science. Where bad science becomes egregious enough, pseudoscience begins" (Kitcher, *Abusing Science: The Case Against Creationism,* Cambridge: MIT Press, 1982).

While fully cognizant of Laudan's reasons for questioning the existence of a firm line of demarcation, I take Kitcher's approach in ultimately deeming the word "pseudoscience" meaningful.

22 *Documents released:* As reported by David Hanners, "Tobacco industry paid thousands to scientists to criticize anti-smoking report," *Saint Paul Pioneer Press,* August 4, 1998. Article includes Stanton Glantz quotation.

23 *you shouldn't defend outcomes:* Interview with John Holdren, May 11, 2004.

3: FROM FDR TO NIXON

25 *one of the most distinguished environmental careers:* For the story of Train's career, see his memoir, *Politics, Pollution, and Pandas: An Environmental Memoir* (Washington: Island Press, 2003). *Nixon may not have cared:* In his memoir, Train writes, "There is no evidence of which I am aware that Nixon had any real personal interest in environmental matters. I certainly never heard him express any. His reaction to these issues was that of a highly political animal" (p. 79).

26 *more middle of the road:* All quotations from my interview with Russell Train, May 6, 2004.
a 1975 White House meeting: Train also describes this encounter in his memoir. See *Politics, Pollution, and Pandas,* pp. 166–171.
after the New York Times *reported:* Andrew C. Revkin and Katherine Q. Seelye, "Report by E.P.A. leaves out data on climate change," *New York Times,* June 19, 2003.
letter published in the Times: Russell Train, "When Politics Trumps Science" (letter to the editor), *New York Times,* June 21, 2003.
I think it's because the White House: Quoted in James Glanz, "Scientists say administration distorts facts," *New York Times,* February 19, 2004.

27 *greatly exaggerated and even mythologized:* For a critical account that nevertheless shows how Bush's famed report fit the ethos of its time, see "Vannevar Bush and the Myth of Creation," Chapter 3 of Daniel S. Greenberg, *Science, Money, and Politics: Political Triumph and Ethical Erosion* (Chicago: University of Chicago Press, 2001).
Soviet launch of Sputnik: For the role of the *Sputnik* crisis in bringing science advice into the White House, see Gregg Herken, *Cardinal Choices: Presidential Science Advising from the Atomic Bomb to SDI* (Stanford: Stanford University Press, 2000). See in particular pp. 98–100; the "technological Pearl Harbor" quotation can be found here. In "Science advice and the presidency: an overview from Roosevelt to Ford," William G. Wells, Jr., notes, "it was not until highly visible political problems crashed about [Eisenhower] that he fully embraced science and technology as instruments of national policy." In *Technology in Society,* vol. 2 (Oxford: Pergamon Press, 1980), pp. 191–220.

28 *apogee of presidential science advising:* Quoted in Chris Mooney, "Political science: the Bush administration snubs its science adviser," *American Prospect,* December 3, 2001.

28 *Wiesner spoke for the president:* Quoted in William G. Wells, Jr., "Science advice and the presidency: an overview from Roosevelt to Ford," in *Technology in Society,* vol 2., pp. 191–220.
cannot be justified solely: Quoted in Herken, *Cardinal Choices,* p. 129. For Herken's account of the space program during the Kennedy years, see pp. 129–130.
were never justified: Interview with W. K. H. Panofsky, June 10, 2004.
in the mid-1950s: The first issue of *National Review* appeared in November 1955, though Buckley began fund-raising for the publication in 1954. See John Judis, *William F. Buckley, Jr., Patron Saint of the Conservatives* (New York: Simon & Schuster, 1988).

29 *runs strong in the conservative movement:* For more on right-wing anti-intellectualism and the Goldwater campaign, see George F. Gilder and Bruce K. Chapman, *The Party That Lost Its Head* (New York: Alfred A. Knopf, 1966).
a group of leading scientists: For a good history of the role of scientists in the 1964 campaign, see Chapter 10 of Daniel S. Greenberg, *Science, Money, and Politics: Political Triumph and Ethical Erosion* (Chicago: University of Chicago Press, 2001). See also Peter J. Kuznick, "Scientists on the stump," *Bulletin of the Atomic Scientists,* November/December 2004, vol. 60, no. 6.
blustery, threatening man: Quoted in Greenberg, *Science, Money, and Politics,* p. 156.
conservative activists who would ultimately achieve: My account of the rise of modern conservatism relies on John Judis, *The Paradox of American Democracy: Elites, Special Interests, and the Betrayal of Public Trust* (New York: Routledge, 2000). See especially Chapter 6, "The Triumph of Conservatism." For Goldwater and the birth of modern conservatism, see also Rick Perlstein's *Before the Storm: Barry Goldwater and the Unmaking of the American Consensus* (New York: Hill and Wang, 2001).

30 *a slew of activist organizations:* See Judis, *The Paradox of American Democracy,* pp. 142–149.
officially launched just days: For the birth of the American Conservative Union in the wake of Goldwater's defeat, see David A. Keene, "Four decades of leadership," *Washington Times,* May 2, 2004.
more genteel conservatives like Buckley: See Alan Crawford, *Thunder on the Right* (New York: Pantheon, 1980), p. 7: "The New Right calls itself the New Right, a designation chosen . . . to distinguish its leadership from what they believe to be the slightly effete conservative leadership of the East Coast, for example, William F. Buckley, Jr., and his *National Review.*"
anti–East Coast, anti-intellectual animus: Crawford, Chapter 6.
the business community's political counterreaction: See generally John Judis, *The Paradox of American Democracy,* and Michael Pertschuk, *The Revolt Against Regulation* (Berkeley: University of California Press, 1982).
a campaign to undermine Carson's science: My account of the backlash against Rachel Carson relies on Frank Graham, Jr., *Since Silent Spring* (Boston: Houghton Mifflin, 1970), especially Chapter 4, "The Counterattack."

31 *a 1963 PSAC study:* President's Science Advisory Committee. 1963. *Use of Pesticides.* The White House, Washington, D.C. For an account of how PSAC shifted the burden of proof onto industry to prove that pesticides were safe, thus backing Carson's concerns, see Naomi Oreskes, "Science and public policy: what's proof got to do with it?" *Environmental Science & Policy* 7 (2004) 369–383.
a fairly thoroughgoing vindication: Quoted in Frank Graham, Jr., *Since Silent Spring* (Boston: Houghton Mifflin, 1970), p. 79.
spurred on science conflicts: On this score, see in particular Sheila Jasanoff, *The Fifth Branch: Science Advisers as Policymakers* (Cambridge: Harvard University Press, 1990). On page 39, Jasanoff writes, "Judging by the volume and intensity of scholarly debate, the rise of social regulation and the resulting transformation of the American administrative process were among the defining political events of the 1970s . . . the cutting adrift of science bearing on policy from its traditional moorings in academic and industry laboratories emerges in retrospect as a major factor in promoting controversy over the technical underpinnings of regulatory decisions. Public debate over the legitimacy of science policy decisions intensified as both the

production and analysis of scientific knowledge were increasingly drawn into public view through governmentally sponsored research, administrative rulemaking, judicial review, and, frequently, media coverage of controversies."

31   *an enormous psychological and social breakpoint:* Interview with Sheila Jasanoff, May 11, 2004.
   Edith Efron's 1984 book: Edith Efron, *The Apocalyptics: How Environmental Politics Controls What We Know About Cancer* (New York: Simon and Schuster, 1984).

32   *judicial review:* See Sheila Jasanoff, *Science at the Bar: Law, Science, and Technology in America* (Cambridge: Harvard University Press, 1995). See especially Chapter 4, "The Technical Discourse of Government."
   Believe it or not: See Pertschuk, *The Revolt Against Regulation,* p. 16: "Businessmen, to an extent that now seems hardly credible, ignored Washington."
   hearkened to the fateful advice: See Judis, pp. 116–119.
   Corporate philanthropy should not be: Irving Kristol, "On corporate philanthropy," in *Two Cheers for Capitalism* (New York: Basic Books, 1978); quotation on page 144.
   Soon businesses turned to funding: This whole paragraph relies on Judis, *The Paradox of American Democracy,* pp. 122–128. See in particular Chapter 1 for the "older generation of foundations."
   frequently in contradiction: In his 1980 book *Thunder on the Right,* Alan Crawford notes of the Heritage Foundation, "It is unusual for a research institution to have a 'staff ideology,' as the Heritage Foundation has. . . . The founder's real interest, in the author's view based on observation of a year or more, appears to be less with balanced public policy research and more with the provision of support for New Right causes" (p. 12).

33   *tens of millions of dollars annually:* Bruce Bimber, *The Politics of Expertise in Congress: The Rise and Fall of the Office of Technology Assessment* (Albany: State University of New York Press, 1996), p. 1.
   were started to counteract: Crawford, *Thunder on the Right,* pp. 25–26.
   a key inspiration: For the definitive account of this story, see Lee Edwards, *The Power of Ideas: The Heritage Foundation at 25 Years* (Ottawa, Illinois: Jameson Books, 1997). See also Judis, *The Paradox of American Democracy,* pp. 122–128.
   conservative sugar daddies: Of interest is Feulner's March 18, 2003, obituary for Coors, published in *National Review Online*: http://www.nationalreview.com/comment/comment-feulner 031803.asp ("Joe Coors, R.I.P.," accessed December 4, 2004). "Joseph Coors, without whom there would be no Heritage Foundation, is dead at the age of 85," Feulner wrote.

34   *the physicist testified against it:* For a good account of Richard Garwin's defiance of Nixon on the SST, and the subsequent sacking of Nixon's science advisors, see Herken, *Cardinal Choices,* Chapter 10, and Greenberg, *Science, Money, and Politics,* Chapter 11.
   less than adequate: Quotation from Herken, *Cardinal Choices,* p. 179.
   also a self-described Republican: Richard Garwin, e-mail communication, November 29, 2004.
   Train angered the Nixon White House: See Train, *Politics, Pollution, and Pandas,* pp. 85–86.
   scientists in their midst as "vipers": Quoted in Herken, *Cardinal Choices,* p. 180.
   he chose to kill the messenger: Jerome Wiesner, "Why we need a tough national science adviser," *Washington Post,* May 24, 1987.
   would rebound somewhat: During the Ford administration, Congress reestablished the position of science adviser, which Nixon had scrapped. Carter's science adviser, geophysicist Frank Press, broadened a role that had traditionally focused on arms-related issues to deliver advice on emerging topics like recombinant DNA and even the greenhouse effect. See Herken, *Cardinal Choices,* Chapters 10 and 11.

4: "CREATION SCIENCE" AND REAGAN'S "DREAM"

35   *moderate former EPA administrator:* See Train, *Politics, Pollution, and Pandas,* pp. 220–222.
36   *I have never understood:* See C. Everett Koop, MD, *Koop: The Memoirs of America's Family Doctor* (New York: HarperCollins, 1992), p. 248.

36 *give us contrary advice:* See Herken, *Cardinal Choices,* p. 200.

weaken the teaching of evolution: Philip J. Hilts, "Evolution on trial again in California; modern version of 'monkey' trial set for California," *New York Times,* March 2, 1981.

teach the "biblical story of creation": See Philip J. Hilts, "Creation vs. evolution: battle resumes in public schools," *Washington Post,* September 13, 1980.

during his 1981 confirmation hearing: Herken, *Cardinal Choices,* p. 201.

ultimately pleading ignorance: John Judis, "Mister Ed: or, Dr. Bennett at the bridge," *New Republic,* April 27, 1987.

judgment of the community: "Bennett seeks local input on teaching creationism," United Press International, September 10, 1986. To be fair, later Bennett took a more defensible position. In their book *The Educated Child: A Parent's Guide from Preschool Through Eighth Grade* (New York: Touchstone, 1999), Bennett and coauthors Chester E. Finn, Jr., and John T. E. Cribb, Jr., write, "First, understand that the theory of evolution is broadly accepted in the scientific community. There is still debate about some of the mechanisms involved, but on the whole, it is considered by scientists to be the soundest explanation for certain patterns and evidence found in nature.

"Second, because it is widely regarded as a central organizing theory of biology, most schools do teach evolution. This is a reasonable position for educators to take. Since the worldwide community of scientists views evolution as one of its most important theories, it would be strange indeed if U.S. science classes ignored it" (pp. 390–391).

37 *Religious America is awakening:* See Philip J. Hilts, "Creation vs. evolution: battle resumes in public schools," *Washington Post,* September 13, 1980.

the marriage started around 1980: Interview with John C. Green, April 26, 2004.

secular conservative intellectuals: Irving Kristol, "Room for Darwin and the Bible," *New York Times,* September 30, 1986.

a necessary political step: For more on the embrace of antievolutionism by neoconservative intellectuals, see Ronald Bailey, "Origin of the specious: why do neoconservatives doubt Darwin?" *Reason,* July 1997.

religious in nature and inspiration: My account of the history and development of "creation science" relies on Ronald L. Numbers, *The Creationists: The Evolution of Scientific Creationism* (Berkeley: University of California Press, 1992). Numbers notes of creationists, "All, it seems, became or remained antievolutionists primarily for biblical reasons. Although creationists increasingly stressed the scientific evidence for their position, one estimated that 'only about five percent of evolutionists-turned-creationists did so on the basis of the overwhelming evidence for creation in the world of nature'" (p. 233).

early American creationists: For examples of early creationist appropriation of science, see Numbers, *The Creationists,* pp. 57–59, on Arthur I. Brown, a creationist surgeon active in the 1920s, and pp. 60–71 on Harry Rimmer, who boasted of his scientific credentials.

William Jennings Bryan: For a discussion of Bryan's attempts to claim scientific credentials, see "William Jennings Bryan, scientist," in James Gilbert, *Redeeming Culture: American Religion in an Age of Science* (Chicago: University of Chicago Press, 1997).

not even creationists could ignore: Interview with Stephen Brush, April 30, 2004.

38 *What they had to do:* Interview with Stephen Brush, April 30, 2004.

with no references to the Bible: Institute for Creation Research, *Scientific Creationism* (Public School Edition), Henry M. Morris, ed. (San Diego: Creation-Life Publishers, 1974). (The quotation comes from page iv of the second printing, 1978.)

comprehensive, coherent, and satisfying world-view: Institute for Creation Research, *Scientific Creationism* (General Edition), Henry M. Morris, ed. (San Diego: Creation-Life Publishers, 1974). (Quotation comes from p. 15 of the second edition, 1985).

purely scientifically: Quoted in Philip J. Hilts, "Creation vs. evolution: battle resumes in public schools," *Washington Post,* September 13, 1980.

a scientific theory only: Fred M. Hechinger, "About education; creationism, politics and public schools," *New York Times,* March 10, 1981.

39 *unworked field:* For more on Price see Numbers, *The Creationists*, Chapter 5.

wind up above younger ones: For a more thorough refutation of creationist arguments, see Philip Kitcher, *Abusing Science: The Case Against Creationism* (Cambridge: MIT Press, 1982). For a critique with a specific focus on geology, see Arthur N. Strahler, *Science and Earth History: The Evolution/Creation Controversy* (Buffalo: Prometheus Books, 1987).

40 *a thorough corruption:* See Mark E. Rushefsky, "The misuse of science in governmental decisionmaking," *Science, Technology, & Human Values*, vol. 9, no. 3 (Summer, 1984), pp. 47–59.

Congress uncovered: See Eliot Marshall, "EPA's troubles reach a crescendo," *Science*, vol. 219, March 25, 1983, p. 1402.

a Nader on toxics: For a detailed account of the "hit list" scandal, see Jonathan Lash, Katherine Gillman, and David Sheridan, *A Season of Spoils: The Story of the Reagan Administration's Attack on the Environment* (New York: Pantheon Books, 1984), pp. 35–40. See also Sheila Jasanoff, *The Fifth Branch: Science Advisers as Policymakers* (Cambridge: Harvard University Press, 1990), p. 89.

Sheila Jasanoff has noted: Sheila Jasanoff, "(No?) accounting for expertise," *Science and Public Policy*, vol. 30, no. 3, June 2003, pp. 157–162.

fed into other controversies: See Associated Press, "List of events which led to Burford resignation," March 9, 1983. The article begins, "The resignation of Environmental Protection Agency Administrator Anne McGill Burford climaxed a series of events which began with a battle over what documents the executive branch would turn over to Congress," and lists as one of these events, "March 1, Rep. John Dingell, D-Mich., warns Reagan in a letter that his Energy and Commerce Committee has uncovered 'evidence of wrongdoing, unethical behavior and potential criminal conduct' at the EPA and urges the president to turn over all subpoenaed documents. Rep. James Scheuer, D-N.Y., releases what he calls a 'hit list' rating scientists advising the EPA. The list contains such descriptions as 'bleeding-heart liberal' and 'snail darter types.' Critics say the agency used the list to purge its advisory committees."

he lightened up: Discussed in Train's memoir, *Politics, Pollution, and Pandas*, pp. 263–267.

41 *There will be no hit lists:* Marjorie Sun, "Ruckelshaus promises EPA cleanup," *Science*, vol. 220, May 20, 1983, p. 801.

exploited scientific uncertainty: My discussion of abuses of scientific uncertainty on "acid rain" relies on a number of sources, ranging from media accounts to scholarly studies to interviews. In particular, I drew substantially on the work of University of Southern Indiana sociologist Stephen C. Zehr, and especially the following papers: "Flexible interpretations of 'acid rain' and the construction of scientific uncertainty in political settings," *Politics and the Life Sciences*, vol. 13, no. 2, August 1994, pp. 205–216, and "The centrality of scientists and the translation of interests in the U.S. acid rain controversy," *Canad. Rev. Soc. & Anth.*, vol. 31, no. 3, 1994, pp. 325–353. Also, for a good account of Reagan and the "acid rain" issue, see Herken, *Cardinal Choices*, p. 204–205.

Mount St. Helens: See Joanne Omang, "Reagan criticizes clean air laws and EPA as obstacles to growth," *Washington Post*, October 9, 1980, and "The 1980 issues; Mr. Reagan v. nature," *Washington Post* (editorial), October 10, 1980.

Ronald Reagan came out: See "The 1980 issues; Mr. Reagan v. nature," *Washington Post*, October 10, 1980, and "The environment and the stump," *New York Times*, October 22, 1980.

"more research" would first be required: See Philip Shabecoff, "Panel of scientists bids U.S. act now to curb acid rain," *New York Times*, June 28, 1983.

magnified uncertainty: See James L. Regens and Robert W. Rycroft, *The Acid Rain Controversy* (Pittsburgh: University of Pittsburgh Press, 1988). On page 48, the authors quote an industry defender who claimed that the "huge acid rain research effort provides ample data showing that the link between SO_2 emissions and the acidity of rain is far weaker than generally supposed, and, further, that the link between acid rain and ecological damage is even weaker, or nearly nonexistent." Similarly, on page 124, the authors quote another industry source: "I cannot overemphasize the importance of knowing what responsibility the industry has for acid deposition. Unless we know that, we cannot judge the efficacy of any control strategy which might be promulgated and, as scientists, we have an obligation to evaluate the extent to which a control strategy will achieve its goal. For acid precipitation, we cannot do that."

42   *They argued about anything:* Interview with Michael Oppenheimer, December 3, 2004.

a number of scientific developments: Eliot Marshall, "Acid rain, a year later; close scrutiny by several technical groups had not made the problem go away, just made the case for regulation stronger," *Science*, vol. 221, July 15, 1983, p. 241.

With two scientific reports: Philip Shabecoff, "Acid rain: new evidence limits political options," *New York Times*, July 5, 1983.

delayed and watered down: Cass Peterson, "White House stalls acid rain report; study stresses stronger action," *Washington Post*, August 18, 1984; Marjorie Sun, "Acid rain report allegedly suppressed," *Science*, vol. 225, September 21, 1984, p. 1374.

catastrophe, which careened: R. Jeffrey Smith, "The knives are out for OSTP; senior White House officials are pushing for its elimination, but science adviser Keyworth says he has Reagan's support," *Science*, vol. 226, December 21, 1984, p. 1399.

Only in 1986 did Reagan: Michael Weisskopf, "Acid rain: pledges but no progress; issue that ensnared Deaver seems to have been shunted aside," *Washington Post*, November 26, 1987.

continued to call for more research: Reuters, "Acid rain problem 'small, not urgent,' U.S. official says," June 21, 1986.

43   *cannot be pinpointed:* Eliot Marshall, "Acid rain, a year later; close scrutiny by several technical groups had not made the problem go away, just made the case for regulation stronger," *Science*, vol. 221, July 15, 1983, p. 241.

The high profile "Star Wars" controversy: My account relies on Frances Fitzgerald, *Way Out There in the Blue: Reagan, Star Wars, and the End of the Cold War* (New York: Simon & Schuster, 2000), as well as on Gregg Herken's *Cardinal Choices*, Chapter 12, and Daniel Greenberg's *Science, Money, and Politics*, Chapter 18. Greenberg describes "Star Wars" as an episode that "demonstrated politics' resistance and, at times, indifference to unsought or politically unpalatable advice from scientists."

actually delivered its very first pamphlet: George C. Marshall Institute, twentieth anniversary (1984–2004) DVD, on file with the author.

on a five-year time scale: My account of the space-based missile defense boosters relies on Fitzgerald, *Way Out There in the Blue*, Chapter 4, "Space Defense Enthusiasts." For the quotation, see p. 135.

drastically oversold: See Fitzgerald, *Way Out There in the Blue*, p. 374.

44   *increasingly at odds:* Herken, *Cardinal Choices*, p. 215.

But the realities of a nuclear era: Interview with Sidney Drell, June 10, 2004.

just as a roof protects: Fitzgerald, *Way Out There in the Blue*, p. 336.

Opinion polls showed: See Fitzgerald, *Way Out There in the Blue*, p. 258.

reports by the Union of Concerned Scientists: Union of Concerned Scientists, *The Fallacy of Star Wars* (New York: Vintage Books, 1984).

the American Physical Society: "Report to the American Physical Society of the Study Group on the Science and Technology of Directed Energy Weapons," in *Reviews of Modern Physics*, vol. 59, no. 3 (part 2), July 1987.

45   *full-fledged "science wars":* For the "science wars" see Fitzgerald, *Way Out There in the Blue*, pp. 246–247.

should not serve as the basis: Directed Energy Missile Defense in Space: A Background Paper (Washington, D.C.: U.S. Congress, Office of Technology Assessment, OTA-BP-ISC-26, April 1984). The quotation comes from Section 10, "Principal Judgments and Observations."

OTA expert review: See Bimber, *The Politics of Expertise in Congress*, pp. 44–45.

a report charging OTA: Michael S. Warner, "Reassessing the Office of Technology Assessment," *Heritage Foundation Reports*, November 7, 1984.

45 *could ignite a new arms buildup:* U.S. Congress, Office of Technology Assessment, *Ballistic Missile Defense Technologies,* OTA-ISC-254 (Washington, DC: U.S. Government Printing Office, September 1985). Quotation on p. 33.

catastrophic failure: U.S. Congress, Office of Technology Assessment, *SDI: Technology, Survivability, and Software,* OTA-ISC-353 (Washington, DC: U.S. Government Printing Office, May 1998). Quotation on p. 5.

withheld three chapters entirely: See Walter Andrews, "Pentagon accused of politics on 'Star Wars' report," UPI, May 7, 1988. In a October 1988 interview with *Technology Review,* OTA director John Gibbons noted, "The three chapters have been thoroughly cleaned. They contain absolutely no surprises for the Soviets. You have to ask why they're denied to the American people." ("How John Gibbons runs through political minefields: life at the OTA," *Technology Review,* October 1988.)

46 *despite the fringe nature of his views:* Daniel Greenberg writes, "In this instance, politics excluded the official advisory system and chose instead to rely on one member, Teller, a politicized scientist who inflamed the Reagan administration's hard-line tendencies" (Greenberg, *Science, Money, and Politics,* p. 282).

took their words as gospel: Interview with W. K. H. Panofsky, June 10, 2004.

rejuvenate the social conservatives: See Kirk Victor, "Not praying together," *National Journal,* October 10, 1987.

Reagan called upon Koop: In his memoir, Koop writes, "Reagan had also embraced a silly idea touted by one of the neophyte right-wingers on the White House staff that the evidence of adverse health effects (presumably mental) of abortion on women that the Surgeon General could pull together would be sufficient to overturn *Roe v. Wade.* Anyone with even a smattering of information—and an idea of how the Supreme Court works—would know this idea was foolish. But the young campus-conservative ideologues heady with what they thought was power did not always live in the real world" (Koop, *Koop: The Memoirs of America's Family Doctor,* pp. 347–348).

47 *a key device for overturning:* For a good explanation of how the Right's agenda works in relation to the question of abortion's alleged health risks, see N. F. Russo and J. E. Denious, "Controlling birth: science, politics, and public policy," *Journal of Social Issues,* vol. 6 (2005), pp. 181–191.

a letter to Reagan declining: Quoted in Warren E. Leary, "Koop challenged on abortion data," *New York Times,* January 15, 1989.

psychological risks from abortion are "minuscule": The Federal Role in Determining the Medical and Psychological Impact of Abortion on Women, 10th Report by the Committee on Government Operations (House Report 101–392), Washington, D.C.: U.S. Government Printing Office, 1989, p. 14.

Conservative Caucus chair Howard Phillips: Quoted in Nancy Shute, "America's M.D.: Surgeon General Koop has been a shot in the arm to the nation's ills," *Chicago Tribune,* June 23, 1989.

it would have been attacked: Quoted in "A surgeon general's parting shots," *Washington Post,* June 20, 1989.

Lee Thomas refused: For Lee Thomas's statements on uncertainty, see Sharon L. Roan, *Ozone Crisis: The 15 Year Evolution of a Sudden Global Emergency* (New York: Wiley, 1989), pp. 146, 152.

48 *assistant to the president:* See D. Allan Bromley, *The President's Scientists: Reminiscences of a White House Science Advisor* (New Haven: Yale University Press, 1994).

5: DEFENSELESS AGAINST THE DUMB

49 *Gingrich predicted:* Newt Gingrich, "We must expand our investment in science," testimony before the Senate Committee on Commerce, Science, and Transportation, May 22, 2002.

Available online at http://www.aei.org/news/newsID.15562/news_detail.asp (accessed December 7, 2004).

50 *much as Gingrich himself did:* In a September 29, 2003 interview Gingrich spokesman Rick Tyler told me, "The Speaker always believed that congressmen and senators were of high enough standing where they could contact scientists working on breakthroughs in the scientific community at any given time, and most scientists are thrilled to brief members on what they're doing, and that the Office of Technology Assessment just provided a barrier and a firewall, to have staff people to go out and talk to the same scientists. And somehow, things got lost in the translation. And for him, personally, it was not a very efficient way to get the right information at the right time."

Which I thought was completely bizarre: Interview with Bob Palmer, April 29, 2004.

51 *a very dumb decision:* House Science Committee Democratic press release, "OTA, Congress's defense against the dumb, closes down; Congress left defenseless," September 29, 1995. Available online at http://www.house.gov/science_democrats/releases/95sep29.htm (accessed December 7, 2004).

How to Revolutionize Washington: Scott Shuger, "How to revolutionize Washington with 140 people," *Washington Monthly,* June 1989.

Congress had created OTA in 1972: My account of OTA's history relies on Bruce Bimber, *The Politics of Expertise in Congress: The Rise and Fall of the Office of Technology Assessment* (Albany: State University of New York Press, 1996).

52 *liberal technocrats:* William Safire, "The Charles River gang returns," *New York Times,* May 26, 1977.

Gibbons promptly shook up: See R. Jeffrey Smith, "Thermidor at OTA," *Science,* vol. 205, August 10, 1979, and "A narrower focus on technology assessment," *Business Week,* August 27, 1979.

Fat City: Donald Lambro, *Fat City* (South Bend, Indiana: Gateway, 1980). OTA is "Program 52" on the chopping block in Lambro's book.

second and occasionally third printings: Rebuttal on file with author.

killing of OTA as "Reagan's revenge": Interview with John Gibbons, June 3, 2004.

53 *consensus body of information:* Interview with Rosina Bierbaum, September 25, 2003.

almost a decade earlier: U.S. Congress, Office of Technology Assessment, *Technologies Underlying Weapons of Mass Destruction,* OTA-BP-ISC-115 (Washington, D.C.: U.S. Government Printing Office, December 1993), p. 78.

it even could have saved a life: "Technology assessment in the war on terrorism and homeland security: the role of OTA," report prepared at the request of Hon. Ernest F. Hollings, chairman, Committee on Commerce, Science, and Transportation, U.S. Senate (Washington: U.S. Government Printing Office, April 2002). On file with author.

left us aghast: From the foreword to Norman J. Vig and Herbert Paschen, *Parliaments and Technology: The Development of Technology Assessment in Europe* (Albany: State University of New York Press, 2000).

sacrificial victim: Interview with Henry Kelly, September 24, 2003.

54 *views on science and technology:* Interview with Roger Herdman, September 25, 2003.

In some cases, it was politicized: Interview with Rick Tyler, September 29, 2003.

took too long to prepare: Interview with Robert Walker, December 2, 2004.

You don't cut the future: Quoted in Jerry Zremski, "Our man against the Newtoids; how one of the richest people in Congress says 'no' to the ultra Right," *Buffalo News (Magazine),* December 10, 1995.

$21.9 million: Figure cited in John E. Yang, "Hill cuts $200 million from its appropriation; lawmakers tentatively agree to forgo their pay raise for 3rd consecutive year," *Washington Post,* September 15, 1995.

sort of symbolic targets: Interview with Amo Houghton, September 25, 2003.

54 *callous treatment:* "An interview with John H. Gibbons," *Cosmos,* 1998, available at http://www.cosmos-club.org/journals/1998/gibbons.html (accessed December 7, 2004).

He was poisoned: John Gibbons, speech at Rice University's Baker Institute for Public Policy, November 2, 2003 (author's notes).

analyses prepared by lobbyists: In *The Politics of Expertise in Congress,* Bruce Bimber notes that "OTA was discarded at the same time that the House leadership strengthened its ties with partisan providers of expertise like the Heritage Foundation" (p. 99).

55 *We constantly found scientists:* Quoted in Jim Dawson, "Legislation to revive OTA focuses on science advice to Congress," *Physics Today,* October 2001. For Gingrich described as a "science geek . . . fascinated with space travel and science fiction," see Shawn Zeller, "Newt's science project," *National Journal,* April 18, 1998.

Scientific Integrity and Public Trust: For a comprehensive (and critical) look at these hearings see George Brown, "Environmental science under siege: fringe science and the 104th Congress," October 23, 1996, at http://www.house.gov/science_democrats/archive/envrpt96.htm (accessed December 8, 2004).

found little substantive support: Brown's investigation into the hearing allegations found "no credible evidence to support the claims that scientists distorted their research to serve political ends."

liberal claptrap: See William K. Stevens, "Trying to stem emissions, U.S. sees its goal fading," *New York Times,* November 28, 1995.

56 *what is the consensus:* Naomi Oreskes, "Science and public policy: what's proof got to do with it?" *Environmental Science & Policy* vol. 7 (2004), pp. 369–383.

Hearings are about trying: Interview with Robert Walker, December 2, 2004.

on the subject of ozone depletion: Committee on Science hearing, Subcommittee on Energy and Environment, "Scientific Integrity and Public Trust: The Science Behind Federal Policies and Mandates, Case Study 1—Stratospheric Ozone: Myths and Realities," September 20, 1995.

the Nobel Prize in chemistry: The prize was awarded to Paul J. Crutzen, Mario Molina, and Sherwood Rowland "for their work in atmospheric chemistry, particularly concerning the formation and decomposition of ozone." See: http://www.nobel.se/chemistry/laureates/1995 (accessed December 8, 2004).

disputing the dangers of acid rain: See, for example, S. Fred Singer, "The answers on acid rain fall on deaf ears," *Wall Street Journal,* March 6, 1990.

57 *There is no scientific consensus:* Quoted in Gary Lee, "Lawmakers move to check CFC phaseout; House Republicans play down links between chemicals and depletion of ozone layer," *Washington Post,* September 21, 1995.

Every good scientist is a skeptic: Interview with James McCarthy, May 11, 2004.

A Dip into the Skeptic Tank: Interview with Bob Palmer, April 29, 2004.

subjected to withering skepticism: My account of the history of the ozone depletion issue relies on Sharon L. Roan, *Ozone Crisis: The 15 Year Evolution of a Sudden Global Emergency* (New York: Wiley, 1989).

CFC manufacturers like DuPont: For DuPont's resistance, see Roan, *Ozone Crisis,* p. 36.

58 *Numerous ozone assessments:* For NASA studies see NASA, *Chlorofluoromethanes and the Stratosphere,* National Aeronautics and Space Administration, Scientific and Technical Information Branch, NASA Reference Publication 1010, 1977; and NASA, *The Stratosphere: Present and Future,* National Aeronautics and Space Administration, Scientific and Technical Information Branch, NASA Reference Publication 1049, 1979. For World Meteorological Organization studies see WMO, *The Stratosphere 1981: Theory and Measurements,* World Meteorological Organization Global Ozone Research and Monitoring Project Report No. 11, 1981; WMO, *Atmospheric Ozone: 1985, Assessment of Our Understanding of the Processes Controlling Its Present Distribution and Change,* World Meteorological Organization Global Ozone Research and Monitoring Project Report No. 16, 1986; WMO, *Report of the International Ozone Trends Panel: 1988,* World Meteorological Organization Global Ozone Research and

Monitoring Project Report No. 18, 1988; WMO, *Scientific Assessment of Stratospheric Ozone: 1989*, World Meteorological Organization Global Ozone Research and Monitoring Project Report No. 20, 1989; WMO, *Scientific Assessment of Ozone Depletion: 1991*, World Meteoro-logical Organization Global Ozone Research and Monitoring Project Report No. 25, 1991; and WMO, *Scientific Assessment of Ozone Depletion: 1994*, World Meteorological Organiza-tion Global Ozone Research and Monitoring Project Report No. 37, 1995.

58 *natural sources like volcanoes:* Dixy Lee Ray (with Lou Guzzo), *Trashing the Planet: How Sci-ence Can Help Us Deal with Acid Rain, Depletion of the Ozone, and Nuclear Waste (Among Other Things)* (Washington: Regnery Gateway, 1990; New York: HarperCollins, 1992), p. 45.

the most footnoted, documented book: Quoted in Gary Taubes, "The ozone backlash; critics of ozone depletion research," *Science*, vol. 260, no. 5114, p. 1580, June 11, 1993.

59 *an international scientific consensus:* Sherwood F. Rowland, "President's lecture: the need for scientific communication with the public," *Science*, vol. 260, no. 5114, p. 1571, June 11, 1993.

the table of scientific inquiry: Dr. John H. Gibbons, "Sound science, sound policy: the ozone story," remarks delivered on September 19, 1995, at the University of Maryland at College Park. It can be read online at http://clinton1.nara.gov/White_House/EOP/OSTP/director/sound_policy.html (accessed December 9, 2004).

a mumbo-jumbo of peer-reviewed documents: Quoted in Ross Gelbspan, *The Heat is On* (New York: Perseus, 1997), p. 65.

my assessment is from reading people: Quoted in William K. Stevens, "GOP bills aim to delay ban on chemicals in ozone dispute," *New York Times*, September 21, 1995.

some primordial soup of mud: Quoted in Francis X. Cline, "Capitol sketchbook; in a bitter cul-tural war, an ardent call to arms," *New York Times*, June 17, 1999.

60 *can be traced back:* My account of the history of climate science relies on Spencer Weart, *The Discovery of Global Warming* (Cambridge: Harvard University Press, 2003). A more expansive version of Weart's history can be read online at http://www.aip.org/history/climate (accessed February 10, 2005).

a rise in media coverage: See McComas and Shanahan, "Telling stories about global climate change," *Communication Research*, Vol. 26, No. 1, February 1999, 30–57.

61 *found his scientific testimony:* See Philip Shabecoff, "Scientist says budget office altered his tes-timony," *New York Times*, May 8, 1989.

forged the Global Climate Coalition: See "Business forms 'climate coalition' to play research and policy role," *Electric Utility Week*, June 26, 1989.

the enormous phenomena that take place in nature: See PR Newswire, "Need for more science and data when assessing effects of climate change, says former Washington governor," Febru-ary 19, 1992. See also PR Newswire, "World's energy policy should not be based on feelings, experts say," February 27, 1992.

thousands of experts: As the IPCC noted in 1995, "the reports of the IPCC and of its Work-ing Groups contain the factual basis of the issue of climate change, gleaned from available ex-pert literature and further carefully reviewed by experts and governments. In total, more than two thousand experts worldwide participate in drafting and reviewing them." (Intergovern-mental Panel on Climate Change, "IPCC second assessment: climate change 1995," available online at http://www.ipcc.ch/pub/sa(E).pdf (accessed December 9, 2004).

62 *The balance of evidence suggests:* Ibid.

bringing unsubstantiated charges: For a discussion of the charges levied against the IPCC, see Paul N. Edwards and Stephen H. Schneider, "Self-Governance and Peer Review in Science-for-Policy: The Case of the IPCC Second Assessment Report," chapter 7 in Clark Miller and Paul N. Ed-wards, eds., *Changing the Atmosphere: Expert Knowledge and Environmental Governance* (Cam-bridge, Massachusetts: MIT Press, 2001). Available online at http://stephenschneider. stanford.edu/Publications/PDF_Papers/ipccpeer.pdf. See also "Special insert: an open letter to

Ben Santer," *University Corporation for Atmospheric Research Quarterly*, Summer 1996, available online at http://www.ucar.edu/communications/quarterly/summer96/insert.html (accessed December 9, 2004). Here, dozens of scientists come to Santer's support, including then–Intergovernmental Panel on Climate Change chairman Bert Bolin.

62 *recipient of substantial energy industry funding:* Michaels's industry ties are documented in Gelbspan, *The Heat is On*; see pp. 40–44. See also Scott Allen, "Global warming debate joined; scientists at hearing doubt threat, activists cite industry ties," *Boston Globe*, November 17, 1995, noting, "Michaels contends that global warming is so minor that the media should 'go find some other issue.' But *Harper's* magazine this week suggests that Michaels has a financial motive to push that view: He has received grants of $115,000 from energy interests, and a coal industry group funds his newsletter. . . . Michaels said he publicly disclosed the research funding from industry and denied that it affects his findings. He said he had testified before Congress about his doubts on global warming 'long before these industry guys knew we existed.'"

soon to go out of style: Quoted in Brown, "Environmental science under siege."

an emphasis on "observed data": Testimony of Patrick J. Michaels, Department of Environmental Sciences, University of Virginia, to the Subcommittee on Energy and the Environment, U.S. House of Representatives, November 16, 1995.

"empirical" evidence: As Brown notes, Michaels's charge "resonates with a corresponding tenet of the Republican Vision for science, which states that 'science programs must seek and be guided by empirically sound data.' That is, model data which do not clearly fit observations must, by definition, be unsound regardless of whatever insights models provide."

attacking a "straw man": Again, see Brown, "Environmental science under siege."

63 *rational basis for action:* Interview with Naomi Oreskes, January 25, 2005.

As Watson noted: Quoted in Brown, "Environmental Science Under Siege."

mathematical representation of physical processes: National Research Council, *Improving the Effectiveness of U.S. Climate Modeling* (Washington, D.C., National Academy Press, 2001).

64 *not one or the other in isolation:* As Brown's report noted, "in reality both observational evidence and theoretical models are essential to constructing an understanding of what is being observed. Neither in isolation is sufficient nor superior from an intellectual standpoint, as suggested by the Republican vision statement. The scientific method, in its purest state, is based on observations, hypotheses, testing of hypotheses, and refining a theoretical construct to explain the phenomenon."

cottage industry: For a telling study on this point, see Aaron M. McCright and Riley E. Dunlap, "Defeating Kyoto: The conservative movement's impact on U.S. climate change policy," *Social Problems*, vol. 50, no. 3, pp. 348–373.

and CO_2 continues to rise: Interview with William Schlesinger, April 23, 2004.

6: JUNKING "SOUND SCIENCE"

65 *pervasive rallying cry:* As Brown notes, "One of the major claims asserted throughout the 104th Congress was that many environmental regulations were not based on 'sound science,' but instead on scare-mongering and gross exaggerations of environmental problems."

George C. Marshall Institute declares: Available at http://www.marshall.org (accessed December 10, 2004).

from climate change to arsenic in drinking water: On climate change, Bush has stated, "When we make decisions, we want to make sure we do so on sound science; not what sounds good, but what is real." George W. Bush, speech on "Clear Skies and global climate change initiatives," February 14, 2002. Available online at http://www.whitehouse.gov/news/releases/2002/02/20020214-5.html (accessed December 10, 2004). On arsenic in drinking water, Bush has stated, "At the very last minute, my predecessor made a decision, and we pulled back his decision so that we can make a decision based upon sound science and what's realistic." Quoted in James Gerstenzang, "Bush defends his stance on environment," *Los Angeles Times*, March 30, 2001.

66 *a 1981 report:* Edward L. Behrens, chairman, American Industrial Health Council Public Affairs Committee, "Ensuring sound science in the development of national chronic health hazards policy (1981 report to the membership)," on file with author. Also available online at http://thecre.com/pdf/20040830_aihc_doc.pdf (accessed December 10, 2004).

challenging agency experts: For the new "scientific" role of the Office of Management and Budget and more specifically the Office of Information and Regulatory Affairs (OIRA) in the Reagan administration, see Sheila Jasanoff, *The Fifth Branch: Science Advisers as Policymakers* (Cambridge, Massachusetts: Harvard University Press, 1990), pp. 190–91. As Jasanoff notes, "Although staffed primarily by economists and policy analysts, OIRA intruded into the scientific aspects of regulatory policy with little hesitation, involving itself between 1983 and 1986 in a series of science policy decisions under the jurisdictions of several federal agencies."

in relation to climate change: "Our responsibility is to maintain the quality of our approach, our commitment to sound science and an open mind to policy options," Bush told a 1990 meeting of the Intergovernmental Panel on Climate Change. See Michael Weisskopf, "Bush pledges research on global warming; speech to U.N.-sponsored panel endorses no proposed remedies," *Washington Post*, February 6, 1990.

became a talking point: See, for example, William K. Stevens, "What really threatens the environment?" *New York Times*, January 29, 1991, as well as Michael Weisskopf, "EPA eases ban on pesticide, finding cancer risk overestimated," *Washington Post*, February 14, 1992.

an Environmental Protection Agency report: U.S. Environmental Protection Agency, Office of Research and Development, Office of Health and Environmental Assessment, Washington, D.C., "Respiratory health effects of passive smoking: lung cancer and other disorders," EPA/600/6-90/006F, 1992. Available online at http://cfpub2.epa.gov/ncea/cfm/recordisplay.cfm?deid=2835 (accessed December 10, 2004). As the study notes, "ETS is a human lung carcinogen, responsible for approximately 3,000 lung cancer deaths annually in U.S. nonsmokers."

high-caliber scientific reports: As summarized in Jonathan M. Samet and Thomas A. Burke, "Turning science into junk: the tobacco industry and passive smoking," *American Journal of Public Health*, vol. 91, no. 11, November 2001. The authors note, "Although the conclusion had been reached as early as 1986 that ETS was a cause of lung cancer, the EPA is a regulatory agency, and its conclusion on ETS had potential implications for tobacco control policy." Later, they add that since 1992, several additional reviews of the evidence have been carried out, and all, except for a review by an industry-sponsored panel, have concluded that passive smoking increases the risk of lung cancer in nonsmokers."

another step in a long process: See Warren E. Leary, "U.S. ties secondhand smoke to cancer," *New York Times*, January 8, 1993.

decided to do something: See Elisa K. Ong and Stanton A. Glantz, "Constructing 'sound science' and 'good epidemiology': tobacco, lawyers, and public relations firms," *American Journal of Public Health*, vol. 91, no. 11, November 2001, pp. 1749–1757, noting, "PM [Philip Morris] began its 'sound science' program in 1993 to stimulate criticism of the 1992 U.S. Environmental Protection Agency (EPA) report, which identified secondhand smoke as a Group A human carcinogen."

67 *1964 U.S. surgeon general's report:* 1964 surgeon general report, "Reducing the health consequences of smoking," available online at http://www.cdc.gov/tobacco/sgr/sgr_1964/sgr64.htm (accessed December 17, 2004).

mobilized to undermine the science: My historical account relies on Stanton Glantz et al, *The Cigarette Papers* (Berkeley: University of California Press, 1996). See, for example, page 173: "After an initial period of uncertainty around the release of the Surgeon General's report . . . the tobacco industry started an aggressive campaign to create controversy about the scientific evidence that smoking is dangerous and to defend the 'right' to smoke." But even earlier, the Tobacco Industry Research Committee (TIRC), founded in 1954, had played a similar role. As Glantz et al. note on page 33, "The industry stated publicly that it was forming TIRC in response to scientific reports suggesting a link between smoking and lung cancer, and that the purpose of TIRC was to fund independent scientific research to determine whether these reports were true. However, the documents show that TIRC was actually formed for public relations purposes, to convince the public that the hazards of smoking had not been definitively proven."

67   *lawyers often funding:* See Lisa Bero, "Tobacco industry manipulation of research," *Public Health Reports*, March–April 2005, vol. 120, noting, "The Master Settlement Agreement between the attorneys general of 46 states and [the tobacco companies] released millions of additional documents to the public. These documents provided an unprecedented look at how tobacco industry lawyers were involved in the design, conduct, and dissemination of tobacco industry-sponsored research." For a more extensive treatment see Lisa Bero et al., "Lawyer control of the tobacco industry's external research program: the Brown and Williamson documents," *Journal of the American Medical Association*, vol. 274, no. 3, July 19, 1995, pp. 241–247. As the paper notes, "The involvement of tobacco industry lawyers in the selection of scientific projects to be funded is in sharp contrast to the industry's public statements about its review process for its external research program. Scientific merit played little role in the selection of external research projects. The results of the projects were used to generate good publicity for the industry, to deflect attention away from tobacco use as a health danger, and to attempt, sometimes surreptitiously, to influence policymakers."

manufacturing uncertainty: See, for example, statement by David Michaels, "Oversight hearing: the impact of science on public policy," House of Representatives Subcommittee on Energy and Mineral Resources, February 4, 2004. On file with author.

1969 Brown & Williamson document: Brown & Williamson, "Smoking and health proposal," document no. 332506, available at http://tobaccodocuments.org/bw (accessed December 17, 2004). In *The Cigarette Papers*, Glantz et al. discuss dating this document on p. 189: "Although the document is undated, the context of the discussion places it around 1969."

epidemiological reports began to appear: For the evolution of scientific understanding on secondhand smoke, see Thomas O. McGarity, "On the prospect of 'Daubertizing' judicial review of risk assessment," *Law and Contemporary Problems*, vol. 66, no. 155, autumn 2003.

once again seeking to sow doubt: Interview with Lisa Bero, June 8, 2004. See also Bero, "Tobacco industry manipulation of research," noting, "The strategies [used by the tobacco industry] have remained remarkably constant since the early 1950s. During the 1950s and 1960s, the tobacco industry focused on refuting data on the adverse effects of active smoking. The industry applied the tools it had developed during this time to refute data on the adverse effects of secondhand smoke exposure from the 1970s through the 1990s."

the Advancement of Sound Science Coalition: Business Wire, "National watchdog organization launched to fight unsound science used for public policy comes to Denver," November 24, 1993.

Internal documents show: See Elisa K. Ong and Stanton A. Glantz, "Constructing 'sound science' and 'good epidemiology': tobacco, lawyers, and public relations firms," *American Journal of Public Health*, vol. 91, no. 11, November 2001, pp. 1749–1757.

Our overriding objective: Quoted in Ong and Glantz, "Constructing 'Sound Science' and 'Good Epidemiology.'" For the original document, see Ellen Merlo, memo to William I. Campbell, February 17, 1993, document no. 2021183916/3930. Available at http://www.pmdocs.com (accessed December 17, 2004).

68   *APCO informed Philip Morris:* Letter from Margery Kraus, president and chief executive officer, APCO Associates, Inc., to Mr. Vic Han, director of communications, Philip Morris, USA,

September 23, 1993, document no. 2024233677/3682. Available at http://www.pmdocs.com (accessed December 18, 2004).

68   *we thought it best to remove:* Note for Ellen Merlo and others from Jack Lenzi, November 15, 1993, document no. 2024233664, available at http://www.pmdocs.com (accessed December 18, 2004).

claimed a wide diversity of members: There is no reason to doubt that TASSC signed up a diverse membership, but in late December 1995, a Philip Morris employee clearly indicated knowledge about the company's role in forming the group. In an exchange on December 1, Philip Morris's John Sorrells asked colleague Karen Daragan, "Chalk it up to ignorance or my relatively short tenure with Big Mo, but I don't know what TASSC is. Can you fill me in?" Daragan replied, "The Advancement of Sound Science Coalition. APCO helped us create the coalition of scientists et al." See John Sorrells, note to Karen Daragan, December 1, 1995, and Karen Daragan, note to John Sorrells, December 1, 1995, documents no. 2048585326B and 2048585326A. Available at http://www.pmdocs.com (accessed December 19, 2004).

according to a February 22, 1994, memo: Memo from John C. Lenzi re: TASSC, February 22, 1994, document no. 2078848225/8226, available at http://www.pmdocs.com (accessed December 18, 2004).

the incomplete science: Business Wire, "National watchdog organization launched to fight unsound science used for public policy comes to Denver," noting, "Specific examples of unsound science in public policy that TASSC highlights include the Alar scare, the poorly researched—and costly—guidelines created for asbestos abatement, the incomplete science used to create ozone repair policy, the decision to move the town of Times Beach, Mo., because of a dioxin scare, and unprecedented regulations to limit radon levels in drinking water (which may include a recommendation to avoid taking showers), to name but a few."

poll of scientists: See PR Newswire, "Scientists say public confidence erodes through government manipulation of research, poll shows," September 29, 1994.

one of its newsletters: The Advancement of Sound Science Coalition, the *Catalyst* (newsletter), summer 1994, document no. 2070270144/0147, available at http://www.pmdocs.com (accessed December 18, 2004). As the newsletter notes, "the EPA's methodology in its secondhand smoke study has been criticized on several fronts. EPA has deemed second-hand smoke a Class A carcinogen, but many experts question the study's methodology."

questioned the scientific credibility of EPA: See, for example, Business Wire, "National watchdog organization launched to fight unsound science used for public policy comes to Denver," November 24, 1993.

a Knight-Ridder news story noted: Heather Dewar, "GOP ready to dilute environmental laws; new buzzwords target property rights," *Houston Chronicle,* November 12, 1994.

69   *five times as much:* See Jane Fritsch, "Tobacco companies pump cash into Republican Party's coffers," *New York Times,* September 13, 1995.

since the early 1980s: See, for example, Jasanoff, *The Fifth Branch: Science Advisers as Policymakers.* On page 185, Jasanoff notes, "The American Industrial Health Council (AIHC), a leading spokesman for the industrial viewpoint, began to lobby strenuously for a science panel to review all agency decisions relating to carcinogenic risk." This was in 1981.

dubious anecdotal examples: For a partial debunking of some "regulatory horror stories," see Edward Flattau, "Mainly imaginary top tens," *Cleveland Plain Dealer,* August 1, 1995 and Frank Clifford, "Does earth still need protection? reformers want to rein in regulation and what they say is bad science, but their targeting of landmark laws may imperil strides made in past quarter-century," *Los Angeles Times,* April 22, 1995. Senator John Glenn (D-OH) delivered a July 14, 1995, speech debunking a slew of such horror stories, which can be read online at http://www.info-pollution.com/glenn.htm (accessed December 13, 2004).

reform bills: For a thorough (if critical) overview of the content of these reforms, see David Vladeck, "Paralysis by analysis: how conservatives plan to kill popular regulation," *American*

Prospect, vol. 6, no. 22, June 23, 1995. See also Robert L. Glicksman and Stephen B. Chapman, "Regulatory reform and (breach of) the Contract with America: improving environmental policy or destroying environmental protection?" *Kansas Journal of Law & Public Policy,* Winter 1996.

70 *potentially stifling scientific adaptability:* Glicksman and Chapman further note that the regulatory reform bills, "by prescribing detailed risk assessment methodology, would codify existing techniques into law and might prevent environmental policymakers from adapting this methodology in response to scientific advances. . . . The freezing in place of current risk assessment methodologies creates the potential for serious misallocations of regulatory effort."

"peer review" panels: Vladeck notes, "Under the bills, regulatory agencies would be required to submit both risk assessments and cost–benefit analyses to peer review panels that could be composed principally of industry scientists. Even a direct conflict of interest is not grounds for disqualification; panelists need only disclose their financial ties to industry."

paralysis by analysis: Vladeck, "Paralysis by analysis."

Dole defended his bill: Robert Dole, "There's no law against common sense; the majority leader makes the case for reforming federal regulations," *Washington Post,* March 5, 1995.

sought to slash funding: As Rep. George Brown (D-CA) noted at the time, "the Science Committee had been actively pursuing a legislative agenda that was unprecedented in its mistrust and hostility toward Federal environmental research programs. For example, the FY96 Budget Resolution, which passed the House in May 1995, recommended a reduction of 20 percent in FY 1996 for Federal environmental research and a steep decline in succeeding years." See Brown, "Environmental science under siege: fringe science and the 104th Congress," October 23, 1996, http://www.house.gov/science_democrats/archive/envrpt96.htm (accessed April 12, 2004). See also Daniel Greenberg, *Science, Money, and Politics: Political Triumph and Ethical Erosion* (Chicago: University of Chicago Press, 2001). See in particular chapter 27, "Science Versus the Budget Cutters."

tracks earthquakes and volcanic eruptions: See Joel Connelly, "Babbitt tries to save the earthquake experts," *Seattle Post-Intelligencer,* February 20, 1995.

focusing much of its energy: Memorandum from APCO Associates to TASSC Supporters, June 9, 1995, noting, "Much of our efforts have surrounded the congressional debate on regulatory reform, since a primary component of the pending legislation deals with controversial provisions on sound science and risk assessment," document no: 2048294354/4356, available at http://www.pmdocs.com (accessed December 18, 2004).

endorsed a regulatory reform proposal: See PR Newswire, "Risk assessment legislation necessary, leader of national science group says," October 13, 1994.

71 *released a study protesting:* See PR Newswire, "Media reports slanted against regulatory reform efforts, study shows," July 7, 1995. The Carruthers quotation comes from this release.

cited in a statement: Robert Dole press release, "Regulatory reform & the media," July 13, 1995.

who unwittingly continue: See, for example, the "Sound Science Initiative," sponsored by the Union of Concerned Scientists, a liberal environmental group: http://www.ucsusa.org/global_environment/invasive_species/page.cfm?pageID=1101 (accessed December 11, 2004).

in an online discussion: Western Caucus, "Sound science and public policy," available online at http://chriscannon.house.gov/wc/meetings/Conference/soundscience.htm (accessed December 11, 2004).

72 *describes the concept:* Interview with Thomas Roskelly, November 1, 2004.

I asked William Kovacs: Interview with William Kovacs, March 18, 2004.

Look, you need to take account: Interview with William O'Keefe, April 13, 2004.

says Robert Walker: Interview with Robert Walker, December 2, 2004.

74 *a strategic memo on the environment:* The document can be read on the Internet at http://www.ewg.org/briefings/luntzmemo/pdf/LuntzResearch_environment.pdf (accessed

December 12, 2004). Its veracity was confirmed by the *New York Times*. See Jennifer Lee, "A call for softer, greener language," *New York Times*, March 2, 2003.

74 *pollster of record:* Luntz is described this way on the website of the Luntz Research Companies. See http://www.luntz.com/PoliticalAnalysis.htm (accessed December 12, 2004).

Luntzspeak: The group has even created the website Luntzspeak.com.

According to a 2001 media analysis: Charles N. Herrick and Dale Jamieson, "Junk science and environmental policy: obscuring public debate with misleading discourse," *Philosophy & Public Policy Quarterly*, vol. 21, no. 2/3 (spring/summer 2001).

75 *In a 2004 lecture:* McGarity's 2004 "Order of the Coif" lecture was later republished as "Our science is sound science and their science junk science: science-based strategies for avoiding accountability and responsibility for risk-producing products and activities," *Kansas Law Review*, June, 2004, 52 Kan. L. Rev. 897.

7: "THE GREATEST HOAX"

77 *Everything's going in the science direction:* Interview with William Kovacs, March 18, 2004.

78 *In a written response to questions:* Written response received from Inhofe's staff on March 24, 2004, via e-mail.

gestapo bureaucracy: Quoted in Eric Pianin, "For Senate committee, a big change; new environment chairman opposes many protections," *Washington Post*, December 30, 2002.

campaign donations: According to the Center for Responsive Politics, in the 2002 election cycle Inhofe's top supporters were the oil and gas ($284,706) and electric utility ($162,213) industries. Information available at http://www.opensecrets.org/politicians/indus.asp?CID=N00005582&cycle=2002 (accessed December 21, 2004).

79 *consistent zero ratings:* The League of Conservation Voters scorecard on Inhofe's voting record can be found at http://www.capwiz.com/lcv/bio/keyvotes/?id=481&congress=1082&lvl=C (accessed December 21, 2004).

You think I'm bad: Quoted in Michael Crowley, "Ill natured," *New Republic*, January 20, 2003.

pledged that on his watch: James Inhofe, statement on "Climate history, science, and mercury emissions," July 29, 2003, available online at http://epw.senate.gov/hearing_statements.cfm?id=212839 (accessed December 21, 2004).

a twelve-thousand-word: James Inhofe, Senate floor speech on "The science of climate change," July 28, 2003. Available online at http://inhofe.senate.gov/pressreleases/climate.htm (accessed December 21, 2004).

80 *U.S. Senate Republican Policy Committee:* See Senate Republican Policy Committee, "The shaky science behind the climate change sense of the Congress resolution," June 2, 2003, available online at http://www.senate.gov/%7Erpc/releases/2003/ev060203.pdf (accessed December 21, 2004). For a critique see John Holdren, comments on the above, June 9, 2003. Available online at http://stephenschneider.stanford.edu/Publications/PDF_Papers/HoldrenRPCClimateComments.pdf (accessed December 21, 2004).

Consensus as strong: Donald Kennedy, "An unfortunate U-turn on carbon," *Science*, vol. 291, p. 2515, March 30, 2001.

in a 2004 interview: Interview with Donald Kennedy, June 22, 2004.

81 *After examining 928 papers:* Naomi Oreskes, "The scientific consensus on climate change," *Science*, vol. 306, no. 5702, p. 1686, December 3, 2004. See also Naomi Oreskes, "Undeniable global warming," *Washington Post*, December 26, 2004. In the latter article, Oreskes describes her methodology in more detail: "The Institute for Scientific Information keeps a database on published scientific articles. . . . We read 928 abstracts published in scientific journals between 1993 and 2003 and listed in the database with the keywords 'global climate change.' Seventy-five percent of the papers either explicitly or implicitly accepted the consensus view. The remaining 25 percent dealt with other facets of the subject, taking no position

on whether current climate change is caused by human activity. None of the papers disagreed with the consensus position. There have been arguments to the contrary, but they are not to be found in scientific literature, which is where scientific debates are properly adjudicated. There, the message is clear and unambiguous."

81 *vanishingly small:* Interview with Naomi Oreskes, January 25, 2005.

In the late '80s, early '90s: Interview with James McCarthy, May 11, 2004.

Major oil companies such as Shell: See *Electric Utility Week,* "Global Climate Coalition restructures; drops individual corporate memberships," March 27, 2000.

became inactive: Interview with Frank Maisano, former spokesman, Global Climate Coalition, January 18, 2005 (originally conducted for *Mother Jones* magazine).

82 *ExxonMobil was donating over a million dollars:* See Jennifer Lee, "Exxon backs groups that question global warming," *New York Times,* May 28, 2003. For ExxonMobil's 2003 giving report see http://www2.exxonmobil.com/Corporate/files/corporate/giving_report.pdf (accessed December 21, 2004). Also on file with the author.

melting of glaciers and sea ice: ACIA, *Impacts of a Warming Arctic: Arctic Climate Impact Assessment* (Cambridge: Cambridge University Press, 2004).

Marshall Institute promptly challenged: See George C. Marshall Institute, "Senate commerce hearing promotes climate alarmism," November 15, 2004, available online at http://www.marshall.org/pdf/materials/270.pdf (accessed March 7, 2005), and James Inhofe press release, "Inhofe questions science behind Arctic report," November 16, 2004. Available online at: http://epw.senate.gov/pressitem.cfm?party=rep&id=228016 (accessed December 22, 2004).

American Petroleum Institute memo: John H. Cushman, "Industrial group plans to battle climate treaty," *New York Times,* April 26, 1998. Environmental Defense has placed the document online at http://www.environmentaldefense.org/documents/3860_GlobalClimateSciencePlanMemo.pdf (accessed December 21, 2004).

83 *Marshall Institute's William O'Keefe:* Interview with William O'Keefe, March 22, 2004.

ExxonMobil took the position: Associated Press, "U.S. scientist voted off international climate panel," April 19, 2002, noting, "A spokesman for Exxon Mobil said Friday that the oil company never lobbied against Watson. The spokesman, Tom Cirigliano, said the company didn't write the unsigned memo but did forward it to the Council on Environmental Quality."

removed from their positions of influence: The memo, obtained by the Natural Resources Defense Council through a Freedom of Information Act request, can be read online at http://www.nrdc.org/media/docs/020403.pdf (accessed December 21, 2004).

84 *senate floor speech:* James Inhofe, Senate floor speech on "The science of climate change," July 28, 2003. Available online at http://inhofe.senate.gov/pressreleases/climate.htm (accessed December 21, 2004).

a board member of Peabody Energy: See PR Newswire, "Peabody Energy (NYSE: BTU) expands independent representation of board of directors," January 16, 2003.

all agree: For the American Meteorological Society, see American Meteorological Society council, "Climate change research: issues for the atmospheric and related sciences," February 9, 2003, noting, "Because human activities are contributing to climate change, we have a collective responsibility to develop and undertake carefully considered response actions," available online at http://www.ametsoc.org/POLICY/climatechangeresearch_2003.html (accessed December 22, 2004). For the American Geophysical Union, see American Geophysical Union Council, "Human impacts on climate," December 2003, noting, "Scientific evidence strongly indicates that natural influences cannot explain the rapid increase in global near-surface temperatures observed during the second half of the 20th century," available online at http://www.agu.org/sci_soc/policy/positions/climate_change.shtml (accessed December 22, 2004).

According to NASA's Goddard Institute: NASA Goddard Institute for Space Studies, "Earth Gets a Warm Feeling All Over," February 8, 2005. Available online at: http://www.giss.nasa.gov/research/news/20050208 (accessed March 12, 2005).

84   *Monty Python skits:* Jerry D. Mahlman interview, March 5, 2004.
85   *It would be astonishing:* Donald Kennedy interview, June 22, 2004.

    a 1998 paper by Wigley: The paper is T. M. L. Wigley, "The Kyoto Protocol: C02, CH4 and climate implications," *Geophysical Research Letters,* vol. 25, no. 13, July 1, 1998, pp. 2285–2299.

    letter to Senators Tom Daschle and Bill Frist: Tom Wigley, letter to Senators Daschle and Frist, July 29, 2003, on file with author. It can also be read online at http://www.climatenetwork. org/uscanweb/csadocs/mccainwigley.doc (accessed August 15, 2004).

86   *a 2001 Nature commentary:* Stephen Schneider, "What is 'dangerous' climate change?" *Nature,* vol. 411, May 3, 2001, available online at http://stephenschneider.stanford.edu/Publications/ PDF_Papers/Nature-SRES-SHS-final.pdf (accessed August 15, 2004).

    Schneider protested: Schneider's response, dated October 26, 2003, can be read online at http://stephenschneider.stanford.edu/Publications/PDF_Papers/McCainQuestionsForSchneider.pdf (accessed August 15, 2004).

    Inhofe has continued to cite: See Chris Mooney, "Earth last," *American Prospect,* May 2004.

    swearing fealty to "sound science": James Inhofe, statement on "Climate history, science, and mercury emissions," July 29, 2003, available online at: http://epw.senate.gov/hearing_ statements.cfm?id=212839 (accessed August 16, 2004). The "sound science" quotation does not appear in this version because Inhofe's spoken language diverged slightly from the prepared statement, but a transcript of the hearing, including the quotation, is on file with the author.

    a highly controversial paper: The paper appeared in two versions: "Proxy climatic and environmental changes of the past 1000 years," Willie Soon and Sallie Baliunas, *Climate Research,* vol. 23, pp. 89–110, January 31, 2003; and then a longer version, "Reconstructing climatic and environmental changes of the past 1000 years: A reappraisal," *Energy & Environment,* vol. 14, nos. 2 & 3, 2003, with three additional authors: Sherwood Idso, Craig Idso, and David Legates.

87   *portentous history-of-science speak:* James Inhofe, statement on "Climate history, science, and mercury emissions," July 29, 2003, available online at: http://epw.senate.gov/hearing_statements. cfm?id=212839 (accessed August 16, 2004).

    unprecedented temperature spikes: For a good overview of the literature, see Richard A. Kerr, "Millennium's hottest decade retains its title, for now," *Science,* February 11, 2005, vol. 307, pp. 828–829.

    neither unusual nor the most extreme: Willie Soon, Senate statement on "climate history and mercury emissions," July 29, 2003, available online at http://epw.senate.gov/hearing_statements. cfm?id=212848 (accessed March 12, 2005). Note that the quotation used in the text does not appear online because Soon's spoken words while giving testimony diverged from the prepared statement. But a transcript of the hearing, including the quotation, is on file with the author.

    partly funded: The Soon and Baliunas "Climate Research" paper lists grants from the American Petroleum Institute as well as the Air Force Office of Scientific Research and NASA.

    Soon has served in the past: In an interview on March 22, 2004, Marshall Institute CEO William O'Keefe noted that Willie Soon had been a "senior scientist" with the institute until resigning that post in 2003. Regarding the group's funding by ExxonMobil, the corporation's 2003 giving report is on file with the author. O'Keefe commented that the ExxonMobil funding was "not the predominant funding" of the institute and that "no money will be accepted by Marshall, and I don't believe any other nonprofit like us, where someone gives money to produce a specific result."

    serves as "science director": Willie Soon is listed as "science director" on the TechCentralStation.com website at http://www.techcentralstation.com/biosoonwillie.html (accessed December 21, 2004). The ExxonMoil 2003 giving report is on file with the author..

    listed as "science director": Soon is described this way in a report by the Center for Science and Public Policy, "EPA mercury MACT rulemaking not justified by science," available at http://ff.org/centers/csspp/pdf/mercurywhitepaper.pdf (accessed December 21, 2004). According to ExxonMobil's 2003 giving report, Frontiers of Freedom received $95,000 for "global climate change outreach," $50,000 for "project support—sound science center," and $50,000 for "global climate change activities." On file with author.

87 *collaborated with the Marshall Institute:* According to William O'Keefe, of the Marshall Institute, Legates has written a publication for the group on climate models, participated in a work group on science in policy, and authored a book chapter for the group.

served as an "adjunct scholar": David Legates is listed as an "adjunct scholar" here: http://www.ncpa.org/pub/ba/ba478 (accessed December 21, 2004). The ExxonMobil 2003 giving report is on file with the author.

88 *his university webpage:* http://www.meteo.psu.edu/~mann/Mann/index.html (accessed March 31, 2006).

have reconstructed climate records: The original paper in question is Michael Mann, Raymond Bradley, and Malcolm Hughes, "Global-scale temperature patterns and climate forcing over the past six centuries," *Nature,* vol. 392, April 23, 1998, pp. 779–787. Later work extended the analysis to cover the last millennium. See, for example, Mann, Bradley, and Hughes, "Northern Hemisphere temperatures during the past millennium: inferences, uncertainties, and limitations," *Geophysical Research Letters,* vol. 26, no. 6, March 15, 1999, pp. 759–762.

Though multiple studies confirm: In his Senate testimony Mann noted, "More than a dozen independent research groups have now reconstructed the average temperature of the Northern Hemisphere in past centuries." See also http://stephenschneider.stanford.edu/Climate/Climate_Science/CliSciFrameset.html, click on "Contrarians." Noted Stanford climatologist Stephen Schneider discusses the Mann debate, notes the other contrarians who have tried to critique Mann's study, and comments, "nearly a dozen other reconstructions of Northern Hemisphere mean temperature by different groups independently confirm the overall existence of a 'hockey-stick'—though I expect that it will be one with a very curvy handle" (accessed March 12, 2005). Finally, for a highly accessible overview of this debate, see Richard A. Kerr, "Millennium's hottest decade retains its title, for now," *Science,* February 11, 2005, vol. 307, pp. 828–829.

hundreds of experts: For instance, the IPCC 2001 Working Group I report, "Climate Change 2001: The Scientific Basis," was the work of 123 "lead authors" and some 516 "contributing authors." And that's just one working group's product. See Intergovernmental Panel on Climate Change, Information Note, "Climate Change 2001: The Scientific Basis," available online at http://www.ipcc.ch/press/pressreleasedraft.htm (accessed March 12, 2005).

89 *prominently challenged in the peer-reviewed literature:* See von Storch et al., "Reconstructing past climate from noisy data," *Science Express,* September 30, 2004.

That's why the federal government turns: Interview with Michael Oppenheimer, March 2, 2004.

There's a whole independent line: Interview with Michael Mann, December 15, 2004.

90 *had resigned to protest deficiencies:* Von Storch has posted a statement on his website on the whole affair: http://w3g.gkss.de/G/Mitarbeiter/storch/CR-problem/cr.2003.htm (accessed August 16, 2004). See also Richard Monastersky, "Storm brews over global warming," *Chronicle of Higher Education,* September 4, 2003; Antonio Regalado, "Debating global warming: global warming skeptics are facing storm clouds," *Wall Street Journal,* July 31, 2003; and Andrew Revkin, "Politics reasserts itself in the debate over climate change and its hazards," *New York Times,* August 5, 2003.

later complained: Monastersky, "Storm brews over global warming."

an article by Mann, Oppenheimer, Wigley: Michael Mann et al., "On past temperatures and anomalous late–20th century warmth," *Eos,* vol. 84, no. 27, July 8, 2003, p. 256. Available online at ftp://holocene.evsc.virginia.edu/pub/mann/eos03.pdf (accessed August 15, 2004). See also Mann et al., "Response to comment 'On past temperatures and anomalous late–20th century warmth,'" *EOS,* vol. 84, no. 44, November 4, 2003, p. 473. Available online at http://holocene.evsc.virginia.edu/MRG/articles/EosReply03.pdf (accessed March 14, 2005).

90 *Otto Kinne agreed with critics:* Otto Kinne, "Climate research: an article unleashed worldwide storms," *Climate Research,* vol. 24, August 5, 2003, pp. 197–198. Available online at http://www.int-res.com/abstracts/cr/v24/n3/CREditorial.pdf (accessed August 15, 2004).
a dissenting EPA memo: Andrew Revkin and Katherine Q. Seelye, "Report by E.P.A. leaves out data on climate change," *New York Times,* June 19, 2003. As the story reports, "White House officials also deleted a reference to a 1999 study showing that global temperatures had risen sharply in the previous decade compared with the last 1,000 years. In its place, administration officials added a reference to a new study, partly financed by the American Petroleum Institute, questioning that conclusion." For the memo itself, see National Wildlife Federation, "EPA ditches climate change in environment report," July 19, 2003, press release. Available online at http://www.nwf.org/news/story.cfm?pageId=E093353A-D994-CA6B-BEC296236ACE6FB0 (accessed December 21, 2004).

91 *greenhouse gases are accumulating:* Committee on the Science of Climate Change, National Research Council, "Climate change science: an analysis of some key questions" (Washington, D.C.: National Academy Press, 2001). Available online at http://www.nap.edu/catalog/10139.html?onpi_webextra6 (accessed August 17, 2004).
Attempted White House edits: Revkin and Seelye, "Report by E.P.A. Leaves Out Data on Climate Change."

92 *In the process of hyping uncertainty:* For an example of another conservative citing this sentence, see Department of Energy secretary Spencer Abraham, "'Sound science,' climate change and policy choices," letter, *Washington Post,* March 7, 2004.
cherry-picked precisely this sentence: George W. Bush and John Kerry, "Bush and Kerry offer their views on science," *Science,* September 16, 2004, available online at http://www.sciencemag.org/cgi/rapidpdf/1104420v1.pdf (accessed September 25, 2004).
I think that's true only because: Interview with Edward Sarachik, May 3, 2004.
We didn't think about how it would sound: Interview with John Wallace, April 27, 2004, followed by e-mail communication, November 22, 2004.

93 *I also spoke with NAS panelist Eric J. Barron:* Interview with Eric Barron, April 20, 2004.
the famed climate science contrarian: For examples of the Right citing Lindzen to question what the NAS found with respect to the IPCC, see Iain Murray, "Our science can beat up your science," *National Review Online,* March 5, 2004; available online at http://www.nationalreview.com/comment/murray200403050931.asp (accessed August 19, 2004). See also Republican Policy Committee, "The shaky science behind the climate change sense of the Congress resolution," June 2, 2003, available online at http://www.senate.gov/~rpc/releases/2003/ev060203.pdf (accessed August 19, 2004).
advised the Bush administration: See Andrew C. Revkin, "After rejecting climate treaty, Bush calls in tutors to give courses and help set one," *New York Times,* April 28, 2001.
I think he's offered a healthy level of caution: Interview with William Schlesinger, April 23, 2004.
between cigarette smoking and lung cancer: Fred Guterl, "The truth about global warming: the forecasts of doom are mostly guesswork, Richard Lindzen argues—and he has Bush's ear," *Newsweek,* July 23, 2001.
putting his own particular spin: Richard S. Lindzen, "The press gets it wrong: our report doesn't support the Kyoto treaty," *Wall Street Journal,* June 11, 2001.

94 *an okay summary:* Interview with Richard Lindzen, April 15, 2004.
I would say there is virtually no difference: Interview with Stephen Schneider, March 30, 2004.

95 *We agreed with the previous findings:* The *NewsHour with Jim Lehrer,* June 7, 2001.
which received a whopping $465,000: Corporate giving report on file with author.
a pathbreaking Clinton-era study: U.S. Global Change Research Program, 2000, *Climate Change Impacts on the United States: The Potential Consequences of Climate Variability and Change* (New York: Cambridge University Press, 2001). Available online at: http://www.usgcrp.gov/usgcrp/nacc/default.htm.

95 *a huge contretemps:* U.S. Department of State, 2002. *U.S. Climate Action Report 2002,* Washington, D.C., May.

But it didn't go unnoticed: Michael MacCracken, who participated in the drafting of the *Climate Action Report,* provided me with this context. Interview with Michael MacCracken, March 9, 2004.

stark shift for the Bush administration: Andrew C. Revkin, "U.S. sees problems in climate change," *New York Times,* June 3, 2002.

put out by the bureaucracy: Andrew C. Revkin, "With White House approval, EPA pollution report omits global warming section," *New York Times,* September 15, 2002.

96 *conservatives first brought a lawsuit: Competitive Enterprise Institute et al. v. William Jefferson Clinton,* October 3, 2000. The lawsuit can be read online at http://www.cei.og/pdf/2300.pdf (accessed August 19, 2004).

rushing to release a junk science report: Statement of Rep. Jo Ann Emerson on the National Assessment, October 5, 2000. Available online at http://www.cei.org/utils/printer.cfm?AID=2535 (accessed August 19, 2004).

Bush administration settled the lawsuit: Rosina Bierbaum letter on the National Assessment, September 6, 2001. Available online at http://www.cei.org/gencon/003,03045.cfm (accessed August 19, 2004).

Emerson had sponsored: Interview with Jeffrey Connor, spokesman for Congresswoman Emerson, November 29, 2004. Emerson also admits authoring the Data Quality Act in a *St. Louis Post Dispatch* article: Andrew Schneider, "EPA warning on asbestos is under attack; lawyers target materials detailing dangers in brakes," October 23, 2003.

provides a new means: Interview with Sidney Shapiro, March 10, 2004.

Inhofe has welcomed the Data Quality Act: Darren Samuelsohn, "Inhofe says Democrats may have the votes to pass CO_2 curbs," *Environment & Energy Daily,* February 12, 2003.

97 *pursued various "data quality" complaints:* See Competitive Enterprise Institute, "Petition to EPA to cease dissemination of *Climate Action Report,*" June 4, 2002, available online at http://www.cei.org/gencon/027,03040.cfm (accessed March 13, 2005), and Competitive Enterprise Institute, Petition to the National Oceanic and Atmospheric Administration regarding the National Assessment, February 19, 2003, available online at http://www.cei.org/pdf/3374.pdf (accessed March 13, 2005).

the very first lawsuit: The lawsuit can be read online at http://www.cei.org/pdf/3595.pdf (accessed August 19, 2004).

In a February 2002 petition: Center for Regulatory Effectiveness, letter to John Marburger, February 11, 2002. Available online at: http://www.thecre.com/quality/20020211_climate-letter.html (accessed February 19, 2005).

by-now familiar misconception: For a discussion of the National Assessment controversy that focuses on misrepresentation and misunderstanding of the nature of climate models, see Charles N. Herrick, "Objectivity versus narrative coherence: science, environmental policy, and the U.S. Data Quality Act," *Environmental Science & Policy,* vol. 7, 2004, pp. 419–433.

Scenarios are plausible alternative futures: National Assessment Synthesis Team, *Climate Change Impacts on the United States: The Potential Consequences of Climate Variability and Change* (Overview), US Global Change Research Program, 400 Virginia Avenue, SW, Suite 750, Washington D.C. 20024. Quotation on page 3.

explains Michael MacCracken: Interview with Michael MacCracken, March 9, 2004.

98 *his committee invited:* Christopher Horner described his appearance that way in an interview, March 22, 2004.

including the controversial **Climate Action Report**: A Senate Democratic staffer who was present at the meeting confirmed that the event involved this report.

highly unusual and a breach: Brian Faler, "Carter to help broker peace on Georgia flag controversy," *Washington Post,* March 9, 2003. (The story also contains mention of the controversy over Horner.)

98   *not there in pursuit of information:* Interview with Christopher Horner, March 22, 2004.
provides a basis for summarizing: Committee on the Science of Climate Change, National Research Council, *Climate Change Science: An Analysis of Some Key Questions* (Washington, D.C.: National Academy Press, 2001). See pages 19–20. Available online at http://www.nap.edu/catalog/10139.html?onpi_webextra6 (accessed August 17, 2004).
important contributions to understanding: Committee to Review the U.S. Climate Change Science Program Strategic Plan, National Research Council, *Implementing Climate and Global Change Research: A Review of the Final U.S. Climate Change Science Program Strategic Plan* (Washington, D.C.: National Academy Press, 2004). Available online at http://www.nap.edu/books/0309088658/html/ (accessed August 19, 2004). Quotation on page 13.
the government website displaying the National Assessment: See http://www.usgcrp.gov/usgcrp/nacc/default.htm (accessed March 13, 2005). See also Competitive Enterprise Institute, "White House acknowledges climate report was not subjected to sound science law," November 6, 2003, available online at http://www.cei.org/gencon/003,03740.cfm (accessed March 13, 2005).

99   *passed any data quality scrutiny:* See November 7, 2003, letter from Michael MacCracken and National Assessment scientists to James Mahoney, director, Climate Change Science Program, protesting the disclaimer. On file with author.
was taken to task: For instance, see *Implementing Climate and Global Change Research*, pp. 29–30: "For the most part, the CCSP's revisions to the strategic plan are quite responsive to comments expressed at the workshop, in written input, and by this committee. One notable exception is the fact that the revised plan does not acknowledge the substantive and procedural contributions of the U.S. National Assessment of the Potential Consequences of Climate Variability and Change (NAST, 2001), a major focus of the Global Change Research Program (GCRP) in the late 1990s. Many participants at the December workshop criticized how the draft strategic plan treated the National Assessment, as did this committee in its first report (NRC, 2003b). The revised plan does not reflect an attempt to address these concerns, and no rationale for this decision has been provided."
See also Andrew C. Revkin, "Bush climate plan rated somewhat improved," *New York Times*, February 19, 2004, noting, "One of the biggest weaknesses in the administration's plan, the panel said, was the absence of any significant reference to existing research examining the potential effects of climate change around the United States. . . . Particularly notable, it said, was the omission of any reference to the National Assessment of the Potential Consequences of Climate Variability and Change. That 2001 report, done at the request of Congress, was compiled by academic and government scientists over several years. . . . White House officials have been continually pressed by industry lobbyists and antiregulatory groups to remove references to the assessment from government documents. . . . The assessment provides 'important contributions,' the advisory panel said, and the independent peer review it went through was exemplary."
says Jerry Mahlman: Interview with Jerry Mahlman, March 5, 2004.
information is normally of value: Interview with Michael MacCracken, March 9, 2004.

100   *Thank you for your letter:* Letter from George W. Bush, governor of Texas, to Michael MacCracken, executive director, National Assessment Coordination Office of the U.S. Global Change Records Program, December 15, 2000. On file with the author.
event that received funding from ExxonMobil: ExxonMobil's 2003 giving report lists a donation of $27,500 to the Annapolis Center for Environmental Quality, Inc., for "general operating support/annual dinner." On file with author.
rational, science-based thinking: U.S. Newswire, "Sen. James Inhofe to be honored by the Annapolis Center for Science-Based Public Policy," April 12, 2004. Press release available online at http://releases.usnewswire.com/GetRelease.asp?id=125–04022004 (accessed August 19, 2004).
said that the event was full: E-mail communication from the Annapolis Center, April 25, 2004.

8: WINE, JAZZ, AND "DATA QUALITY"

104   *favorite bad guy:* E-mail communication from Rena Steinzor, October 4, 2004.

105   *It's hard not to like the guy:* Interview with Sidney Shapiro, March 10, 2004.

regulatory policy "nerd": Quotations of Jim Tozzi in this chapter are based on a series of personal interviews conducted on March 17, March 24, and April 12, 2004.

paralysis by analysis: See, for example, David Vladeck, "Paralysis by analysis: how conservatives plan to kill popular regulation," *American Prospect,* vol. 6, no. 22, June 23, 1995.

Rep. Jo Ann Emerson: Jeffrey Connor, spokesman for Congresswoman Emerson, November 29, 2004, confirms her role in the Data Quality Act's origins. For more on the provenance of the act, see "The 411 on 515," James T. O'Reilly, *Administrative Law Review,* Spring 2002.

106   *says William Kovacs:* Interview with William Kovacs, March 18, 2004.

mid-2004 analyses: See Rick Weiss, "'Data quality' law is nemesis of regulation," *Washington Post,* August 16, 2004. ("A *Washington Post* analysis of government records indicates that in the first 20 months since the act was fully implemented, it has been used predominantly by industry. Setting aside the many Data Quality Act petitions filed to correct narrow typographical or factual errors in government publications or Web sites, the analysis found 39 petitions with potentially broad economic, policy or regulatory impact. Of those, 32 were filed by regulated industries, business or trade organizations or their lobbyists.") See also OMB Watch, "The reality of data quality act's first year: a correction of OMB's report to Congress," July 2004, available online at http://www.ombwatch.org/info/dataqualityreport.pdf (accessed September 15, 2004). ("OMB accurately states that a wide range of stakeholders have filed information quality challenges, attempting to dismiss fears that these challenges would be dominated by industry. But OMB fails to disclose that 72 percent of the challenges—nearly three quarters—were from industry.")

"sound science" law: See, for example, Chris Horner, "Who's afraid of sound science," September 17, 2004, available online at http://www.thecre.com/quality/20040920_quality.html (accessed September 28, 2004).

contacts at OMB are second to none: Nicoli, D., memo to Howard Liebengood [Re: WRO Retention of Jim Tozzi], February 17, 1998. Document number 2078294554, available at http://www.pmdocs.com (accessed September 15, 2004).

107   *secretary of the army's office:* Interview with Jim Tozzi, March 17, 2004. See also Dan Davidson, "Nixon's 'nerd' turns regulations watchdog," *Federal Times,* November 11, 2002. Available online at http://www.thecre.com/pdf/20021111_fedtimes-tozzi.pdf (accessed September 15, 2004).

black hole: For a reference to OIRA's reputation as a "black hole," see Douglas Jehl, "Regulations czar prefers new path," *New York Times,* March 25, 2001.

I don't want to leave fingerprints: Quoted in Peter Behr, "If there's a new rule, Jim Tozzi has read it; Office of Management and Budget," *Washington Post,* July 10, 1981.

108   *innocent bystanders:* A detailed discussion of one chapter in tobacco's war on the science of secondhand smoke can be found in Thomas O. McGarity, "On the prospect of 'Daubertizing' judicial review of risk assessment," *Law and Contemporary Problems,* vol. 66, p. 155, autumn 2003.

difficult or impossible to prove: Elisa Ong and Stanton Glantz, "Constructing 'sound science' and 'good epidemiology': tobacco, lawyers, and public relations firms," *American Journal of Public Health,* vol. 91, no. 11, November 2001.

As one 1994 Philip Morris memo explained: Letter from Matthew Winokur to David W. Bushong re: Sound science issues, May 10, 1994. Document number 2078741714/1717, available at http://www.pmdocs.com (accessed December 24, 2004).

Tobacco interests were trying: Again, see Ong and Glantz, "Constructing 'sound science' and 'good epidemiology.'"

108 *Many studies demonstrate:* Jonathan Samet and Thomas Burke, "Turning science into junk: the tobacco industry and passive smoking," *Am. J. Public Health,* vol. 91, no. 11, 2001, pp. 1742–1744.

regulation of nearly any environmental toxin: Interview with Stanton Glantz, March 31, 2004.

while receiving donations from Philip Morris: Luisa F. Martinez letter to Thorne Auchter, Federal Focus, July 18, 1995 (describing a $250,000 grant). Document number 2046379787, available at http://www.pmdocs.com (accessed September 16, 2004). See also Thorne Auchter letter to Tony Andrade, vice president, World Wide Regulatory Affairs, Philip Morris, Inc., December 9, 1994 (requesting a $200,000 contribution; letter stamped "OK to Pay"). Document number 2074282703, available at http://www.pmdocs.com (accessed September 16, 2004).

London Principles: The "London Principles" can be accessed online at http://www.fedfocus. org/science/london-principles.html (accessed September 16, 2004). The scientists involved in drafting the principles are listed here: http://www.fedfocus.org/science/london-panel.html (accessed September 16, 2004).

109 *the realm of "weak association":* Matthew Winokur, memo on "Epi Guidelines," August 24, 1994. Document numbers 2028381624 and 2028381625, available at http://www.pmdocs.com (accessed September 20, 2004).

Tozzi worked to support: Memo from J. Bolland and T. Borelli to T. Collamore, February 17, 1993. Document numbers 2046597149–7150, available at http://www.pmdocs.com (accessed September 16, 2004).

A signed 1994 Philip Morris contract: Signed Philip Morris contract for Jim Tozzi, August 8, 1994. Document numbers 2029377061–63, available at http://www.pmdocs.com (accessed September 16, 2004).

Tozzi worked on the so-called Shelby amendment: See memo from Jim Tozzi to Matthew Winokur on "data access," October 12, 1998, document numbers 2065231124–1127; letter from Jim Tozzi to Matthew Winokur, October 21, 1998, document numbers 2065231111–1112; and memo from Jim Tozzi to Matthew Winokur, November 20, 1998, document number 2065231059. All available at http://www.pmdocs.com (accessed September 16, 2004). Tozzi confirmed in interviews that he worked for Philip Morris on "data access."

the one-sentence amendment: For background on the Shelby amendment, see Philip J. Hilts, "A law opening research data sets off debate," *New York Times,* July 31, 1999.

Though subsequently limited in scope: For a discussion of the Clinton administration's weakening of the Shelby amendment, see James T. O'Reilly, "The 411 on 515," *Administrative Law Review,* Spring 2002.

the "daughter" of Shelby: Ibid.

Tozzi circulated a "data quality" proposal: For comments on a data quality proposal from the Center for Regulatory Effectiveness by various individuals with Philip Morris, see letter from Jim Tozzi to Robert Elves, PM USA, January 24, 2000, document number 2072826845; letter from Jim Tozzi to Robert Elves, February 29, 2000 (commenting on a CRE draft regulation that "defines the four key Data Quality terms [i.e., quality, objectivity, utility, and integrity]"), document numbers 2072826830–6831; e-mail from Robert Elves to Robin D. Kinser and others, March 17, 2000, document number 2072826818; and comments on CRE "data quality" draft, February 23, 2000, document 2072826836. All available at http://www.pmdocs.com (accessed September 16, 2004).

110 *it's in the data:* Interview with Gary Bass, October 20, 2004.

an EPA risk assessment document: Environmental Protection Agency, *Reregistration Eligibility Science Chapter for Atrazine,* April 22, 2002 (available online at http://www.epa.gov/oppsrrd1/ reregistration/atrazine/efed_redchap_22apr02.pdf [accessed March 7, 2005]).

The groups' petition objected: Kansas Corn Growers Association, Triazine Network, and Center for Regulatory Effectiveness, petition to the U.S. EPA on atrazine, November 25, 2002.

Available online at http://www.thecre.com/pdf/petition-atrazine2B.pdf (accessed September 21, 2004).

111 *Multinational Business Services received:* Lobbying documents dated August 21, 2002 and February 25, 2003 on file with author.

has also received some financial support: E-mail communication from Jere White, Triazine Network, March 10, 2005. See also Goldie Blumenstyk, "The price of research: a Berkeley scientist says a corporate sponsor tried to bury his unwelcome findings and then buy his silence," *Chronicle of Higher Education,* October 31, 2003.

You can't make a necessary connection: Interview with Jim Tozzi, February 17, 2005.

Syngenta spokeswoman Sherry Ford added: Sherry Ford, e-mail response to questions received March 23, 2005.

112 *a charge EcoRisk denies:* Interview with Tyrone Hayes, December 13, 2004. The charges (and EcoRisk's response) have been reported previously in Blumenstyk, "The price of research." See also Rick Weiss, "'Data quality' law is nemesis of regulation," *Washington Post,* August 16, 2004.

Syngenta's Sherry Ford: Sherry Ford, e-mail response to questions received March 23, 2005.

The big article: Tyrone Hayes et al., "Hermaphroditic, demasculinized frogs after exposure to the herbicide atrazine at low ecologically relevant doses," *Proceedings of the National Academy of Sciences,* vol. 99 no. 8, April 16, 2002.

chemical castration effect: Interview with Tyrone Hayes, December 13, 2004.

113 *in a series of* FoxNews.com *columns:* See Steven Milloy, "Frog Study Leaps to Conclusions," *FoxNews.com,* April 19, 2002, available online at http://www.foxnews.com/story/0,2933, 50669,00.html (accessed September 22, 2004); "Freaky-Frog Fraud," *FoxNews.com,* November 8, 2002, available online at http://www.foxnews.com/story/0,2933,69497,00.html (accessed September 22, 2004); and "Freaky Frogs Not Linked With Herbicide, Says EPA," *FoxNews.com,* Thursday, June 12, 2003, available at http://www.foxnews.com/story/ 0,2933,89222,00.html (accessed September 22, 2004).

Milloy never even bothered: E-mail communication from Tyrone Hayes, September 25, 2004.

Syngenta's atrazine website: See http://www.syngentacropprotection-us.com/prod/herbicide/ atrazine/index.asp, accessed September 23, 2004, and on file with author.

After his new work emerged: Tyrone Hayes et al., "Atrazine-induced hermaphroditism at 0.1 PPB in American frogs (*Rana pipiens*): laboratory and field evidence," *Environmental Health Perspectives,* October 23, 2002.

unable to reproduce: For the original press release describing EcoRisk scientists' inability to replicate Hayes's results, see PR Newswire, "Frog research on atrazine casts doubt on earlier studies," June 20, 2002. The press release can also be read online at http://www.farmassist.com/ prod/herbicide/atrazine/index.asp?nav=Ecorisk (accessed March 27, 2005).

114 *according to an EcoRisk press release:* Ibid.

Hayes has scathingly critiqued: Tyrone B. Hayes, "There is no denying this: defusing the confusion about atrazine," *Bioscience,* December 2004, Vol. 54 No 12.

the European Union has moved to phase out: See Commission Decision, March 10, 2004, "concerning the non-inclusion of atrazine in Annex I to Council Directive 91/414/EEC and the withdrawal of authorisations for plant protection products containing this active substance," which notes, "Member states shall ensure that . . . authorisations for plant protection products containing atrazine are withdrawn by 10 September 2004; from 16 March 2004 no authorizations for plant protection products containing atrazine are granted or renewed under the derogation provided for in Article 8(2) of Directive 91/414/EEC," but also adds that that some authorizations for "plant protection products containing atrazine" may remain in effect until 30 June 2007.

Nevertheless, the U.S. EPA concluded: U.S. EPA, January 2003 Atrazine IRED (Interim Registration Eligibility Decision), p. 73, noting, "The Agency's ecological risk assessment does not suggest that endocrine disruption, or potential effects on endocrine-mediated pathways, be regarded as a regulatory endpoint at this time. Nor does the Agency have evidence to state that there is no reliable evidence that atrazine causes endocrine effects in the environment. Based on the existing uncertainties in the available database, atrazine should be subject to more definitive testing once the appropriate testing protocols have been established." Available online at http://www.epa.gov/oppsrrd1/REDs/atrazine_ired.pdf (accessed March 8, 2005).

114 *notes Sean Moulton:* Interview with Sean Moulton, October 20, 2004.

both targeted the National Toxicology Program: See Center for Regulatory Effectiveness et al., petition to National Toxicology Program on carcinogens, June 28, 2004. Available online at http://thecre.com/pdf/20040630_dqa_petition.pdf (accessed September 25, 2004). See also Center for Regulatory Effectiveness et al., petition to the National Toxicology Program on carcinogens, July 16, 2004. Available online at http://aspe.hhs.gov/infoquality/request& response/18a.shtml (accessed 'March 7, 2005).

115 *by a chemical company:* See petition to U.S. EPA from the Chemical Products Corporation, October 29, 2002, available online at http://www.epa.gov/quality/informationguidelines/documents/2293.pdf (accessed September 21, 2004).

by a law firm: Petition to U.S. EPA from Morgan, Lewis & Bockius, August 19, 2003. Available online at http://www.epa.gov/quality/informationguidelines/documents/12467.pdf (accessed September 21, 2004). See also Andrew Schneider, "EPA warning on asbestos is under attack; lawyers target materials detailing dangers in brakes," *St. Louis Post-Dispatch*, October 26, 2003, reporting, "The international law firm of Morgan, Lewis & Bockius has petitioned the Environmental Protection Agency to stop distributing warning booklets, posters and videotapes that give mechanics guidance on the need to protect themselves from asbestos. . . . The firm refused repeated requests to identify its client in the effort to stop the booklets, but it has represented at least one major asbestos firm and two insurance companies involved in asbestos litigation."

by paint manufacturers: See petition to U.S. EPA from Willkie Farr & Gallagher, LLP on behalf of the National Paint and Coatings Association and the Sherwin-Williams Company, June 2, 2004. Available online at http://www.epa.gov/quality/informationguidelines/documents/04020.pdf (accessed September 21, 2004). See also Center for Progressive Regulation letter to U.S. EPA administrator Michael Leavitt and OIRA administrator John Graham, objecting to paint manufacturers' complaint, August 3, 2004. Available online at http://www.progressiveregulation.org/articles/Paint_DQA_0804.pdf (accessed September 21, 2004).

logging interests challenged: W.K. Olsen & Associates et al., petition to U.S. Forest Service on Northern Goshawk analyses, January 17, 2003. Available online at http://www.fs.fed.us/qoi/documents/2003/01/rm–217-goshawk-petition.pdf (accessed September 21, 2004).

including the Environmental Working Group: Environmental Working Group, petition to the U.S. Food and Drug Administration on mercury in fish and shellfish, December 22, 2003. Available online at http://www.ewg.org/issues_content/mercury/20031222/FDA_DQAchallenge.pdf (accessed September 27, 2004).

Americans for Free Access: See Americans for Free Access, "Government must correct medical marijuana information, petition says," *U.S. Newswire,* October 1, 2004. Available online at http://releases.usnewswire.com/GetRelease.asp?id=37292 (accessed October 4, 2004).

My musician friends: E-mail communication, October 1, 2004.

116 *He has drafted sample legislation:* See OMB Watch, "Analysis of state level data quality and access legislation," March 24, 2003. Available online at http://www.ombwatch.org/article/articleview/1393/1/1 (accessed September 27, 2004).

to the World Health Organization: Jim Tozzi, letter to the World Health Organization and Food and Agriculture Organization on U.S. data quality standards, September 8, 2003. Available online at http://www.thecre.com/pdf/20030908__who.pdf (accessed September 27, 2004).

and to universities: Jim Tozzi, letter to Jane Buck, president of American Association of University Professors, August 6, 2003. Available online at http://www.thecre.com/pdf/universityDQltrBuck.pdf (accessed September 27, 2004).

116 *tried using the act preemptively:* William G. Kelly, Jr., letter to EPA concerning NRDC's comments on biosolids, February 27, 2003. Available online http://www.thecre.com/pdf/20030310_biosolids.pdf (accessed September 27, 2004).

an attempt to quell scientific debate: Center for Progressive Regulation, letter to EPA administrator Christine Todd Whitman and OIRA administrator John Graham, May 19, 2003. Available online at http://www.progressiveregulation.org/perspectives/DQA_Letter_CRE.PDF (accessed September 27, 2004).

a strategic test case: Salt Institute and U.S. Chamber of Commerce, lawsuit filed against Tommy G. Thompson, March 31, 2004. On file with author.

denied a right to judicial review: Salt Institute and Chamber of Commerce of the United States of America v. Tommy G. Thompson, final decision available at http://thecre.com/pdf/20041122_dqa_decision.pdf (accessed December 25, 2004).

117 *conservative legal scholars:* See Alan Charles Raul and Julia Zampa Dwyer, "'Regulatory Daubert': a proposal to enhance judicial review of agency science by incorporating *Daubert* principles into administrative law," *Law and Contemporary Problems,* vol. 66, no. 4, autumn 2003. For a critique of Raul and Dwyer, see Thomas O. McGarity, "On the prospect of 'Daubertizing' judicial review of risk assessment," *Law and Contemporary Problems,* vol. 66, no. 4, autumn 2003.

corpuscular: See McGarity, "On the Prospect of 'Daubertizing' Judicial Review of Risk Assessment," p. 155.

Hayes has noted: Hayes, "There Is No Denying This: Defusing the Confusion about Atrazine."

a group funded in part by industry: For funding information on the Harvard Center for Risk Analysis, see http://www.hcra.harvard.edu/funding.html (accessed March 28, 2005).

Graham also served: See PR Newswire, "Science watchdog group celebrates third anniversary with renewed commitment to exposing use of junk science," December 3, 1996.

he once even commented: Angela Antonelli, Hon. David McIntosh, Fred Smith, and Dr. John Graham, "Making regulatory reform a reality," Heritage Lecture #559, January 31, 1996. Available online at http://www.heritage.org/Research/GovernmentReform/HL559.cfm (accessed December 25, 2004).

118 *data quality guidelines:* See Office of Management and Budget, "Guidelines for ensuring and maximizing the quality, objectivity, utility, and integrity of information disseminated by federal agencies," *Federal Register,* vol. 67, no. 36, February 22, 2002. Available online at http://www.whitehouse.gov/omb/fedreg/reproducible2.pdf (accessed February 22, 2005).

Graham proposed using the act's thin language: Office of Management and Budget, "Proposed bulletin on peer review and information quality," *Federal Register,* vol. 68, no. 178, September 15, 2003, pp. 54023–54029. Available online at http://www.whitehouse.gov/omb/fedreg/030915.pdf (accessed September 27, 2004).

119 *different goals and constraints in mind:* For a more thorough discussion see Sheila Jasanoff, *The Fifth Branch: Science Advisers as Policymakers* (Cambridge: Harvard University Press, 1990), pp. 76–83.

Scientific heavyweights like the American Public Health Association: American Public Health Association, comments on OMB peer review proposal, December 11, 2003. Available online at http://www.progressiveregulation.org/articles/peer/APHA_PR_Comments.pdf (accessed September 27, 2004).

Federation of American Societies for Experimental Biology: Association of American Medical Colleges and Federation of American Societies for Experimental Biology, comments on OMB peer review proposal, December 4, 2003. Available online at http://www.progressiveregulation.org/articles/peer/AAMC_PR_Comments.pdf (accessed September 27, 2004).

also announced worries: American Association for the Advancement of Science, comments on OMB peer review proposal, December 12, 2003. Available online at http://www.progressiveregulation.org/articles/peer/AAS_PR_Comments.pdf (accessed September 27, 2004).

119 *wolf in sheep's clothing:* Henry Waxman et al., letter to Joshua Bolten, OMB administrator, on OMB peer review proposal, December 15, 2003. Available online at http://www.progressive regulation.org/articles/peer/Waxman_PR_comments.pdf (accessed September 27, 2004).

"peer review" system: Office of Management and Budget, "Proposed Bulletin on Peer Review and Information Quality."

120 *revised "peer review" bulletin:* Office of Management and Budget, "Revised information quality bulletin for peer review," April 15, 2004. Available online at http://www.whitehouse. gov/omb/inforeg/peer_review041404.pdf (accessed September 27, 2004).

American Association for the Advancement of Science: American Association for the Advancement of Science, resolution on the OMB proposed peer review bulletin, March 9, 2004. Available online at http://archives.aaas.org/docs/resolutions.php?doc_id=434 (accessed September 27, 2004).

The "Bad Science" Fiction: Wendy Wagner, "The 'bad science' fiction: reclaiming the debate over the role of science in public health and environmental regulation," *Law and Contemporary Problems,* vol. 66, no. 4, autumn 2003.

121 *"peer review" critic Sidney Shapiro:* Sidney Shapiro, comments at "An early review of OMB's final peer review bulletin," American Bar Association, Wednesday, February 23, 2005. Author's notes.

says David Vladeck: Interview with David Vladeck, December 1, 2004.

finalized the peer review plan: See Office of Management and Budget, "Final information quality bulletin for peer review," December 15, 2004, available online at http://www.whitehouse. gov/omb/inforeg/peer2004/peer_bulletin.pdf (accessed December 26, 2004).

highly influential scientific assessments: John Graham, comments at "An early review of OMB's final peer review bulletin," American Bar Association, Wednesday, February 23, 2005. Author's notes.

122 *sometimes you get the monkey:* Quoted in Rick Weiss, "'Data quality' law is nemesis of regulation," *Washington Post,* August 16, 2004.

9: EATING AWAY AT SCIENCE

125 **World Health Organization and Food and Agriculture Organization (WHO/FAO) report:** For a description of the WHO/FAO "expert consultation" process, see Nishida et al., "The Joint WHO/FAO Expert Consultation on diet, nutrition and the prevention of chronic diseases: process, product and policy implications," *Public Health Nutrition* 7(1A), 245–250. The ultimate report that arose from the process in 2003, "Diet, nutrition, and the prevention of chronic diseases," can be read online at http://www.who.int/hpr/NPH/docs/who_fao_ expert_report.pdf (accessed August 25, 2004).

recommendation to "eat less": The U.S. food industry's long history of resisting "eat less" messages is chronicled in Marion Nestle, *Food Politics: How the Food Industry Influences Nutrition and Health* (Berkeley: University of California Press, 2002).

126 *asking congressional allies to block:* Sugar Association letter to Gro Harlem Brundtland, director general, World Health Organization, April 14, 2003. Food activists acquired a copy of this letter, and have put it online here: http://www.commercialalert.org/sugarthreat.pdf (accessed August 26, 2004). The sugar industry's letter to the WHO was also reported on in the *Washington Post.* See Juliet Eilperin, "U.S. sugar industry targets new study," *Washington Post,* April 23, 2003.

removed from the WHO's website: Letter from food industry groups to Heath and Human Services secretary Tommy Thompson, March 20, 2003. Available online at http://www.sugar.org/ newsroom/CoalitionletterHHSWHO.pdf (accessed August 26, 2004).

cease further promotion: The Craig/Breaux letter, dated March 28, 2003, was included as an attachment with the Sugar Association's letter of April 14, 2003.

Briscoe held a press event: The press release announcing the conference call can be read online at http://www.sugar.org/newsroom/WHOdiscussion.pdf (accessed August 25, 2004).

126 *speaking out of both sides of her mouth:* Two separate sources agree on this quotation. The first is science writer David Appell's firsthand account of the press conference, April 27, 2003. On file with author. The second is Stephen Clapp, "WHO report continues to draw fire from sugar lobby," *Food Chemical News,* April 28, 2003, vol. 45, no. 11, which also contains the quotation.

simply arbitrary and capricious: Steven Milloy, "World Health Baloney," *FOXNews.com,* April 25, 2003. Available online at http://www.foxnews.com/story/0,2933,85104,00.html (accessed November 22, 2004).

on his website, JunkScience.com: You can find Milloy's "junk scientist" roster at http://www.junkscience.com/roster.html, and his discussion of Kumanyika at http://www.junkscience.com/news/diet-trends-part2.html (both accessed November 22, 2004).

127 *introduction to the report:* Institute of Medicine, *Dietary Reference Intakes for Energy, Carbohydrate, Fiber, Fat, Fatty Acids, Cholesterol, Protein and Amino Acids (Macronutrients)* (Washington, D.C.: National Academies Press, 2002). Accessible online at http://www.nap.edu/books/0309085373/html (accessed August 26, 2004).

I did not determine: E-mail from Shiriki Kumanyika, February 15, 2004.

reporters wrongly read the IOM study: The IOM report suggested 25 percent as a "maximum intake level" for consumption of added sugars to prevent nutrient losses. But this was not a dietary recommendation. Alas, due to "inartfully worded statements" contained in a draft version of the report, the IOM study was widely cited as providing such a recommendation. See March 19, 2003, letter from Michael F. Jacobson, executive director, Center for Science in the Public Interest, to Ms. Paula Trumbo, Food and Nutrition Board. On file with author.

interpretations suggesting: Letter from Harvey V. Fineberg to Tommy Thompson, April 15, 2003. On file with author.

128 *turned their attention to the obesity epidemic:* The shift of the global disease burden from infectious diseases to incommunicable chronic conditions is discussed in the WHO's "Global strategy on diet, physical activity and health," May 22, 2004. Available online at http://www.who.int/gb/ebwha/pdf_files/WHA57/A57_R17-en.pdf (accessed August 29, 2004).

Obesity rates: The American Obesity Association defines as obese a person having a body mass index (BMI) of 30 or more, where BMI is calculated by dividing weight in pounds by height in inches squared and multiplying the result by 704.5 [BMI = weight ÷ (height)^2 × 704.5]. An individual with a BMI between 25 and 29.9 is considered overweight. For more detail, see http://www.obesity.org/subs/fastfacts/obesity_what2.shtml (accessed March 29, 2005).

American women on average consumed: "Trends in intake of energy and macronutrients—United States, 1971–2000," *Morbidity and Mortality Weekly Report,* vol. 53, no. 4, February 6, 2004, pp. 80–82, online at http://www.cdc.gov/mmwr/preview/mmwrhtml/mm5304a3.htm (accessed August 29, 2004).

fifteen percent of adolescents and children: For American Obesity Association figures on child obesity, see http://www.obesity.org/subs/childhood/prevalence.shtml (accessed August 29, 2004).

reached $75 billion in 2003: Finkelstein et al., "State-level estimates of annual medical expenditures attributable to obesity," *Obesity Research,* vol. 12 (2004), pp. 18–24.

American public remains ambivalent: This is the conclusion reached in Regina G. Lawrence, "Framing obesity: the evolution of news discourse on a public health issue," *Harvard Journal of Press/Politics,* vol. 9, no. 3, 2004, pp. 56–75.

toxic food environment: The phrase "toxic food environment" comes from Kelly Brownell and Katherine Battle Horgen, *Food Fight: The Inside Story of the Food Industry, America's Obesity Crisis, and What We Can Do About It* (New York: McGraw-Hill, 2004).

129 *larding of America:* For a discussion of these trends, see Kelly Brownell, "Fast food and obesity in children," *Pediatrics,* vol. 113, no. 1, January 2004, p. 132.

129 *plausibly could increase risk for obesity:* Bowman et al., "Effects of fast-food consumption on energy intake and diet quality among children in a national household survey," *Pediatrics,* vol. 113, no. 1 (January 2004).

the role of soft drinks: Interview with Derek Yach, September 22, 2004.

130 *associated with obesity in children:* David Ludwig et al., "Relation between consumption of sugar-sweetened drinks and childhood obesity; a prospective, observational analysis," *Lancet,* vol. 357, February 17, 2001, pp. 505–508.

an 83 percent increased risk: Matthias B. Schulze et al., "Sugar-sweetened beverages, weight gain, and incidence of type 2 diabetes in young and middle-aged women," *Journal of the American Medical Association,* vol. 292, no. 8, August 25, 2004, pp. 927–934.

that there's a relationship: Interview with Kelly Brownell, February 20, 2004.

soft drinks are important contributors: Interview with David Ludwig, February 24, 2004.

Obesity is a complex problem: American Beverage Association, "No scientific evidence to link obesity and HFCS; culprits are too many calories and not enough exercise," March 25, 2004, available online at http://www.nsda.org/pressroom/032504hfcsresponse.asp (accessed March 1, 2005).

All foods can be a part: National Restaurant Association talking points on "Nutrition and Healthy Lifestyles," available online at http://www.restaurant.org/government/state/nutrition/resources/nra_20040208_talkingpoints_health.pdf (accessed August 30, 2004).

rather a lack of activity: Quoted in Mary Clare Jalonick, Martin Kady II, and Amol Sharma, "Dueling science: critics cite other statistics," *Congressional Quarterly Weekly,* March 19, 2004.

131 *a flip and chatty rebuttal:* Todd G. Buchholz, "Burgers, fries, and lawyers: the beef behind obesity lawsuits," conducted for the U.S. Chamber of Commerce, July 2, 2003. Available online at http://www.legalreformnow.com/resources/burgers.pdf (accessed August 29, 2004).

on its board of directors: For the Chamber of Commerce board of directors, see http://www.uschamber.com/about/board/all.htm (accessed February 14, 2004).

at least back to the lead industry: See David Rosner and Gerald Markowitz, "Industry challenges to the principle of prevention in public health: the precautionary principle in historical perspective," *Public Health Reports,* vol. 117, no. 6, November 1, 2002.

calling on governments worldwide: The final global strategy, adopted by the World Health Assembly on May 22, 2004, can be read online at http://www.who.int/gb/ebwha/pdf_files/WHA57/A57_R17-en.pdf (accessed September 2, 2004).

draft version of the global strategy: World Health Organization Executive Board, "Draft global strategy on diet, physical activity, and health," November 27, 2003, on file with the author.

George H. W. Bush's godson: Confirmed in interview with HHS Spokesman William Pierce, February 23, 2004.

132 *sent a missive:* William R. Steiger, letter to Lee Jong-wook, with attached analysis and annexes, January 5, 2004, on file with the author.

the Grocery Manufacturers of America: For Grocery Manufacturers of America arguments emphasizing the role of the individual, rather than government, in controlling weight, and criticisms of the WHO/FAO expert consultation, see Stephen Clapp, "WHO outlines process for global anti-obesity strategy," *Food Chemical News,* August 26, 2002. See also Rob Stein, "U.S. says it will contest WHO plan to fight obesity," *Washington Post,* January 16, 2004.

executives from Coca Cola and Nestlé USA: During the presidential race, the Bush campaign published a list of "Rangers" and "Pioneers," on file with author, which listed Fanjul and Coker, as well as Barclay Resler, Coca-Cola's vice president of government affairs ("ranger"); and Joe Weller, chairman and CEO of Nestlé USA ("pioneer").

If they're mirroring us: Interview with Bill Pierce, February 23, 2004.

the collective thinking: Interview with William Steiger, March 8, 2005.

133 *changes in the environment:* Statement by William Dietz on the CDC's role in combating the obesity epidemic, Senate Committee on Health, Education, Labor, and Pensions, May 21,

2002. Dietz's testimony can be read online at http://www.hhs.gov/asl/testify/t020521a.html (accessed September 2, 2004).

133   *not just this technical document:* Interview with Kaare Norum, February 10, 2004.

134   *somebody in the food business:* Interview with Jim Tozzi, March 17, 2004.

challenged whether: Center for Regulatory Effectiveness, request for correction of information contained in a World Health Organization report, September 8, 2003, available online at http://www.thecre.com/pdf/20030908_correction.pdf (accessed September 2, 2004).

Tozzi's group even e-mailed: E-mail communication from Carlos Camargo, February 5, 2005.

Department of Agriculture's food pyramid: Interview with Marion Nestle, February 11, 2004. See also Kelly Brownell and Marion Nestle, "The sweet and lowdown on sugar," *New York Times,* January 23, 2004.

advised the global health body: Jim Tozzi letter to WHO director-general Lee Jong-wook and FAO director general Jacques Diouf, September 8, 2003. Available online at http://www.thecre.com/pdf/20030908__who.pdf (accessed September 6, 2004).

As implemented by the U.S. Department of Health and Human Services: For HHS Data Quality guidelines, see http://www.thecre.com/pdf/20021026_hhs-dqfinal.pdf (accessed September 3, 2004).

135   *it omitted, as a political compromise:* See John Zarocostas, "WHA adopts landmark global strategy on diet and health," *Lancet,* vol. 363, May 29, 2004.

136   *an August 2003 press release:* See joint press release, "HHS, USDA designate experts to the Dietary Guidelines Advisory Committee," August 11, 2003, noting that "In order to prepare the *Guidelines* for release in 2005—the sixth edition—the designees will examine the new Dietary Reference Intakes by the Institute of Medicine; the World Health Organization report on Diet, Nutrition, and the Prevention of Chronic Diseases; and other recent scientific research." Available online at http://www.health.gov/dietaryguidelines/dga2005/pressrelease.htm (accessed December 26, 2004).

choose carbohydrates wisely: 2005 Dietary Guidelines Advisory Committee Report, quotation from Executive Summary. Available online at http://www.health.gov/dietaryguidelines/dga2005/report (accessed March 28, 2005).

moderate your intake of sugar: See Marian Burros, "Added sugars, less urgency? Fine print and the dietary guidelines," *New York Times,* August 25, 2004.

Deep in the fine print: As the report stated, "Compared with individuals who consume small amounts of foods and beverages that are high in added sugars, those who consume large amounts tend to consume more calories but smaller amounts of micronutrients. Although more research is needed, available prospective studies suggest a positive association between the consumption of sugar-sweetened beverages and weight gain. A reduced intake of added sugars (especially sugar-sweetened beverages) may be helpful in achieving recommended intakes of nutrients and in weight control." See Section 5: Carbohydrates, available online at http://www.health.gov/dietaryguidelines/dga2005/report/HTML/D5_Carbs.htm (accessed March 28, 2005).

according to committee member Carlos Camargo: Interview with Carlos Camargo, February 2, 2005.

in the section relevant to sugar: The section on carbohydrates can be read online at http://www.health.gov/dietaryguidelines/dga2005/report/HTML/D5_Carbs.htm, while the section on fats can be read online at http://www.health.gov/dietaryguidelines/dga2005/report/HTML/D4_Fats.htm (accessed March 28, 2005).

137   *two U.S. government employees:* The individuals were Dr. Deborah A. Galuska, of the CDC's Division of Nutrition and Physical Activity, and Dr. Arthur Schatzkin, head of the nutritional epidemiology branch of the NIH's National Cancer Institute. See "Diet, Nutrition, and the Prevention of Chronic Diseases," online at http://www.who.int/hpr/NPH/docs/who_fao_expert_report.pdf (accessed August 25, 2004).

137   *noted the paper:* David Brown, "Ideas on WHO delegates at odds," *Washington Post,* July 21, 2004.

    advocate U.S. Government policies: William Steiger letter to Denis G. Aitken, assistant director-general, World Health Organization, April 15, 2004. On file with author.

138   *under Steiger:* Jocelyn Kaiser, "The man behind the memos," *Science,* September 10, 2004, vol. 305.

    in an outraged letter to Health and Human Services' Tommy Thompson: Henry Waxman letter to HHS secretary Tommy Thompson, June 24, 2004. Online at http://www.house.gov/reform/min/pdfs_108_2/pdfs_inves/pdf_health_hhs_who_policy_june_24_let.pdf (accessed November 22, 2004). The letter also contains, as an attachment, the text of William Steiger's April 15, 2004, letter to the WHO.

    from such luminaries as D. A. Henderson: For Henderson's criticism, see Tom Hamburger, "White House tries to rein in scientists," *Los Angeles Times,* June 26, 2004.

    Jeffrey Koplan: Jeffrey Koplan, "No place in medicine for politics," *Atlanta Journal-Constitution* (op-ed piece), July 12, 2004.

    It seems abundantly clear: Roger Pielke, Jr., "Henry Waxman, HHS, and a Bush Administration Misuse of Science," June 28, 2004, online at http://sciencepolicy.colorado.edu/prometheus/archives/science_policy_general/index.html#000111 (accessed November 22, 2004).

139   *Pregnant mothers run the gravest risk:* Guy Gugliotta, "Mercury threat to fetus raised," *Washington Post,* February 6, 2004.

    nearly all of the nation's rivers and lakes: See Michael Janofsky, "E.P.A. says mercury taints fish across the U.S.," *New York Times,* August 25, 2004.

140   *market-based "cap and trade" system:* Eric Pianin, "EPA led mercury policy shift; agency scuttled task force that advised tough approach," *Washington Post,* December 20, 2003.

    A July 2003 hearing: The hearing occurred on July 29, 2003, and was entitled "Climate history, science, and health effects of mercury emissions." See http://epw.senate.gov/hearing_statements.cfm?id=212852 (accessed September 6, 2004).

    would do little to reduce that pollution: Leonard Levin, "New perspectives on mercury in the human environment," testimony before the Committee on Environment and Public Works, July 29, 2003. Available online at http://epw.senate.gov/108th/Levin_072903.pdf (accessed February 26, 2005). As Levin argued, "a drop of nearly half in utility mercury emissions results in a drop of 3% (on average) in mercury depositing to the ground, and a drop of less than one-tenth of a percent in the number of children 'at risk.'"

    did not find harmful effects: As the National Academy of Sciences commented on the studies, "Results from the three large epidemiological studies—the Seychelles, Faroe Islands, and New Zealand studies—have added substantially to the body of knowledge on brain development following long-term exposure to small amounts of MeHg. Each of the studies was well designed and carefully conducted, and each examined prenatal MeHg exposures within the range of the general U.S. population exposures. In the Faroe Islands and New Zealand studies, MeHg exposure was associated with poor neurodevelopmental outcomes, but no relation with outcome was seen in the Seychelles study." National Research Council, Committee on Life Sciences, *Toxicological Effects of Methylmercury* (Washington, D.C.: National Academies Press, 2000), p. 5. Available online at http://www.nap.edu/openbook/0309071402/html (accessed February 16, 2005).

141   *At least eight studies:* Statement of Deborah C. Rice, Ph.D., Maine Department of Environmental Protection, Augusta, Maine, Health Effects of Methylmercury with Particular Reference to the U.S. Population. Hearing by the Senate Committee on Environment and Public Works, July 29, 2003. Rice's testimony can be read online at http://epw.senate.gov/hearing_statements.cfm?id=212850 (accessed September 6, 2004).

    We do not believe: Statement by the University of Rochester Research Team Studying the Developmental Effects of Methylmercury read by Dr. Gary Myers to the U.S. Senate Committee on Environment and Public Works, July 29, 2003. Myers's testimony can be read online at http://epw.senate.gov/hearing_statements.cfm?id=212851 (accessed September 6, 2004).

141 *By bringing Dr. Myers in:* Interview with Deborah Rice, March 12, 2004.

are not harmful: U.S. Chamber of Commerce, "Reality check: straight talk about mercury." Readable here: http://www.uschamber.com/NR/rdonlyres/emaxvxmr64ymbj2e34womfomacim6dlmucjt3ghz6eovdsjsrttbziscxfbslcosr5sclvnjjfzkbl/mercurybook.pdf (accessed September 6, 2004).

EPRI has contributed $486,000: Joint Institute for Food Safety and Applied Nutrition, 1998–1999 annual report. On file with author.

some related studies: Interview with Gary Myers, March 11, 2004.

142 *received $195,000 in total funding:* 2003 ExxonMobil Worldwide Giving Report, on file with author.

the first year of its existence: The Center for Science and Public Policy opened its doors in March 2003. See press release, "Frontiers opens 'center for science and public policy,'" March 18, 2003, available online at http://ff.org/centers/csspp/press/12220040244.html (accessed December 28, 2004).

not justified by science: Center for Science and Public Policy, "EPA mercury MACT rule not justified by science," available online at http://ff.org/centers/csspp/pdf/mercurywhitepaper.pdf (accessed September 6, 2004).

The Mercury Scare: Wall Street Journal (editorial), "The mercury scare," April 8, 2004.

the true "gold standard": As the NAS put it, "The committee concludes that, given the strengths of the Faroe Islands study, it is the most appropriate study for deriving an RfD [reference dose]."

editorial prompted a letter: Philippe Grandjean, "Mercury exposure risks in prenatal development" (letter to the editor), *Wall Street Journal,* April 15, 2004.

Grandjean told me: Interview with Philippe Grandjean, March 23, 2004.

143 *Soon has made industry-friendly scientific arguments:* See Ian Ith, "Skeptic scorns mercury risk; science in error, physicist says; environmentalists question his motives," *Seattle Times,* July 29, 2004, noting, "Soon contends that less than 1 percent of the world's mercury comes from American power plants, with the vast majority coming from natural sources such as volcanic eruptions, supernovas in space and forest fires."

70 percent of mercury inputs: Schuster et al., "Atmospheric mercury deposition during the last 270 years: a glacial ice core record of natural and anthropogenic sources," *Environmental Science and Technology,* vol. 36, no. 11, 2002, pp. 2303–2310.

available information indicates: United Nations Environment Programme—Chemicals, "Global mercury assessment," December 2002. Available online at http://www.chem.unep.ch/mercury/Report/Final%20report/final-assessment-report–25nov02.pdf (accessed December 27, 2004).

China now accounts: Interview with Nicola Pirrone, March 4, 2005.

144 *from sources within this country:* U.S. EPA, "Regulatory Finding on Emissions of Hazardous Air Pollutants From Electric Utility Steam-Generating Units," *Federal Register:* December 20, 2000 (Volume 65, Number 245), 79825–79831, reading "Based on modeling . . . the EPA estimates that roughly 60 percent of the total mercury deposited in the U.S. comes from U.S. anthropogenic air emission sources; the percentage is estimated to be even higher in certain regions (e.g., northeast U.S.)." Available online at http://www.epa.gov/fedrgstr/EPA-AIR/2000/December/Day–20/a32395.htm (accessed September 8, 2004).

40 percent or more: Interview with Nicola Pirrone, March 4, 2005.

current, peer-reviewed scientific literature: Richard W. Pombo and Jim Gibbons, "Mercury in Perspective: Fact and Fiction About the Debate Over Mercury," February 16, 2005, available online at http://resourcescommittee.house.gov/Press/reports/mercury_in_perspective.pdf (accessed February 27, 2005).

144 *there is a plausible link:* See U.S. EPA, "Regulatory Finding on Emissions of Hazardous Air Pollutants From Electric Utility Steam-Generating Units," *Federal Register:* December 20, 2000 (Volume 65, Number 245), 79825–79831. The Pombo and Gibbons report quoted the EPA as follows: "It is acknowledged that there are uncertainties regarding the extent of the risks due to electric utility mercury emissions. For example, there is no quantification of how much of the methylmercury in fish consumed by the U.S. population is due to electric utility emissions relative to other mercury sources (e.g., natural and other anthropogenic sources)." The preceding two sentences (which were not quoted) read, "There is a plausible link between emissions of mercury from anthropogenic sources (including coal-fired electric utility steam generating units) and methylmercury in fish. Therefore, mercury emissions from electric utility steam generating units are considered a threat to public health and the environment." The next two sentences (which were also not quoted) read, "Nonetheless, the available information indicates that mercury emissions from electric utility steam generating units comprise a substantial portion of the environmental loadings and are a threat to public health and the environment. The EPA believes that it is not necessary to quantify the amount of mercury in fish due to electric utility steam generating unit emissions relative to other sources for the purposes of this finding."

10: FISHY SCIENCE

147 *shut off irrigation water:* For a discussion of acreage, as well as general background on the Klamath dispute, see Holly Doremus and A. Dan Tarlock, "Fish, farms, and the clash of cultures in the Klamath Basin," *Ecology Law Quarterly,* p. 284.
148 *The shutoff marked the first time:* See ibid., introduction.
  It triggered immense anger: Republican congressman Greg Walden, who represents Klamath Falls, describes the consequences of the water shutoff for the community here: http://www.klamathbasincrisis.org/esa/esawaldentestfs020404.htm (accessed July 19, 2004).
  even engaged in civil disobedience: See Michael Milstein, "Farmers defy feds, escalate fight," *Oregonian,* July 14, 2001; Bettina Boxall, "Officials cut off flow of water to farmers; drought: releases into Klamath Basin are halted to protect fish. Protesters vow not to give up," *Los Angeles Times,* August 24, 2001.
  guarding the canal: As reported in the *Oregonian,* "Guarding Klamath waters costs the U.S. $750,000," November 1, 2001.
  Endangered Species Data Quality Act of 2004: The press release on the bill's passage, dated July 21, 2004, can be read here: http://resourcescommittee.house.gov/Press/releases/2004/0721esabills.htm (accessed December 27, 2004). See also Greg Walden press release, "Walden bipartisan ESA modernization bill approved by House Resources Committee," July 21, 2004, noting the bill's name change. Available online at http://www.house.gov/walden/press/releases/2004/pr_040721_esabill.html (accessed March 18, 2005).
149 *there wasn't strong scientific evidence:* The National Academy of Sciences' final and interim reports are titled "Scientific Evaluation of Biological Opinions on Endangered Species and Threatened Fishes in the Klamath River Basin: Interim Report" (2002), and "Endangered and Threatened Fishes in the Klamath River Basin: Causes of Decline and Strategies for Recovery" (2004), both from the National Academies Press. On the Web at http://www.nap.edu/catalog/10296.html?onpi_newsdoc020402 and http://www.nap.edu/catalog/10838.html (accessed July 19, 2004). The review arose from a joint request by the Departments of the Interior and Commerce. As part of the Department of the Interior, the Fish and Wildlife Service (FWS) has responsibility for preserving endangered and threatened terrestrial and freshwater species, including the two sucker species. As part of the Department of Commerce, the National Marine Fisheries Service of the National Oceanic and Atmospheric Administration—sometimes called NOAA Fisheries—has responsibility for marine and anadromous species, such as coho salmon, which swim upriver from the ocean to spawn.

149 *hundreds of wildlife scientists:* Over four hundred biology and wildlife experts have signed a statement, circulated by Earthjustice, Defenders of Wildlife, and other environmental groups, criticizing H.R. 1662, Greg Walden's "sound science" Endangered Species Act reform bill. Letter from scientists to members of Congress, July 2004, expressing concern about legislation that would change endangered species law, H.R. 1662/S. 2009 and H.R. 2933. Statement and list of signatories on file with author.

two scientists from the NAS's Klamath committee: Peter B. Moyle and Jeffrey F. Mount, "Endangered Species Act at work in Klamath basin," *Sacramento Bee,* December 28, 2003.

Who knows, or can say: Quoted in *Tennessee Valley Authority v. Hill,* 4437 U.S. 153, 180 (1978).

150 *notoriously affirmed a lower court's ruling:* Ibid.

Congress quickly reined in: Interview with Holly Doremus, February 19, 2004.

the dam was a highly dubious project: A good discussion of the "economic infeasibility" of the Tellico project can be found in John H. Gibbons, Holly Gwin, William Chandler, "The efficacy of federal environmental legislation: the TVA experience with endangered species and clean air," *Utah Law Review,* vol. 1979, no. 4.

pit bull of environmental laws: For an example of the ESA being referred to this way, see the May 27, 1999, Senate testimony of William R. Murray, of the American Forest & Paper Association, available online at http://epw.senate.gov/107th/mur_5–27.htm (accessed July 21, 2004).

It's the most effective, substantive law: Holly Doremus interview, February 19, 2004.

151 *waffle-stomping socialists:* See Frank Clifford, "GOP divided on environmental deregulation," *Los Angeles Times,* September 25, 1995.

without significant amendment since 1982: See introduction to J. B. Ruhl, "The battle over Endangered Species Act methodology," *Environmental Law,* vol. 34, no. 2 (2004), pp. 555–603.

increasing number of science-based legal challenges: See Holly Doremus, "The Purposes, Effects, and Future of the Endangered Species Act's Best Available Science Mandate," *Environmental Law,* vol. 34, no. 2, pp. 397–450.

brand new legal battleground: Interview with Dennis Murphy, February 26, 2004.

many of the same political actors: Consider House Committee on Resources chair Richard Pombo's 1996 book *This Land Is Our Land: How to End the War on Private Property,* coauthored with Joseph Farah (New York: St. Martin's). Pombo strongly suggests that the ESA might be a violation of the Fifth Amendment's right to property, and writes that the act must be "comprehensively rewritten to restore it to its original intent" (p. 186). Pombo also describes his idea of what such a rewrite would entail: voluntary measures, compensation for loss of private property, and overall, a "dramatic and fundamental reform of the existing law by recognizing that the key to protecting threatened or endangered species is through incentives and rewards, not threats and fines" (p. 188).

best scientific and commercial data available: The text of the act can be read here: http://endangered.fws.gov/esa.html (accessed July 20, 2004).

In a 1995 report: National Research Council, Committee on Life Sciences, *Science and the Endangered Species Act* (Washington, D.C.: National Academies Press, 1995), p. 12. Available online at http://books.nap.edu/books/0309052912/html/index.html (accessed February 22, 2005).

generally due to inadequate budgets: Carroll, R., et al. 1996. "Strengthening the use of science in achieving the goals of the Endangered Species Act: an assessment by the Ecological Society of America," *Ecological Applications,* vol. 6, no. 1, pp. 1–11. Available online at http://www.esa.org/pao/esaPositions/Papers/StrentheningUSAGESA.php (accessed February 22, 2005).

A 2003 report from the Government Accountability Office: Endangered Species: Fish and Wildlife Uses Best Available Science to Make Listing Decisions, but Additional Guidance Needed for Critical Habitat Designations. U.S. General Accounting Office, August 2003 (GAO-03-803). PDF version: www.gao.gov/cgi-bin/getrpt?GAO-03-803. Text version of the study available here: http://www.gao.gov/atext/d03803.txt (accessed July 21, 2004).

152   *some species tend to dominate others:* Richard Pombo and Joseph Farah, *This Land Is Our Land: How to End the War on Private Property.* (New York: St Martin's Press, 1996). The quotation comes from pp. 111–112.

A look at Walden's proposed "sound science" bill: The version of the bill discussed here passed the House Committee on Resources by a 25–16 vote on July 21, 2004, during the 108th Congress. As the hardcover edition of this book went to press, Walden had announced that he was "working on updating his sound science legislation for this current session of Congress [the 109th]." US Fed News, "Rep. Walden cosponsors bipartisan critical habitat legislation to update Endangered Species Act," March 15, 2005.

These sympathies shine through: See Committee on Resources, "Report Together with Dissenting Views to Accompany H.R. 1662," November 19, 2004. Available online at: http://frwebgate.access.gpo.gov/cgi-bin/getdoc.cgi?dbname=108_cong_reports&docid=f:hr785.108.pdf (accessed March 19, 2005).

153   *We shouldn't confuse peer review:* My distinction between research science and regulatory science relies on comments submitted by Sheila Jasanoff to the Office of Management and Budget in 2003 regarding the office's "peer review" proposal. The comments can be read here: http://www.whitehouse.gov/omb/inforeg/2003iq/159.pdf (accessed July 22, 2004).

constituting peer review panels could be difficult: Congressional Research Service, "The Endangered Species Act and 'sound science,'" July 16, 2002 (report code RS21264).

154   *you might as well shut down:* Interview with Gordon Orians, February 18, 2004.

tries to spread smallpox: Interview with Stuart Pimm, April 26, 2004.

155   *the Bush administration has itself endorsed:* See Testimony of Craig Manson, assistant secretary for fish and wildlife and parks, U.S. Department of the Interior, before the House Resources Committee, regarding H.R. 4840, the "Sound Science for Endangered Species Act Planning Act of 2002." This is an earlier version of H.R. 1662, Greg Walden's "sound science" bill. Manson stated, "The Administration supports H.R. 4840 with modifications to address our concerns. We believe that, if implemented, this legislation will broaden opportunities for scientific input and assure additional public involvement in Endangered Species Act implementation. We also believe it will also improve the U.S. Fish and Wildlife Service's (Service) decision-making process and result in increased public confidence in the Service's decisions." Available online at http://www.doi.gov/ocl/2002/hr4840.htm (accessed December 14, 2004).

marked the first time ever: As noted in Ruhl, "The battle over Endangered Species Act methodology."

produced after just four months: The timing is clear from the following Department of the Interior press release: "National Academy to review scientific decisions, needs of aquatic endangered species in Klamath Basin project," October 2, 2001. Available online at http://www.doi.gov/news/011002.html (accessed July 29, 2004). The interim report's release was reported on by the *Washington Post* on February 4, 2002 (Michael Grunwald, "Scientific report roils a salmon war").

156   *argued that the ESA was unconstitutional:* Elizabeth Shogren, "The new administration; Senate panel OKs Norton for Interior," *Los Angeles Times,* January 25, 2001.

for this year and future years: "Interior Secretary Norton responds to National Academy of Sciences draft interim report," Department of the Interior press release, February 4, 2004. Available online at http://www.doi.gov/news/020225c.html (accessed July 29, 2004). (The date on the file is incorrect.)

I challenge anyone: Walden's testimony was delivered on February 4, 2004, before the Subcommittee on Energy and Mineral Resources of the Committee on Resources. It is available at http://resourcescommittee.house.gov/archives/108/testimony/2004/gregwalden.pdf (accessed July 29, 2004).

minimizing risk to the species: National Research Council, "Endangered and Threatened Fishes in the Klamath River Basin," p. 10.

And they just have no appreciation: Interview with J. B. Ruhl, February 16, 2004.

156 *a $685,000 budget:* E-mail communication from Dr. James Reisa, director, Board on Environmental Studies and Toxicology, National Academy of Sciences, March 5, 2004.

157 *The agencies are generally justified:* Interview with Peter Moyle, August 26, 2004.

Moyle has also set the record straight: Quoted in Michael Milstein, "Scientists critical of Klamath water ban," *Oregonian,* February 4, 2002; Sharon Levy, "Turbulence in the Klamath River Basin," *BioScience,* April 1, 2003.

(ESA) is working as intended: Peter B. Moyle and Jeffrey F. Mount, "Endangered Species Act at work in Klamath Basin," *Sacramento Bee,* December 28, 2003.

more rigorous, thorough, and defensible: Michael S. Cooperman and Douglas F. Markle, "The Endangered Species Act and the National Research Council's interim judgment in the Klamath Basin," *Fisheries,* March 2003, vol. 28, no. 3.

158 *perfect storm:* Interview with Douglas Markle, August 2, 2004.

flow is the only controllable factor: "September 2002 Klamath River fish kill: final analysis of contributing factors and impacts," Northern California North Coast Region, California Department of Fish and Game, July 2004. Available online at http://www.dfg.ca.gov/html/krfishkill-2004.pdf (accessed August 2, 2004).

from a very high level: Kelly's official statement on file with author. It can also be read online at http://www.peer.org/docs/noaa/kellynarrative.pdf (accessed February 22, 2005).

While I was out there: Interview with Mike Kelly, August 26, 2004.

159 *In a recently published article:* Ruhl, "The battle over Endangered Species Act methodology."

160 *God created the earth:* These quotations come from the author's notes taken while attending the event, but a very similar description of Doolittle's words appeared in the *North Coast Journal,* July 29, 2004 (Hank Sims, "Upstream, downstream: Klamath politics and the great national divide), available online at: http://www.northcoastjournal.com/072904/cover0729.html (accessed March 19, 2005).

at a cost of some $700 million: For this figure, see White House Press Release, "Fact Sheet: President Bush highlights salmon recovery successes," August 22, 2003. Available online at: http://www.whitehouse.gov/news/releases/2003/08/20030822-1.html (accessed March 20, 2005).

161 *went public with their views:* For a newspaper account of this controversy, see Kenneth R. Weiss, "Action to protect salmon urged; scientists say their advice was dropped from a report to the U.S. fisheries service," *Los Angeles Times,* March 26, 2004.

The hatcheries debate: For a historical look at the Pacific salmon saga and the myriad drawbacks of hatcheries, see Jim Lichatowich, *Salmon Without Rivers: A History of the Pacific Salmon Crisis* (Washington, D.C.: Island Press, 1999).

fisheries agency indulged in "junk science": The "junk science" quotation comes from the Pacific Legal Foundation's discussion of the *Alsea v. Evans* case on its website, available at http://www.pacificlegal.org/view_PLFCaseDetail.asp?iID=132&sSubIndex=Protecting+Humanity+from+the+Endangered+Species+Act&iParentID=14&sParentName=Making+Environmental+Policy+More+Humane (accessed August 3, 2004).

In a September 2001 ruling: Alsea Valley Alliance v. Evans, 161 F. Supp. 2d 1154 (D. Or. 2001).

one of the most groundbreaking: Pacific Legal Foundation press release, "Ninth Circuit, once again, upholds decision ordering federal fisheries agency to treat hatchery fish equally," June 9, 2004. Available online at http://www.pacificlegal.org/view_PLFNews.asp?iID=247&sTitle=Ninth+Circuit,+Once+Again,+Upholds+Decision+Ordering+Federal+Fisheries+Agency+to+Treat+Hatchery+Fish+Equally+ (accessed August 3, 2004).

162 *cannot maintain wild salmon populations:* Ransom Myers et al., "Hatcheries and endangered salmon," *Science,* vol. 303, March 26, 2004.

The paper revealed: Timothy Egan, "Shift on salmon reignites fight on species law," *New York Times,* May 9, 2004.

162   *the spotted owl:* For Rutzick's role in spotted owl litigation, see David G. Savage, "Court rules for loggers in spotted owl habitats," *Los Angeles Times*, March 26, 1992.

he left in early 2005: Erik Robinson, "Senior salmon adviser to Bush resigns; Rutzick passed over for top regional post," *Columbian*, February 15, 2005.

will bring the runs back: Mark C. Rutzick, "New ESA decision good news for coho, communities, and the rule of law," November, 2001, available online at http://www.dougtimber.org/news/read_article.php?article_id=167 (accessed August 3, 2004).

Rutzick's argument is "simply not true": Interview with Ransom Myers, August 3, 2004.

163   *radical environmentalists:* See Weiss, "Action to protect salmon urged."

there's very little science: Interview with Robert Paine, August 9, 2004.

the policy: Department of Commerce, National Oceanic and Atmospheric Administration, "Policy on the Consideration of Hatchery-Origin Fish in Endangered Species Act Listing Determinations for Pacific Salmon and Steelhead," *Federal Register*, vol. 70, no. 123, June 28, 2005. Available online at: http://www.nwr.noaa.gov/Publications/FR-Notices/2005/upload/70FR37204.pdf (accessed April 4, 2006).

Just as natural habitat provides: Quoted in Michael Milstein and Joe Rojas-Burke, "Policy will put hatchery fish in salmon count," *Oregonian*, April 29, 2004.

Judge Hogan challenged the listing: See Kenneth Weiss, "U.S. erred on salmon listing, judge rules," *Los Angeles Times*, January 13, 2005.

PLF has also filed a sweeping lawsuit: Pacific Legal Foundation, "PLF files Lawsuit Challenging 16 Salmon ESA Listings Throughout the West," December 13, 2005, available online at: http://www.pacificlegal.org/view_PLFNews.asp?iID=311&sTitle=PLF+Files+Lawsuit+Challenging+16+Salmon+ESA+Listings+Throughout+the+West (accessed April 4, 2006).

164   *Florida panther:* See "Scientific Integrity in Policymaking: Further investigation of the Bush administration's abuse of science," Union of Concerned Scientists, July 2004. Available online at: http://www.ucsusa.org/global_environment/rsi/page.cfm?pageID=1641 (accessed February 22, 2005). In addition to the Florida panther, the report discusses abuses of science with respect to trumpeter swans. The discussion of abuses with respect to bull trout would not fit my criteria, since the issue involves economic analyses rather than scientific ones.

sage grouse: See Felicity Barringer, "Interior aide and biologists clashed over protecting bird," *New York Times*, December 5, 2004, noting that "The scientific opinions of a Bush administration appointee at the Interior Department with no background in wildlife biology were provided as part of the source material for the panel of Fish and Wildlife Service biologists and managers who recommended against giving the greater sage grouse protection under the Endangered Species Act."

ordered the agency's scientists to change: See Michael Milstein, "Bush officials order rewrite of protected seabird report," *Oregonian*, September 2, 2004, and Jeff Barnard, "Bush administration moves to change protection for old growth bird," Associated Press, September 1, 2004.

directed, for nonscientific reasons: Union of Concerned Scientists and Public Employees for Environmental Responsibility, "U.S. Fish and Wildlife Service Survey," available online at http://www.ucsusa.org/global_environment/rsi/page.cfm?pageID=1601 (accessed February 22, 2005).

165   *It is a logical fallacy:* "Craig's List: An interview with Bush's point person on species and parks," *Grist* (conducted by Amanda Griscom), April 15, 2004. Available online at: http://www.gristmagazine.com/maindish/manson041504.asp (accessed November 18, 2004).

an overwhelming body of evidence: Kai Chan, letter in *Grist*, April 30, 2004. The letter can be read online at http://www.gristmagazine.com/letters/letters043004.asp (accessed November 18, 2004).

the highest in the entire fossil record: National Research Council, *Science and the Endangered Species Act.* Quotation from page 13.

11: "CREATION SCIENCE" 2.0

169 *We serve the greatest Scientist:* W. David Hager, "Standing in the gap," Asbury College Tape Ministry, October 29, 2004. Audio CD on file with the author.

170 *the unofficial Republican magazine:* George F. Gilder and Bruce K. Chapman, *The Party That Lost Its Head: The Republican Collapse and Imperatives for Revival* (New York: Alfred A. Knopf, 1966). Quotation on page 57.

fading world of the class-B movie: Ibid. Quotation on p. 5.

172 *Rockefeller Republican:* Quotation from a lecture delivered by Larson at an October 15–17, 2004, conference at Case Western Reserve University entitled "Evolution and God: 150 years of love and war between science and religion." Larson's talk, entitled "The origins of the Discovery Institute," was delivered on October 16 and included a profile of Bruce Chapmàn, whom Larson knows personally. I was present at the lecture. The presentation of the Discovery Institute's history in this chapter draws both on Larson's talk and a November 1, 2004, interview with Larson.

173 *theory in crisis:* Bruce Chapman, "How should schools teach evolution? Don't forget weaknesses in theory," *Dallas Morning News,* September 21, 2003.

under the Reagan banner: John Herbers, "Working profile: aide focuses on 'ability to look forward,'" *New York Times,* July 20, 1983.

like the guest list: Herb Robinson, "Ideas from the region's best and brightest," *Seattle Times,* September 27, 1991.

174 *Discovery Institute's "No. 1 project":* Larry Witham, "Contesting science's anti-religious bias; Seattle institute's scholars promote 'intelligent design' theory of universe," *Washington Times,* December 29, 1999.

the Darwinist materialist paradigm: George Gilder, "Biocosm," *Wired,* October 2004. Available online at http://www.wired.com/wired/archive/12.10/evolution.html?pg=5 (accessed October 21, 2004).

Ruckelshaus told me: Personal communication, November 10, 2004, at "Science Communications and the News Media," University of Washington, Seattle, November 8–10, 2004.

a marvel of evolutionary adaptation: William Ruckelshaus, "Preventing a Catastrophe in Cascadia: Can we save the salmon?" Discovery Institute, June 21, 2000. Available online at http://www.discovery.org/scripts/viewDB/index.php?command=view&id=260 (accessed March 15, 2005).

serious and scholarly: John Herbers, "Man in the news: chief of census bureau," *New York Times,* July 14, 1981.

175 *Darwin read (and was impressed by) Paley:* See Charles Darwin, ed. Nora Barlow, *The Autobiography of Charles Darwin, 1809–1882* (New York: Norton, 1958), pp. 58–59.

To suppose that the eye: Charles Darwin, *The Origin of Species,* Penguin World's Classics edition (Oxford: Oxford University Press, 1996), p. 152.

176 *in the idiom of information theory:* William A. Dembski, "Signs of Intelligence: A Primer on the Discernment of Intelligent Design," pp. 171–192 of William A. Dembski and James M. Kushiner, eds., *Signs of Intelligence: Understanding Intelligent Design* (Grand Rapids, Michigan: Brazos Press, 2001). Quotation on page 192.

in this quixotic enterprise: The best documentation of the scientific failure of intelligent design is Barbara Forrest and Paul R. Gross, *Creationism's Trojan Horse: The Wedge of Intelligent Design* (Oxford: Oxford University Press, 2004). See especially Chapter 3, "Searching for the Science."

teach the controversy: For an example of a Discovery Institute representative using this phrase, see Stephen C. Meyer, "Teach the controversy," *Cincinnati Enquirer,* March 30, 2002. Available online at http://www.discovery.org/scripts/viewDB/index.php?program=CSC&command=view&id=1134 (accessed October 5, 2004).

176 *First Amendment legal strategies:* See, for example, David K. DeWolf, Stephen C. Meyer, Mark E. DeForrest, *Intelligent Design in Public School Science Curricula: A Casebook* (Richardson, Texas, Foundation for Thought and Ethics, 1999). For a more scholarly version see DeWolf, Meyer, and DeForrest, "Teaching the origins controversy: science, or religion, or speech?," *Utah Law Review,* 2000, vol. 39, no. 1.

critically analyze five different aspects: State of Ohio, Department of Education, "Critical Analysis of Evolution—Grade 10," March 2004. Available online at: http://www.ode.state.oh. us/academic_content_standards/sciencesboe/pdf_setA/L10-H23_Critical_Analysis_of_ Evolution_Mar_SBOE_changes.pdf (accessed October 12, 2004).

that scientific experts have rejected: See Patricia Princehouse, Remarks to the Ohio Board of Education, July 13, 2004, on file with author.

opposed by the National Academy of Sciences: Bruce Alberts, president of the National Academy of Sciences, letter to the Ohio Board of Education, February 9, 2004. On file with author.

seek to debunk radioisotope dating: See, for example, John D. Morris, "Can radioisotope dating be trusted?" *Back to Genesis* 104, no. 104b (Institute for Creation Research), August 1997. Available online at http://www.icr.org/pubs/btg-b/btg-104b.htm (accessed October 4, 2004).

177 *Robert T. Pennock defines creationism:* See Robert T. Pennock, "Creationism and intelligent design," *Annual Review of Genomics and Human Genetics,* vol. 4, 2003, pp. 143–163.

Seventy-two Nobel laureates signed: Amicus curiae brief of 72 Nobel laureates et al. in *Edwards v. Aguillard,* 1986. Available online at http://www.talkorigins.org/faqs/edwards-v-aguillard/ amicus1.html (accessed March 16, 2005).

better credentials than their creationist predecessors: Francis J. Beckwith, "Science and religion twenty years after *McLean v. Arkansas:* evolution, public education, and the new challenge of intelligent design," *Harvard Journal of Law and Public Policy,* vol. 26, no. 2, 2003.

178 *observes Harvard's Steven Pinker:* Interview with Steven Pinker, May 11, 2004.

the small Kansas town of Pratt: Scott Stephens, "Kansas town lighted fuse of Ohio feud over origins," *Cleveland Plain Dealer,* August 12, 2002.

continued its quest: Barbara Forrest and Glenn Branch, "Wedging creationism into the academy," *Academe,* vol. 91, no. 1, January–February 2005.

to represent a scientific innovation: See, for example, Peter Slevin, "Battle on teaching evolution sharpens," *Washington Post,* March 14, 2005, quoting ID proponent John Calvert: "The thing that excites me is we really are in a revolution of scientific thought."

an intriguing evaluation: See "The Supreme Court decision and its meaning," *Impact,* no. 170 (Institute for Creation Research), August 1987, available online at http://www.icr.org/ pubs/imp/imp-170.htm (accessed October 5, 2004).

pioneered in the wake: Glenn Branch, "The intelligent design controversy," *Seed,* 2004, spring, 9, pp. 19–21.

179 *the so-called Wedge Document:* The Discovery Institute eventually acknowledged the "Wedge Document" as its own and offered an explanation of what it is. See Discovery Institute, "The 'Wedge Document': So What?'" available online at http://www.discovery.org/scripts/viewDB/ filesDB-download.php?id=109 (accessed October 7, 2004).

180 *engaging in ad hominem attacks:* I know this from firsthand experience. See Chris Mooney, "Survival of the slickest: how anti-evolutionists are mutating their message," *American Prospect,* December 2, 2002. Available online at http://www.prospect.org/print/V13/22/ mooney-c.html (accessed October 5, 2004).

was even previously named: See National Center for Science Education, "Evolving banners at the Discovery Institute," August 29, 2002. Available online at http://www.ncseweb.org/resources/ articles/4116_evolving_banners_at_the_discov_8_29_2002.asp (accessed October 7, 2004).

placed on the Web without permission: Mooney, "Survival of the Slickest."

181 *creationism historian Ronald Numbers:* Interview with Ronald Numbers, October 18, 2004.

181 *This article imagines:* Nancy Pearcey, "Phillip Johnson was right: the unhappy evolution of Darwinism," *World*, February 24, 2001. Available online at http://www.discovery.org/scripts/viewDB/index.php?command=view&id=598 (accessed October 21, 2004).

rivals those of Newton and Einstein: Michael Behe, *Darwin's Black Box: The Biochemical Challenge to Evolution* (New York: The Free Press, 1996), pp. 232–233.

it doesn't exist: Quoted in Randy Dotinga, "A who's who of players in the battle of biology class," *Christian Science Monitor,* December 7, 2004.

182 *at the advanced age of 38:* Stephen Goode, "Johnson challenges advocates of evolution," *Insight on the News*, October 25, 1999. Available online at http://www.arn.org/docs/johnson/insightprofile1099.htm (accessed March 21, 2005). See also Barbara Forrest and Paul Gross, *Creationism's Trojan Horse: The Wedge of Intelligent Design* (Oxford: Oxford University Press, 2003), p. 17.

devote my life to destroying Darwin: See Jonathan Wells, "Darwinism: Why I Went for a Second Ph.D.," available online at http://www.tparents.org/library/unification/talks/wells/DARWIN.htm (accessed October 7, 2004).

to mobilize a new generation of scholars: Jeff Robinson, "Dembski to head seminary's new science & theology center," *BP News,* September 16, 2004. Available online at http://www.bpnews.net/bpnews.asp?ID=19115 (accessed October 5, 2004).

183 *background in Republican politics:* Meyer "advised U.S. Rep. George Nethercutt [a Republican] in his 1994 campaign to unseat House Speaker Thomas Foley," according to his by-line in a 1996 article. See Stephen C. Meyer, "Dole must put Clinton's liberalism in the spotlight," *Chicago Tribune*, October 2, 1996.

the person who brought ID to DI: Larson lecture, October 16, 2004, "Origins of the Discovery Institute."

All those who become associated: Palm Beach Atlantic University, "Guiding principles," available online at http://www.pba.edu/CD/guiding.htm (accessed October 5, 2004).

mainstream, moderate evangelical activities: Interview with Edward Larson, November 1, 2004.

Howard F. Ahmanson, Jr.: The Discovery Institute's board of directors can be seen at http://www.discovery.org/fellows (accessed March 16, 2005). Ahmanson's funding of the Institute has been widely reported. For a few examples, see Michael D. Lemonick, Noah Isackson, Jeffrey Ressner, "Stealth attack on evolution; who is behind the movement to give equal time to Darwin's critics, and what do they really want?" *Time*, January 31, 2005; Peter Slevin, "Battle on teaching evolution sharpens," *Washington Post*, March 14, 2004 (noting, "Meyer said the institute accepts money from such wealthy conservatives as Howard Ahmanson Jr."); and Teresa Watanabe, "Enlisting science to find the fingerprints of a creator," *Los Angeles Times*, March 25, 2001 (noting, "Primarily funded by evangelical Christians—particularly the wealthy Ahmanson family of Irvine—the institute's $1 million annual program has produced 25 books, a stream of conferences, and more than 100 fellowships for doctoral and postdoctoral research. Fieldstead & Co., which is owned by Howard and Roberta Ahmanson, has pledged $2.8 million through 2003 to support the intelligent design program").

Tennessee-based Maclellan Foundation: According to the *Los Angeles Times*, the Maclellan Foundation contributed $350,000 to the Discovery Institute in the hope that it would prove that "evolution was not the process by which we were created." See Teresa Watanabe, "Enlisting science to find the fingerprints of a creator," *Los Angeles Times*, March 25, 2001.

184 *fulfillment of the Great Commission:* The Maclellan Foundation, "About Us," available online at: http://www.maclellanfdn.org/about/home.asp (accessed January 2, 2005).

at a 2004 conference: The event was titled "Intelligent design conference: dispelling the myth of Darwinism," and took place June 24–26, 2004, in Highlands, North Carolina. The speakers can be viewed online at http://www.idconference.org/html/speakers.html (accessed October 7, 2004).

In a 2002 resolution: AAAS Board Resolution on Intelligent Design Theory, October 18, 2002. Available online at http://www.aaas.org/news/releases/2002/1106id2.shtml (accessed October 7, 2004).

184 *literature searches have failed:* George W. Gilchrist, The elusive scientific basis of intelligent design theory. *Reports of the National Center for Science Education*, vol. 17, no.3, pp. 14–15. See also Barbara Forrest and Paul Gross, *Creationism's Trojan Horse: The Wedge of Intelligent Design*, Oxford: Oxford University Press, 2003, Chapter 3.

Brown University's Kenneth Miller: Kenneth Miller, "Answering the biochemical argument from design," available online at http://www.millerandlevine.com/km/evol/design1/article.html (accessed December 29, 2004).

185 *Meyer argued that evolutionary theory:* Stephen Meyer, "The origin of biological information and the higher taxonomic categories," *Proceedings of the Biological Society of Washington,* vol. 117, no. 2, August 4, 2004, pp. 213–239.

Even the Discovery Institute acknowledged: See Discovery Institute, "Meyer responds to errors in *Chronicle of Higher Education* article," September 20, 2004, noting, "My piece was merely the first peer-reviewed article to advocate intelligent design openly in a science journal." Available online at http://www.discovery.org/scripts/viewDB/index.php?command=view&id=2207&program=CSC%20Responses (accessed February 26, 2005).

though the group claims: Discovery Institute, "Media backgrounder: intelligent design article sparks controversy," September 7, 2004.

considerable media attention: See, for example, Richard Monastersky, "Society disowns paper attacking Darwinism," *Chronicle of Higher Education,* September 24, 2004.

demanding an explanation: As reported in Ronald Jenner, "The tainting of *Proc. Biol. Soc. Wash.*," *Palaeontology Newsletter,* 2004, no. 57, pp. 10–17.

A Scientific Dissent from Darwinism: Discovery Institute, "A scientific dissent from Darwinism," September 24, 2001. Available online at http://www.reviewevolution.com/press/Darwin Ad.pdf (accessed October 7, 2004).

186 *Baraminology Study Group:* Baraminology Study Group, website at: http://www.bryancore. org/bsg/opbsg (accessed October 7, 2004).

hosted by Bryan College: The Baraminology Study Group insists that it is not "at" Bryan College, nor is it "associated with" Bryan College, other than its website hosting. See http://www. bryancore.org/bsg/clarifications.html (accessed March 17, 2005).

did not describe himself as an evolutionist: As Sternberg states, "Although it is irritating to have to respond to ad hominem arguments rather than arguments on the issues, I will state for the record that I do not accept the claims of young-earth creationism. Rather, I am a process structuralist." Quoted on http://www.rsternberg.net (accessed February 24, 2005). Elsewhere, he explains that "Structuralist analysis is generally ahistorical, systems-oriented, and non-evolutionary (not anti-evolutionary)." Quoted on http://www.rsternberg.net/Structuralism. htm (accessed February 24, 2005).

his resignation preceded: As Sternberg states, "In October of 2003 I resigned as managing editor of the *Proceedings*; after almost two years I was tiring of my editorial responsibilities and eager to have more time for my own research and writing. At that time, however, no new managing editor could be found, and so without withdrawing my letter of resignation I agreed to continue on as managing editor until such time as the Council could find my replacement. That happened in May 2004, when Dr. Richard Banks agreed to replace me after the issue Volume 117–3 and a major "bulletin" that was nearly complete (both are currently in press). So as planned for some time, Dr. Banks has recently taken over as managing editor of the *Proceedings*. This transition had nothing to do with the publication of the Meyer paper." Quoted on http://www.rsternberg.net (accessed February 24, 2005).

contrary to typical editorial practices: Council of the Biological Society of Washington, statement regarding the publication of the paper by Stephen C. Meyer in vol. 117, no. 2 of the *Proceedings,* date unclear. Available online at http://www.biolsocwash.org/id_statement.pdf (accessed February 24, 2005).

as managing editor: Quoted at http://www.rsternberg.net (accessed February 24, 2005).

186 *does not meet the scientific standards:* Council of the Biological Society of Washington, Statement regarding the publication of the paper by Stephen C. Meyer in Volume 117(2) of the *Proceedings,* October 2004.

gag rule on science: Discovery Institute, "Darwinists impose gag rule on science," September 8, 2004. Available online at http://www.discovery.org/scripts/viewDB/index.php?command=view&id=2194&program=CSC-News&callingPage=discoMainPage (accessed October 12, 2004).

187 *three of Meyer's scientific critics:* Alan Gishlick, Nick Matzke, and Wesley R. Elsberry, "Meyer's hopeless monster." Review of Meyer, Stephen C. 2004. The origin of biological information and the higher taxonomic categories. *Proceedings of the Biological Society of Washington,* vol. 117, no. 2, pp. 213–239," August 24, 2004. Available online at: http://www.pandasthumb. org/pt-archives/000430.html (accessed October 21, 2004).

Another expert who commented: Roger Thomas, e-mail communication, January 12, 2005.

Rather than continuing to trust: Jenner, "The tainting of *Proc. Biol. Soc. Wash.*"

188 *stunning successes of the theory of evolution:* For a good overview of the many diverse sources of evidence in favor of the theory of evolution, see "Science and Creationism: A View from the National Academy of Sciences," Chapter on "Evidence Supporting Biological Evolution," available online at http://books.nap.edu/html/creationism/evidence.html (accessed March 17, 2005).

Barbara Forrest has noted: Barbara Forrest, "Neo-creationism: where's the harm in that?" Remarks delivered at Case Western Reserve University, October 16, 2004.

189 *James Dobson's Focus on the Family:* See, for example, Lee Strobel, "The case for a creator," *Breakaway (A Focus on the Family Magazine for Teen Guys),* October 2004. Available online at http://www.family.org/teenguys/breakmag/departments/a0033928.html (accessed October 21, 2004).

Phyllis Schlafly's Eagle Forum: See, for example, "Missouri bill would require teaching alternatives to evolution," *Education Reporter,* February 2004. Available online at http://www.eagleforum. org/educate/2004/feb04/evolution.html (accessed October 21, 2004).

Concerned Women for America: See, for example, Tanya L. Green, "Advocates of intelligent design to get hearing in March," Concerned Women for America, February 14, 2002. Available online at http://www.cwfa.org/articledisplay.asp?id=920&department=CWA&categoryid=education (accessed October 21, 2004).

Kennedy's Coral Ridge Ministries: See, for example, "Intelligent design theory challenges Darwin in schools," Coral Ridge Ministries *Impact* Newsletter, August 2003. Available online at http://www.coralridge.org/impact/2003_Aug_Pg1.htm (accessed October 21, 2004).

American Family Association: See, for example, Jim Brown and Jenni Parker, "Scientist: Darwinists trying to squelch intelligent design debate," *AgapePress,* September 14, 2004. Available online at http://headlines.agapepress.org/archive/9/afa/142004a.asp (accessed October 21, 2004).

Alliance Defense Fund: See Michael Moore, "Legal defense organization says it may aid Darby school board," *Missoulian,* February 16, 2004. Available online at http://www.missoulian. com/articles/2004/02/16/news/local/znews01.txt (accessed October 21, 2004).

ID movement progenitor Phillip Johnson: As noted in Larry Witham, "Senate bill tackles evolution debate; advises states to allow academic openness on concept," *Washington Times,* June 18, 2001.

not on a philosophically neutral assessment: See Phillip E. Johnson, "The Unraveling of Scientific Materialism," *First Things,* November 1997, vol. 77, pp. 22–25.

the Darwinian monopoly: Quoted in American Geological Institute, "Special Update: Evolution Opponents on the Offensive in Senate, House," June 19, 2001. Available online at http://www.agiweb.org/gap/legis107/evolution_update0601.html (accessed March 17, 2005).

should be allowed to learn: Rick Santorum, "Illiberal education in Ohio schools," *Washington Times,* March 14, 2002.

190 *insist is a scientific theory:* See Discovery Institute press release, "Congress urges teaching of diverse views on evolution, but Darwinists try to deny it," December 28, 2001, quoting Discovery's John West, "The reality is that Congress has clearly voiced its support for teaching 'the

full range of scientific views that exist' about evolution. That includes both scientific criticisms of natural selection and random mutation as the mechanism for evolution as well as scientific alternatives to Darwinism such as intelligent design theory." Available online at http://www.discovery.org/scripts/viewDB/index.php?programs=CRSC&command=view&id =1121 (accessed March 17, 2005).

190 *ID is not science:* See Chris Mooney, "Intelligent Denials: Bush's science adviser said one important thing about politicized science in a recent appearance. But only one," *American Prospect Online,* February 22, 2005. Available online at: http://www.prospect.org/web/page.ww?section=root&name=ViewWeb&articleId=9216 (accessed March 17, 2005).

saw significant antievolution activity: E-mail communication from Glenn Branch, of the National Center for Science Education, February 23, 2005.

found eight state parties: Glenn Branch, "Evolution and the Elections," *Reports of the National Center for Science Education,* Nov/Dec 2004, 24 (6): pp. 4–9.

12: STEMMING RESEARCH

195 *These advocates claim:* See, for example, the U.S. Conference of Catholic Bishops, "Scientific experts agree: embryonic stem cells are unnecessary for medical progress," available online at http://www.usccb.org/prolife/issues/bioethic/fact401.htm (accessed March 25, 2005).

196 *it will take at least five years:* Peter van Etten, comments at James A. Baker III Institute for Public Policy, Rice University, November 21, 2004. I was present at the event.

biologist James Thomson: Thomson et al., "Embryonic stem cell lines derived from human blastocysts," *Science,* vol. 282, November 6, 1998.

197 *bypass four current "bottlenecks":* Lawrence S. B. Goldstein, "What are embryonic stem cells?," speech at Rice University's James A. Baker III Institute for Public Policy, November 20, 2004.

198 *violating respect for nascent human life:* Leon Kass, "Playing politics with the sick," *Washington Post,* October 8, 2004.

we must first exhaust: Statement by Tommy G. Thompson, secretary of health and human services, on the President's embryonic stem cell policy, August 8, 2004. Available online at http://www.os.hhs.gov/news/press/2004pres/20040808.html (accessed November 4, 2004).

199 *from a purely scientific perspective:* Rick Weiss, "Bush's stem cell policy reiterated, but some see shift," *Washington Post,* May 16, 2004.

laptop analogy: Interview with Evan Snyder, June 8, 2004.

200 *potentially unsuitable for transplantation:* See Rick Weiss, "Approved stem cells' potential questioned," *Washington Post,* October 29, 2004, reporting on Martin et al., "Human embryonic stem cells express immunogenic nonhuman sialic acid," *Nature Medicine,* vol. 11, no. 2, February 2005, pp. 228–232.

pathologist Irving Weissman explained: Interview with Irving Weissman, June 9, 2004.

201 *which the FDA:* On the FDA's ban, see Louis Guenin, "Stem cells, cloning, and regulation," *Mayo Clinic Proceedings,* vol. 80, 2005, pp. 241–250.

responsible for the lives: U.S. Senate Committee on Commerce, Science, and Transportation, science, technology, and space hearing, "Adult stem cell research," July 14, 2004. I attended the hearing, and the Weissman quotation comes from my notes taken at the event.

No human therapies of any kind: Eric Cohen, "Inflated promise, distorted facts," *National Review Online,* May 25, 2004. Available online at http://nationalreview.com/comment/cohen200405251335.asp (accessed November 3, 2004).

202 *It's a phony argument:* Interview with Paul Berg, June 9, 2004.

junk science argument: Michael Reagan, "I'm with my dad on stem cell research," *Human Events Online,* June 23, 2004. Available online at http://www.humaneventsonline.com/article.php?id=4286 (accessed November 4, 2004).

stem cell transplantation therapies: On this question, conservatives frequently cite the following article: Rick Weiss, "Stem cells an unlikely therapy for Alzheimer's," *Washington Post,* June

10, 2004. But note that Weiss uses the word "therapy" in the title. The body of his article explains that basic research on Alzheimer's using embryonic stem cells still holds considerable promise. To wit: "The key, said Harvard stem cell researcher George Daley, is not to get 'preoccupied with stem cells as cellular therapies.' Their real value for Alzheimer's will be as laboratory tools to explore basic questions of biology, Daley said."

203 *Christopher Reeve will get up:* See "Frist knocks Edwards over stem cell comment," *CNN.com,* October 12, 2004, available online at http://www.cnn.com/2004/ALLPOLITICS/10/12/edwards.stem.cell (accessed November 30, 2004).

204 *likened to "creation science":* See Louis Guenin, testimony before the House Committee on Energy and Commerce, June 20, 2001, noting of adult stem cell research, "We have learned from encounters with such ventures as 'creation science,' which purportedly refutes the theory of evolution, that we must be sceptical when nonscientist advocates offer purported analyses of scientific data to reinforce conclusions that they have already reached on nonscientific grounds."

conservative bioethics writer Wesley Smith: See, for example, Wesley J. Smith, "The 'wrong' cure: adult stem cells get the shaft," *National Review Online,* September 9, 2004. Available online at http://www.nationalreview.com/smithw/smith200409090835.asp (accessed November 3, 2004).

Wouldn't it be wise: Second presidential debate, Washington University in St. Louis, October 8, 2004. The questioner's name was Elizabeth Long. Her full question was, "Senator Kerry, thousands of people have already been cured or treated by the use of adult stem cells or umbilical cord stem cells. However, no one has been cured by using embryonic stem cells. Wouldn't it be wise to use stem cells obtained without the destruction of an embryo?"

as great, if not greater, potential: David A. Prentice, "Case study: the case against federal funding of human embryonic stem cell research," *Yale Journal of Health Policy, Law and Ethics,* Fall, 2001.

of both the National Institutes of Health: National Institutes of Health, "Stem cell basics: what are the similarities and differences between embryonic and adult stem cells?" describing limits to these cells' biomedical potential. Available online at http://stemcells.nih.gov/info/basics/basics5.asp (accessed November 4, 2004). As the NIH states, "Adult stem cells are generally limited to differentiating into different cell types of their tissue of origin. However, some evidence suggests that adult stem cell plasticity may exist, increasing the number of cell types a given adult stem cell can become."

205 *research on all types of stem cells:* International Society for Stem Cell Research, Board of Directors, Letter to President Bush, June 22, 2004. Available online at http://www.isscr.org/media/documents/Bushpolicy02.pdf (accessed March 26, 2005).

adult stem cell work that we did: Interview with David Prentice, July 13, 2004.

Prentice has served: Prentice listed these qualifications on his Indiana State University faculty website. Printout on file with author.

a surprising ability for transformation: See David Prentice, "Adult stem cells," Appendix K in *Monitoring Stem Cell Research,* President's Council on Bioethics, January 2004.

206 *Weissman should know:* Interview with Irving Weissman, June 9, 2004.

it may happen: Interview with Elizabeth Blackburn, December 13, 2004.

they don't grow nearly as robustly: Interview with Elizabeth Blackburn, December 13, 2004.

no strong scientific opinion: Lawrence S. B. Goldstein, "What are embryonic stem cells?" speech at Rice University's James A. Baker III Institute for Public Policy, November 20, 2004.

207 *multipotent adult progenitor cells:* Y. Jiang et al., "Pluripotency of mesenchymal stem cells derived from adult marrow," *Nature,* vol. 418, July 4, 2002.

They have put words in my mouth: Quoted in Gareth Cook, "From adult stem cells comes debate," *Boston Globe,* November 1, 2004.

At a July 2004 hearing: U.S. Senate Committee on Commerce, Science, and Transportation, science, technology, and space hearing, "Adult stem cell research," July 14, 2004.

who had previously downplayed: "A forum for dialogue: stem cell research: the science and the ethics," Loyola University Stritch School of Medicine, October 27, 2004. According to a summary of the event online, available at http://il-cha.org/summary2.php (accessed November 4, 2004), Peduzzi-Nelson's presentation went as follows: "Dr. Peduzzi-Nelson spoke on the science involved in stem cell research and human cloning. She discussed the different types of stem cells being used in research and explained the basics of how stem cells are extracted as well as the process involved in cloning. Dr. Peduzzi-Nelson noted that some of the disadvantages to embryonic stem cells include uncontrolled growth and the likelihood of rejection. In contrast, with adult stem cells, even though there is a slower rate of growth, the overgrowth and rejection problems can be avoided. Dr. Peduzzi-Nelson also discussed pre-clinical and clinical trials that have shown the use of embryonic stem cells has not been successful, while clinical trials with the use of a patient's own neural stem cells have resulted in improvements to the patient."

207 *there is no doubt that President Reagan:* Testimony of Dr. Jean D. Peduzzi-Nelson before the Senate Commerce Subcommittee on Science, Technology and Space, July 14, 2004. On file with author.

that she was under oath: In addition to the author, two media outlets covered this exchange. See Mark Sherman, "Senate hearing turns contentious over embryonic stem cell research," Associated Press, July 14, 2004, and Kristen Philipkoski, "Senator pushes adult stem cells," *Wired News,* July 15, 2004, available online at http://www.wired.com/news/medtech/0,1286,64221,00.html (accessed November 5, 2004).

208 *with a vast biomedical potential:* David Prentice, testimony, "Embryonic stem cell research: exploring the controversy," Senate Committee on Commerce, Science, and Transportation, Subcommittee on Science, Technology and Space, September 29, 2004. Available online at http://commerce.senate.gov/hearings/testimony.cfm?id=1323&wit_id=3857 (accessed November 4, 2004).

thoroughly moral at the same time: Family Research Council, "FRC Endorses Real Science and Real Cures," September 29, 2004, available online at http://www.frc.org/get.cfm?i=PR04I07&f=PR04I07&t=e (accessed March 26, 2005).

209 *council's membership was revealed:* See White House press release, "President names members of bioethics council," January 16, 2002. Available online at http://www.whitehouse.gov/news/releases/2002/01/20020116–9.html (accessed December 30, 2004).

she began to criticize: Interview with Elizabeth Blackburn, December 13, 2004. For Blackburn's criticism of the council's *Beyond Therapy* report, see Chris Mooney, "Half empty," *SAGE Crossroads,* November 24, 2003, available online at http://www.sagecrossroads.net/Default.aspx?tabid=28&newsType=ArticleView&articleId=39 (accessed March 29, 2005). For her criticism of *Monitoring Stem Cell Research,* see the President's Council on Bioethics, January 15, 2004, meeting transcript, "Session 1: Stem Cells: Council's Report to the President," available online at http://www.bioethics.gov/transcripts/jan04/session1.html (accessed March 29, 2005).

three individuals: See Rick Weiss, "Bush ejects two from bioethics council," *Washington Post,* February 28, 2004.

210 *good news on the pro-life front:* Family Research Council, Tony Perkins's Washington update, "Good news on the pro-life front," March 1, 2004. Available online at http://www.frc.org/get.cfm?i=WU04C01 (accessed October 3, 2004).

over 170 bioethicists signed: Jeffrey Brainard, "A new kind of bioethics," *Chronicle of Higher Education,* May 21, 2004.

malicious and false: Leon Kass, "We don't play politics with science," *Washington Post,* March 3, 2004.

Blackburn offered strong cautions: See the President's Council on Bioethics, January 15, 2004, meeting transcript, "Session 1: Stem Cells: Council's Report to the President," available online at http://www.bioethics.gov/transcripts/jan04/session1.html (accessed March 29, 2005).

capabilities of embryonic versus adult stem cells: See Elizabeth Blackburn, "A 'full range' of bioethical views just got narrower," *Washington Post,* March 7, 2004.

210 *Blackburn published a scientific takedown:* Elizabeth Blackburn, Janet Rowley, "Reason as our guide," *PLoS Biology,* vol. 2, no. 4, April 2004, pp. 420–422.

absolutely destructive practices: Quoted in Paul Elias, "Scientists rally around stem cell advocate sacked by Bush team," Associated Press, March 19, 2004.

211 *These charges are utterly baseless:* Written response received from Leon Kass, March 20, 2005.

212 *who also testified:* See President's Council on Bioethics, "Session 3: Stem Cell Research: Recent Scientific and Clinical Developments," July 24, 2003. Available online at http://www.bioethics.gov/transcripts/july03/session3.html (accessed November 5, 2004).

ignited debate about the relative merits: President's Council on Bioethics, *Monitoring Stem Cell Research,* January 2004, Chapter 1. Available online at http://www.bioethics.gov/reports/stemcell/chapter1.html (accessed November 5, 2004).

reads as a political document: President's Council on Bioethics, *Monitoring Stem Cell Research,* January 2004, Chapter 2. Available online at http://www.bioethics.gov/reports/stemcell/chapter2.html (accessed November 5, 2004).

Bush himself had drawn no such distinction: When I asked Leon Kass about this passage, and specifically, why the President's Council on Bioethics had added the parenthetical remark, given that Bush had drawn no such distinction in his August 2001 speech, he replied, "Technically, a stem cell *line* is a purified and well-characterized stable population. As our report makes clear, the derivation of stem cells begins with a 'preparation' of cells that must then be turned into a stable line. To be eligible for federal funding, the embryo-destroying derivation of stem cells had to have occurred prior to the President's decision. It takes time to convert the *eligible* population of cells into *available* stem cell *lines*." All of this is true and properly nuanced. But as noted in the text, it wasn't what the Bush administration was saying at the time of the president's announcement.

diverse, robust, and viable for research: Laura Meckler, "Thompson says critics missing the point on stem cell debate," Associated Press, September 7, 2001.

Only after considerable stonewalling: Ceci Connolly and Justin Gillis, "Thompson: stem cell work viable; most lines unproven, HHS chief concedes," *Washington Post,* September 6, 2001.

213 *a slate of individuals:* I am indebted to Timothy Noah, of *Slate,* who compiled the quotations from the new members given here. See http://slate.msn.com/id/2096848 (accessed March 30, 2005).

one human being in the service of others: Council on Biotechnology Policy, "The sanctity of life in a brave new world: A manifesto on biotechnology and human dignity," available online at http://www.pfm.org/AM/Template.cfm?Section=Biotechnology_Policy&template=/CM/HTMLDisplay.cfm&ContentID=2222 (accessed March 30, 2005).

cloning is an evil: Diana Schaub, review of "Human Cloning and Human Dignity: The Report of the President's Council on Bioethics," *First Things,* January 2003.

Berry College political scientist Peter Lawler: See Peter Augustine Lawler, "The right choice: Hadley Arkes on natural rights from the Declaration to Roe v. Wade," *Weekly Standard,* November 25, 2002, in which Lawler writes, "Unless we become clear as a nation that abortion is wrong, women will—I predict—eventually find themselves compelled to submit to therapeutic abortions of genetically defective babies and then to do whatever is required to enhance their children genetically. . . . We will not be able to protect the genuine reproductive freedom of women—their right to have and love their own babies—unless there is a pro-life consensus embodied in our law. Those who believe the effective regulation of biotechnological development can be morally neutral about abortion are simply wrong."

13: SEXED-UP SCIENCE

217 *I decided to become a biochemist:* Joel Brind, "Reading the data: defining a link between abortion and breast cancer," *Physician,* July/August 2000. Available online at http://www.family.org/physmag/issues/a0012382.cfm (accessed November 6, 2004).

217 *he even cofounded a think tank:* The webpage for the Breast Cancer Prevention Institute that describes its history can be found at http://www.bcpinstitute.org/history.htm (accessed November 6, 2004).

218 *Tuskegee syphilis scandal:* For Brind's role in the Coalition on Abortion/Breast Cancer, see http://www.abortionbreastcancer.com/About_Us.htm. For the group's use of the Tuskegee syphilis scandal analogy, see http://www.abortionbreastcancer.com/tuskegee.htm (both accessed November 6, 2004).

National Library of Medicine's PubMed database: A search of the PubMed database for articles by "Brind J" conducted on March 24, 2005, yielded thirty-two results going back to the year 1973. I examined all the results and found ten that were "on point"; that is, they were actually about the relationship between abortion and breast cancer. These I analyzed further.

Of the ten, eight were explicitly listed, under "publication types," as "comment" and/or "letter." The page numbers provided by PubMed back this up: none was longer than a few pages. Of the remaining two publications, one was Brind J. et al., "Correcting the record on abortion and breast cancer," *Breast J.*, 1999 May; 5 (3) 215–216. Again very brief, I suspected that this was a letter as well, and sure enough, Brind himself classifies this as a letter in a list of publications on the website of the Breast Cancer Prevention Institute, available online at http://www.bcpinstitute.org/printabc_word.doc (accessed March 24, 2005).

The final publication was Brind J. et al., "Induced abortion as an independent risk factor for breast cancer; a comprehensive review and meta-analysis," *J. Epidemiol Community Health*, 1996 October; 50 (5): 481–496. In the abstract, this article is described as an analysis of "all 28 published reports which include specific data on induced abortion and breast cancer incidence."

in pro-life newsletters: Brind's faculty webpage lists numerous articles published in *National Right to Life News*. Available online at http://www.baruch.cuny.edu/wsas/departments/natural_science/faculty/brind.html (accessed November 6, 2004). Brind also publishes a newsletter, the "Abortion Breast Cancer Quarterly Update." See Joel Brind, "Reading the data: defining a link between abortion and breast cancer," *Physician*, July/August 2000. Available online at http://www.family.org/physmag/issues/a0012382.cfm (accessed November 6, 2004).

Rep. David Weldon: See David Weldon, "Dear Colleague" letter on abortion and breast cancer, available online at http://www.abortionbreastcancer.com/weldon_letter.htm (accessed November 6, 2004).

several states have even passed laws: Scott Gold, "Texas OKs disputed abortion legislation," *Los Angeles Times*, May 22, 2003, and Shankar Vedantam, "Abortion link to breast cancer discounted; scientists compared 53 studies and based findings on better designed research," *Washington Post*, March 26, 2004.

full-term pregnancy: See National Cancer Institute, "Pregnancy and breast cancer risk," available online at http://cis.nci.nih.gov/fact/3_77.htm (accessed November 6, 2004).

discounted the ABC link: Melbye et al., "Induced abortion and the risk of breast cancer," *New England Journal of Medicine*, vol. 336, no. 2, January 1997, pp. 81–85.

Harvard epidemiologist Karin Michels: Interview with Karin Michels, June 25, 2004.

adds Lynn Rosenberg: Interview with Lynn Rosenberg, June 28, 2004.

alleging methodological flaws: See Rick Weiss, "Study disputes breast cancer, abortion link; no added risk seen in early-term procedures," *Washington Post*, January 9, 1997.

over one hundred experts: Number cited in Gold, "Texas OKs Disputed Abortion Legislation."

is not associated with an increase: National Cancer Institute, "Summary report: early reproductive events and breast cancer workshop," March 2003. Available online at http://www.cancer.gov/cancerinfo/ere-workshop-report (accessed November 6, 2004).

Brind alone dissented: The National Cancer Institute acknowledged the existence of a minority dissent from its workshop report. See "Minority dissenting comment regarding early reproductive events and breast cancer workshop," March 25, 2003. Available online at http://cancer.gov/cancer_information/doc.aspx?viewid=15e3f2d5–5cdd–4697-a2ba-f3388d732642 (accessed November 6, 2004). But the NCI did not name the dissenter. On the website of the Coalition on Abortion/Breast Cancer, however, Brind's March 10, 2003, document "Early reproductive events and breast cancer: a minority report," can be accessed. Available at http://www.abortion breastcancer.com/minorityreport.htm (accessed November 6, 2004).

219 *studies on the issue were inconclusive:* See Adam Clymer, "Critics say government deleted sexual material from web sites to push abstinence," *New York Times,* November 26, 2002, and Adam Clymer, "U.S. revises sex information, and a fight goes on," *New York Times,* December 27, 2002.

220 *notes Michael Cromartie:* Interview with Michael Cromartie, June 30, 2004.

 loudly protested his appointment: See Bernard Weintraub, "Reagan Nominee for surgeon general runs into obstacles on Capitol Hill," *New York Times,* April 7, 1981.

 I am the nation's surgeon general: Quoted in Nancy Shute, "America's M.D.: Surgeon General Koop has been a shot in the arm to the nation's ills," *Chicago Tribune,* June 23, 1989.

221 *describing his theological worldview:* Interview with David Reardon, July 12, 2004.

 Conservatives widely applauded the book: For a description of the book and a list of endorsements, see http://www.afterabortion.org/awsnm.html (accessed November 14, 2004).

 Elliot Institute for Social Sciences Research: Interview with David Reardon, July 12, 2004.

 now calls himself "Dr. David C. Reardon": See, for example, David Reardon et al., "Psychiatric admissions of low-income women following abortion and childbirth," *Canadian Medical Association Journal,* vol. 168, no. 10, May 13, 2003.

 diploma mills and other unaccredited schools: Government Accountability Office (formerly General Accounting Office), "Diploma mills: federal employees have obtained degrees from diploma mills and other unaccredited schools, some at government expense," May 11, 2004. Available online at http://www.gao.gov/new.items/d04771t.pdf (accessed November 15, 2004).

 Council for Higher Education Accreditation: Search of the Council for Higher Education Accreditation's "Database of Institutions Accredited by Recognized United States Accrediting Organizations," May 6, 2005. Database available online at http://www.chea.org/ (accessed May 6, 2005).

 By e-mail: David C. Reardon, e-mail communication, July 12, 2004.

 Joe Pitts, of Pennsylvania, sponsored a bill: Megan Fromm, "Study of post-abortion depression proposed," *Washington Times,* June 22, 2004.

 a safe medical procedure: American Psychological Association, "Briefing paper on the impact of abortion on women: what does the psychological research say?" March 2004. Available online at http://www.apa.org/ppo/issues/womenabortfacts.html (accessed November 14, 2004).

222 *Reardon's work fails to overturn:* Interviews with Nancy Adler, June 24, 2004, Nancy Filipe Russo, June 28, 2004, Brenda Major, June 24, 2004. All three, along with Henry David, coauthored a major paper in *Science,* published on April 6, 1990, finding that "Severe negative reactions after abortions are rare and can best be understood in the framework of coping with a normal life stress." See Nancy Adler et al., "Psychological responses after abortion," *Science,* vol. 248, no. 4951, April 6, 1990, p. 41.

 a far stronger and less defensible claim: See, for example, the Elliot Institute, "A list of major psychological sequelae of abortion," 1997, available online at http://www.afterabortion.info/ psychol.html (accessed March 23, 2005).

 a 2003 Reardon study: David C. Reardon et al., "Psychiatric admissions of low-income women following abortion and childbirth," *Canadian Medical Association Journal,* vol. 168, no. 10, May 13, 2003.

 As Major noted: Brenda Major, "Psychological implications of abortion—highly charged and rife with misleading research," *Canadian Medical Association Journal,* vol. 168, no. 10, May 13, 2003.

 in well-designed studies: Interview with Nancy Felipe Russo, June 28, 2004; e-mail follow-up, September 3, 2004.

223 *proposed legislation:* Elliot Institute, "Proposals for legislation," available online at http://www. afterabortion.org/leg/index.html (accessed November 15, 2004).

223 *Nothing good comes from evil:* David Reardon, "A defense of the neglected rhetorical strategy (NRS)," *Ethics & Medicine,* Volume 18, no. 2, summer 2002.

224 *safe sex:* For instance, a Medical Institute press release heralding George W. Bush's January 20, 2004, state of the union address noted, "Research makes clear that the 'safe sex' or 'safer sex' approach—built as it is upon a reliance on condoms and the disrespectful belief that children will have sex anyway—is a failure." Available online at http://www.medinstitute.org/Bush.htm (accessed November 15, 2004).

that the Medical Institute deems acceptable: See Medical Institute, "Scientific review of condom effectiveness research reveals condoms provide inadequate risk reduction for sexually transmitted diseases," October 16, 2002. Available online at http://www.medinstitute.org/media/Monograph.htm (accessed November 15, 2004).

McIlhaney has also served: Centers for Disease Control and Prevention, "Secretary Thompson appoints nine to CDC advisory committee," February 20, 2003. Available online at http://www.cdc.gov/od/oc/media/pressrel/r030220d.htm (accessed November 15, 2004).

We're a medical, scientific organization: Interview with Joe S. McIlhaney, July 14, 2004.

Douglas Kirby: Interview with Douglas Kirby, June 25, 2004.

rarely willing to bankroll: E-mail communication from Douglas Kirby, July 29, 2004.

no studies: In fact, there is some evidence suggesting that the "abstinence only" approach fails in its objective of preventing the spread of sexually transmitted diseases. In a March 2005 study published in the *Journal of Adolescent Health,* Yale sociologist Hannah Bruckner and Columbia sociologist Peter Bearman found that teenagers who took so-called Virginity Pledges—promising to abstain from sex until marriage—delayed having sex for longer but did not have correspondingly diminished STD infection rates. As it happened, most pledgers didn't actually keep their oaths all the way to marriage, and those that broke them were less likely to use condoms the first time they had sex. Moreover, the minority of pledgers who actually managed to abstain from vaginal sex until marriage were more likely to try out oral or anal sex than to abstain from intercourse entirely. The study is Hannah Bruckner and Peter Bearman, "After the promise: the STD consequences of adolescent virginity pledges," *Journal of Adolescent Health,* vol. 36, 2005, pp. 271–278.

225 *No program was able to demonstrate:* Hauser D. *Five Years of Abstinence-Only-Until-Marriage Education: Assessing the Impact* (Washington, D.C.: Advocates for Youth, 2004). Available online at http://www.advocatesforyouth.org/publications/stateevaluations.pdf (accessed December 30, 2004).

Programs That Work had identified: Congressman Henry Waxman, "The effectiveness of abstinence-only education," available online at http://democrats.reform.house.gov/features/politics_and_science/example_abstinence.htm (accessed November 15, 2004).

report on federally supported abstinence education: U.S. House of Representatives, Committee on Government Reform (Minority), "The content of federally funded abstinence-only education programs," December 2004, available online at http://www.democrats.reform.house.gov/Documents/20041201102153-50247.pdf (accessed December 30, 2004).

sweat and tears: See Charles Babington, "Viewing Videotape, Frist disputes Fla. doctors' diagnosis of Schiavo," *Washington Post,* March 19, 2005.

226 *"safe enough":* The Medical Institute, "Do condoms make sex safe enough?" See http://www.medinstitute.org/products/index.htm (accessed November 15, 2004).

should protect against these other diseases as well: National Institute of Allergy and Infectious Diseases, *Workshop Summary: Scientific Evidence on Condom Effectiveness for Sexually Transmitted Disease (STD) Prevention,* July 21, 2001. Quotation on page 11. See also Willard Cates, Jr., "The NIH condom report: the glass is 90% full," *Family Planning Perspectives,* vol. 33, no. 5, September/October 2001.

The answer is yes: Interview with Ward Cates, July 27, 2004.

to downplay their effectiveness: See "Condom Effectiveness" on the website Politics & Science: Investigating the State of Science Under the Bush Administration, presented by Representative

Henry Waxman. Available online at http://democrats.reform.house.gov/features/politics_and_ science/example_condoms.htm (accessed December 30, 2004).

226 *information about how to use condoms properly:* For the original fact sheet, see http://www. democrats.reform.house.gov/Documents/20040817143856-95300.pdf (accessed December 30, 2004). For its revised version, see http://www.democrats.reform.house.gov/Documents/ 20040817143928-82727.pdf (accessed December 30, 2004).

227 *for the lie that it is:* Tom Coburn press release, "Safe sex myth exposed by scientific report; condoms do not prevent most STDs," July 19, 2001. On file with author.

he has taken a leadership role: Interview with Heather Boonstra, December 2, 2004.

the truth of condom ineffectiveness: Quoted in Chris Bull, "Bush's abstinence man: Tom Coburn, head of the president's AIDS advisory council, preaches his preference for 'just say no' over condom-based HIV education," *Advocate,* May 28, 2002.

conspirators in unnecessary deaths: "Anti-condom claims dangerous," editorial, *Atlanta Journal-Constitution,* August 22, 2003.

228 *happens to serve on the advisory board:* The Medical Institute, "Executive & board members," available at http://www.medinstitute.org/about/members.htm (accessed November 15, 2004).

served on the Physicians Resource Council: As noted in Amy Fagan, "Pro-choice groups object to doctor: say he mixes religion and medicine, unfit for FDA panel," *Washington Times,* October 17, 2002.

chain e-mails: Following George W. Bush's reelection in 2004, a sign-on e-mail began to circulate calling for a protest against Hager's nomination as chair of the FDA's Reproductive Health Drugs Advisory Committee. Several women I know, including my sister, received the anti-Hager call to arms. I then had to explain to them that the missive perpetuated a number of falsehoods. In fact, Hager had already been sitting on the FDA committee since late 2002. Although media reports initially suggested that he might serve as the group's chairman (see Karen Tumulty, "Jesus and the FDA," *Time,* October 5, 2002), he did not wind up in that role, perhaps because of the controversy that these media reports engendered.

The recycled e-mail seemed to reflect a revealing bout of left-wing panic following the 2004 election. Newspapers widely reported that George W. Bush had won the race thanks to unprecedented turnout among conservative Christian voters. And to the liberal mind, empowering someone like Hager, even in a minor FDA advisory role, symbolized the consequences of allowing religion to dominate politics and trump science.

For a further dissection of the Hager rumor, see the urban legend tracking website *Snopes.com:* http://www.snopes.com/politics/bush/hager.asp (accessed December 30, 2004).

abortion pill RU–486: Karen Tumulty, "Jesus and the FDA," *Time,* October 5, 2002.

for promoting promiscuity: See Barbara Isaacs, "Pill still fills the bill," *Lexington Herald Leader,* Mary 23, 2000, quoting Hager: "It altered sexual activity in this country. . . . It was a more convenient way for young people to be sexually active outside of marriage."

who seek to live for the Lord: Baker Publishing Group, "About Us," available online at http://www.bakerpublishinggroup.com/ME2/dirsect.asp?sid=DB7502D634414EA0BD09C A2A58EA1D03&nm=About+Us (accessed March 22, 2005).

spiritual ramifications of this problem: W. David Hager and Linda Carruth Hager, *Stress and the Woman's Body* (Grand Rapids, Michigan: Fleming H. Revell, 1996), p. 64.

joy in your life: Hager and Hager, *Stress and the Woman's Body,* p. 11.

the healing process may take a long time: W. David Hager, *As Jesus Cared for Women: Restoring Women Then and Now* (Grand Rapids, Michigan: Fleming H. Revell, 1998). From p. 30: "Today we might say that demonic possession is a result of persistent sin or experimentation with the occult so that Satan is allowed to exert his control in the person's life. Or in some case the evil spirit may be present as a result of generational sin." From p. 35: "Although love and devotion are usually behind the pursuit of health for a family member, they are not enough in themselves. Faith must also be present for healing to occur." Later, Hager notes that "Even with divine intervention, the healing process may take a long time." From p. 43: "Remarkable miracles of conception still occur." From p. 57: "Vivian had great faith—more than her doctor. I was astounded when I realized that she was miraculously healed." And so forth.

229   *that protested his appointment:* For a list of groups that opposed Hager, see http://www.aauw.org/about/newsroom/press_releases/nomination.cfm (accessed March 22, 2005).

  the safest product that we have seen: Food and Drug Administration, Nonprescription Drugs Advisory Committee in Joint Session with the Advisory Committee for Reproductive Health Drugs, meeting transcript, December 16, 2003, p. 363. Available online at http://www.fda.gov/ohrms/dockets/ac/03/transcripts/4015T1.pdf (accessed December 30, 2004).

  personally refuses to prescribe: On Hager's refusal to prescribe emergency contraception see Barbara Isaacs, "'Morning-after' pill offers women emergency contraceptive alternative," *Houston Chronicle,* July 21, 1999. Confirmed in David Hager interview, June 22, 2004.

  acknowledges such a possibility: Barr Pharmaceuticals, "About Plan B," available online at http://www.g02planb.com/section/about (accessed November 15, 2004).

  three million each year: Stanley K. Henshaw, "Unintended pregnancy in the United States," *Family Planning Perspectives,* vol. 30, no. 1, January/February 1998.

  fifty-one thousand abortions: For the one hundred thousand unintended pregnancies figure, see New York State Comptroller press release, "Improved access to emergency contraception could prevent unwanted pregnancies and abortions and save New York $450 million annually," November 6, 2003, available online at http://www.osc.state.ny.us/press/releases/nov03/110603.htm (accessed March 23, 2005). For the fifty-one thousand abortions figure, see Rachel K. Jones et al., "Contraceptive use among U.S. women having abortions in 2000–2001," *Perspectives on Sexual and Reproductive Health,* vol. 34, no. 6, November/December 2002; e-mail communication from Stanley Henshaw, March 22, 2005.

  I think that was not a good call: Interview with Donald Kennedy, June 22, 2004.

230   *Hager raised questions:* See FDA meeting transcript, p. 137.

  on multiple occasions: Ibid., pp. 352, 365, 370, 372.

  appeared to seize on this interlude: Letter from Steven Galson, acting director, Center for Drug Evaluation and Research, to Barr Pharmaceuticals, May 6, 2004. Available online at http://www.fda.gov/cder/drug/infopage/planB/planB_NALetter.pdf (accessed December 30, 2004). The FDA itself admits that having Galson sign was unusual. See http://www.fda.gov/cder/drug/infopage/planB/planBQandA.htm (accessed December 30, 2004).

  The opinion I wrote: W. David Hager, "Standing in the gap," Asbury College Tape Ministry, October 29, 2004. Audio CD on file with author.

231   *there's no data either:* Interview with James Trussell, June 19, 2004.

  Concerned Women for America: See Concerned Women for America, comments to the FDA Advisory Committee on Reproductive Drugs, December 16, 2003. Available online at http://www.cwfa.org/images/content/ww-maptest.pdf (accessed December 30, 2004).

  conservative members of Congress: Letter from David Weldon and colleagues to George W. Bush, January 9, 2004, on file with the author.

  Yet that wasn't the case: Interview with James Trussell, June 19, 2004.

  in a recent randomized study: Tina R. Raine et al., "Direct access to emergency contraception through pharmacies and effect on unintended pregnancy and STIs: a randomized controlled trial," *JAMA,* vol. 293, no. 1, January 5, 2005, noting, "Access to EC did not have a detrimental effect on contraceptive use or sexual behavior. . . . Our study supports the hypothesis that behavior is not influenced by access to EC and that women who have increased access to EC do not have more unprotected intercourse."

  there's no evidence that this drug: Interview with Alastair Wood, May 26, 2004.

231   *we don't have that information:* Interview with W. David Hager, June 22, 2004.
       Obviously, you can't go advertising: See FDA meeting transcript, p. 301.

14: BUSH LEAGUE SCIENCE

238   *had signed a statement:* Union of Concerned Scientists and signatories, "Restoring scientific
       integrity in policymaking," February 18, 2004. Available online at http://www.ucsusa.org/
       global_environment/rsi/page.cfm?pageID=1320 (accessed October 2, 2004).
239   *We almost wistfully think:* Interview with John Gibbons, June 3, 2004.
       Forty-nine Nobel laureates: Union of Concerned Scientists, "RSI Signatories," available online
       at http://www.ucsusa.org/scientific_integrity/interference/prominent-statement-signatories.
       html (accessed April 9, 2006).
       palpable unease: Interview with Kevin Knobloch, Peter Frumhoff, and Suzanne Shaw, of the
       Union of Concerned Scientists, May 10, 2004.
240   *took over a year to fill:* President Bush nominated Dr. Elias Zerhouni as National Institutes of
       Health director and Dr. Richard Carmona as surgeon general on March 26, 2002, well over a
       year into his presidency. See Earl Lane, "Bush picks top docs; Harlem native nominated to be
       U.S. surgeon general," *Newsday*, March 27, 2002. Meanwhile, Mark McClellan went uncon-
       firmed as Food and Drug Administration commissioner until October 17, 2002. See Christo-
       pher Lee, "Bush slow to fill top federal posts; Brookings study cites growing snags in process;
       White House challenges findings," *Washington Post*, October 18, 2002, noting, "The Food
       and Drug Administration went nearly 20 months without a nominee for commissioner until
       Bush tapped health policy adviser Mark McClellan last month."
       just eight days after the president's inauguration: See National Academy of Sciences, Commit-
       tee on Science, Engineering, and Public Policy, Science and Technology in the National Inter-
       est, *Ensuring the Best Presidential and Federal Advisory Committee Science and Technology
       Appointments* (Washington: National Academies Press, 2004), p. 68.
       that title was never offered: Chris Mooney, "Political science," *American Prospect* vol. 12, no.
       21, December 3, 2001.
       voiced serious concerns: Ibid.
241   *one for science and one for technology:* The nominees were Kathie Olsen for science policy and
       Richard Russell for technology policy. For information on their confirmation hearing, see Jef-
       frey Mervis, "Senate puts the heat on science nominees," *Science*, July 26, 2002.
       sure sign of declining influence: See Andrew Lawler, "Marburger shakes up White House of-
       fice," *Science*, November 2, 2001.
       says Harvard's John Holdren: Interview with John Holdren, May 11, 2004.
       revealed a "broad restructuring": Rick Weiss, "HHS seeks science advice to match Bush views,"
       Washington Post, September 17, 2002.
       science moles: Steven Milloy, "Lingering infestation of science moles," *Washington Times*, May
       30, 2001.
       he didn't end up on the panel: For the Miller case, see Aaron Zitner, "Advisors put under a mi-
       croscope; the Bush team is going to great lengths to vet members of scientific panels. Creden-
       tials, not ideology, should be the focus, critics say," *Los Angeles Times*, December 23, 2002. See
       also Ken Silverstein, "Bush's new political science: when it comes to pubic-health appointments,
       the administration has its own litmus test," *Mother Jones*, November/December 2002.
       hundreds of which advise: For this figure, I rely on the National Academy of Sciences, Com-
       mittee on Science, Engineering, and Public Policy, Science and Technology in the National In-
       terest, *Ensuring the Best Presidential and Federal Advisory Committee Science and Technology
       Appointments* (Washington: National Academies Press, 2004), p. 37: "According to the GSA
       Committee Management Secretariat, in 2004 there were 967 federal advisory committees.
       Half of them have a major scientific and technical component as measured by their charters
       or the number of scientists, engineers, and health professionals who are members."

242 *editorial entitled "An Epidemic of Politics":* Donald Kennedy, "An epidemic of politics," *Science,* vol. 299, no. 5607, January 31, 2003, p. 625.

to appease religious conservatives: See Adam Clymer, "U.S. revises sex information, and a fight goes on," *New York Times,* December 27, 2002, as well as an earlier article by the same author, "Critics say government deleted sexual material from web sites to push abstinence," *New York Times,* November 26, 2002.

a short report: "Weird science: the Interior Department's manipulation of science for political purposes," Democratic Staff, Committee on Resources, December 17, 2002. Online at http://resourcescommittee.house.gov/democrats/hot2002/weirdscience.pdf (accessed December 11, 2004).

a far more sweeping report: U.S. House of Representatives, Committee on Government Reform, Minority Staff, "Politics and science in the Bush Administration," August 2003. Available online at http://democrats.reform.house.gov/features/politics_and_science/pdfs/pdf_politics_and_science_rep.pdf (accessed December 12, 2004).

243 *had crossed a new line:* The information in this paragraph relies on my interviews with representatives of the Union of Concerned Scientists on May 10, 2004, with John Holdren, a UCS statement signatory, on May 11, 2004, and with Kurt Gottfried on February 7, 2005.

D. Allan Bromley, a UCS critic: Interview with D. Allan Bromley, April 26, 2004.

The pattern was shockingly similar: Interview with James McCarthy, May 11, 2004.

posted an online story: James Glanz, "Scientists accuse White House of distorting facts," *New York Times,* February 18, 2004.

the most e-mailed article: E-mail communication from Suzanne Shaw, director of communications at the Union of Concerned Scientists, May 12, 2004.

conspiracy report: For editorials, see "Uses and abuses of science," *New York Times,* February 23, 2004, and "A political load on science," *The Los Angeles Times,* February 20, 2004. For Marburger's denunciation of a "conspiracy report," see Dan Vergano, "White House manipulates science, leaders in field say," *USA Today,* February 19, 2004.

in response to written questions: Written response from John Marburger, December 22, 2004.

a big pattern of disrespect: Quoted in James Glanz, "Scientists say administration distorts facts," *New York Times,* February 19, 2004.

244 *with three individuals expected:* See Rick Weiss, "Bush ejects two from bioethics council; changes renew criticism that the president puts politics ahead of science," *Washington Post,* February 28, 2004.

arrogant and dismissive response: In written questions, I asked bioethics council chair Leon Kass whether he maintained that the group's controversial membership change had been settled on prior to the release of the Union of Concerned Scientists report on the politicization of science. He responded on March 20, 2005, "I do. Absolutely so. The review of membership was conducted in the fall of 2003, looking ahead to our second term that began in January 2004." The UCS report hit the headlines in February 2004.

protesting Blackburn's ouster: Jeffrey Brainard, "A new kind of bioethics: eschewing the academic mainstream, Bush panel focuses on technology's dangers," *Chronicle of Higher Education,* May 21, 2004.

The question is, what's the big picture?: Diane Rehm Show, March 4, 2004. Audio available online at http://www.wamu.org/programs/dr/04/03/04.php (accessed December 11, 2004).

245 *politically motivated statement:* Quoted in James Glanz, "Scientists say administration distorts facts," *New York Times,* February 19, 2004.

press conference at the National Press Club: The Marshall Institute press conference occurred on March 23, 2004. I was in attendance.

released a detailed rebuttal: Statement of the Honorable John H. Marburger, III on Scientific Integrity in the Bush Administration, April 2, 2004. Available online at http://www.ostp.gov/html/ucs/ResponsetoCongressonUCSDocumentApril2004.pdf (accessed December 11, 2004).

245 *Union of Concerned Scientists had published:* Union of Concerned Scientists, "Scientific Integrity in Policymaking: An Investigation into the Bush Administration's Misuse of Science," March 2004. Available online at http://www.ucsusa.org/documents/RSI_final_fullreport.pdf (accessed December 12, 2004).

an absolutely bloodless speech: John Marburger, speech at Rice University's Baker Institute for Public Policy, November 2, 2003 (author's notes).

You have to remember: Written response from John Marburger, December 22, 2004.

held a private meeting: Interview with W. K. H. Panofsky, June 10, 2004.

246 *thorough investigation into all the allegations:* Written response from John Marburger, December 22, 2004.

says Harvard's Holdren: Interview with John Holdren, May 11, 2004.

Rose Garden speech: As the president put it, "There is a natural greenhouse effect that contributes to warming. Greenhouse gases trap heat, and thus warm the earth because they prevent a significant proportion of infrared radiation from escaping into space. Concentration of greenhouse gases, especially CO_2, have increased substantially since the beginning of the industrial revolution. And the National Academy of Sciences indicate that the increase is due in large part to human activity.

"Yet, the Academy's report tells us that we do not know how much effect natural fluctuations in climate may have had on warming. We do not know how much our climate could, or will change in the future. We do not know how fast change will occur, or even how some of our actions could impact it." Speech by President Bush on climate change, June 11, 2001, available online at http://www.whitehouse.gov/news/releases/2001/06/20010611-2.html (accessed December 12, 2004).

247 *Bush went on to quote passages:* George W. Bush and John Kerry, "Bush and Kerry offer their views on science," *Science,* September 16, 2004. Available online at http://www.sciencemag.org/cgi/rapidpdf/1104420v1.pdf (accessed December 12, 2004).

248 *repeatedly resisted even mild language:* Juliet Eilperin, "U.S. wants no warming proposal; administration aims to prevent Arctic Council suggestions," *Washington Post,* November 4, 2004.

The State Department, much like Marburger: Richard Boucher (State Department spokesman), "The U.S. commitment to a climate study," *Washington Post* (letters section), November 10, 2004.

first reported on by the New York Times: Andrew C. Revkin and Katherine Q. Seelye, "Report by E.P.A. leaves out data on climate change," *New York Times,* June 19, 2003. A number of other news outlets also reported on the memo later that same day, and their reports indicate that they had obtained the same document. See CNBC, *The News with Brian Williams,* June 19, 2003 (7 p.m. ET), CBS Evening News, "President being taken to task for cynical changing of a major report on global warming," June 19, 2003 (6:30 p.m. ET), and ABC News, *World News Tonight with Peter Jennings,* "Global warming Bush administration keeps truth a secret," June 19, 2003 (6:30 p.m. ET).

no longer accurately represents: See National Wildlife Federation, "EPA ditches climate change in environment report," July 19, 2003, press release. Available online at http://www.nwf.org/news/story.cfm?pageId=E093353A-D994-CA6B-BEC296236ACE6FB0 (accessed December 21, 2004).

249 *I can state categorically:* Russell Train, "When politics trumps science," letter to the editor, *New York Times,* June 21, 2003.

Bush administration had supported: See testimony of Craig Manson, assistant secretary for fish and wildlife and parks, U.S. Department of the Interior, before the House Resources Committee, regarding H.R. 4840, the "Sound Science for Endangered Species Act Planning Act of 2002." This is an earlier version of H.R. 1662, Greg Walden's "sound science" bill. Manson stated, "The Administration supports H.R. 4840 with modifications to address our concerns."

We believe that if implemented, this legislation will broaden opportunities for scientific input and assure additional public involvement in Endangered Species Act implementation. We also believe it will also improve the U.S. Fish and Wildlife Service's (Service) decision-making process and result in increased public confidence in the Service's decisions." Available online at http://www.doi.gov/ocl/2002/hr4840.htm (accessed December 14, 2004).

250 *biologists have overwhelmingly opposed:* Over four hundred biology and wildlife experts have signed a statement, circulated by Earthjustice, Defenders of Wildlife, and other environmental groups, criticizing H.R. 1662, Greg Walden's "sound science" Endangered Species Act reform bill. Statement and list of signatories on file with author.

The possible relationship between abortion and breast cancer: National Cancer Institute, "Early reproductive events and breast cancer," fact sheet posted on November 25, 2002, available online at http://www.democrats.reform.house.gov/Documents/20040817143807-43596.pdf (accessed December 26, 2004).

political creation of scientific uncertainty: Letter from Henry Waxman and members of Congress to Health and Human Services secretary Tommy Thompson, December 18, 2002. Available online at http://www.democrats.reform.house.gov/Documents/20040817143418-92080.pdf (accessed December 26, 2004).

a previous and much more thorough fact sheet: National Cancer Institute, "Abortion and breast cancer," date reviewed, March 6, 2002, available online at http://www.democrats.reform.house.gov/Documents/20040817143732-39165.pdf (accessed December 26, 2004).

apparently in response: My account relies on Adam Clymer, "U.S. revises sex information, and a fight goes on," *New York Times,* December 27, 2002, as well as an earlier article by the same author, "Critics say government deleted sexual material from web sites to push abstinence," *New York Times,* November 26, 2002.

induced abortion is not associated: National Cancer Institute, "Summary report: early reproductive events and breast cancer workshop," March 2003. Available online at http://www.cancer.gov/cancerinfo/ere-workshop-report (accessed November 6, 2004).

251 *a new fact sheet appeared:* See Andrew von Eschenbach, letter to Rep. Henry Waxman, April 25, 2003, available online at http://www.democrats.reform.house.gov/Documents/2004 0817144235-83520.pdf (accessed December 26, 2004).

Miller had complained about the interview: See, for example, Ken Silverstein, "Bush's new political science," *Mother Jones,* November/December 2002.

I received a telephone call: E-mail communication from William R. Miller, February 15, 2005.

252 *other scientists emerged:* For example, the Union of Concerned Scientists in a later report noted many more such cases. See "Scientific integrity in policy making: further investigation of the Bush administration's misuses of science," July 2004, pp. 25–31. Available online at http://www.ucsusa.org/documents/Scientific_Integrity_in_Policy_Making_July_2004.pdf (accessed December 17, 2004).

253 *riddled with distortion, inaccuracies, and omissions:* Christopher Marquis, "Bush misuses science data, report says," *New York Times,* August 8, 2003.

I'd be particularly interested: Interview with Henry Waxman, May 17, 2004.

254 *A 2003 book published by the George C. Marshall Institute:* Michael Gough (ed.), *Politicizing Science: The Alchemy of Policymaking* (Stanford: Hoover Institution Press, 2003).

do not increase the use of illegal drugs: Institute of Medicine, *Preventing HIV Transmission: The Role of Sterile Needles and Bleach* (Washington, D.C.: National Academies Press, 1995). Quotation on page 6.

255 *We have concluded that needle-exchange:* See Lauran Neergaard, "U.S. won't fund needle exchanges," Associated Press, April 20, 1998.

In an extraordinary February 2005 editorial: "Deadly ignorance," *The Washington Post* (editorial), February 27, 2005.

257 *want to use it for political purposes:* Interview with Thomas Murray, October 6, 2003.
we create our own reality: Ron Suskind, "Without a doubt," *New York Times Magazine,* 17 October 2004.

258 *Health and Human Services spokesman William Pierce:* See Rick Weiss, "HHS seeks science advice to match Bush views," *Washington Post,* September 17, 2002. The article says that Pierce "defended [Health and Human Services secretary] Thompson's prerogative to hear preferentially from experts who share the president's philosophical sensibilities."
I think it's an appropriate question: Quoted in David Brown, "Panel debates politics' role in scientists' appointments," *Washington Post,* July 22, 2004.
such as hair color or height: See National Academy of Sciences, Committee on Science, Engineering, and Public Policy, Science and Technology in the National Interest: *Ensuring the Best Presidential and Federal Advisory Committee Science and Technology Appointments* (Washington: National Academies Press, 2004), p. 5.

259 *push back at some point in the future:* Robert Walker, response to a question at September 30, 2004, American Association for the Advancement of Science forum on science and technology policy in the 2004 presidential campaign. Webcast available at http://www.aaas.org/news/press_room/election (accessed February 12, 2005).
you reduce your credibility: Interview with Robert Walker, December 2, 2004.
Forty-eight Nobel laureates: Peter Agre et al., "An open letter to the American people" (Kerry endorsement letter from forty-eight Nobel laureates), June 21, 2004. On file with author.
actually on the government payroll: See Andrew C. Revkin, "NASA expert criticizes Bush on global warming policy," *New York Times,* October 26, 2004.

260 *What if we have a president who believes in science:* John Kerry, 2004 Democratic national convention speech, July 29, 2004.
Bush is being political with science: Gregg Easterbrook, "Politics and science do mix: claims that Bush misuses research are hypocritical," *Los Angeles Times,* April 6, 2004.

EPILOGUE: WHAT WE CAN DO

263 *and proposed helpful solutions:* Federation of American Scientists, "Flying blind: the rise, fall, and possible resurrection of science policy advice in the United States," December 2004.
but you can't just pick up the phone: Interview with David Guston, September 16, 2003.
authors of a 2003 collection: M. Granger Morgan and Jon M. Peha (eds.), *Science and Technology Advice for Congress* (Washington, D.C.: Resources for the Future, 2003).

264 *made it possible for science:* Interview with Rush Holt, April 29, 2004.
Proposed legislation: The legislation, introduced in the 109th Congress, was H.R. 839.

265 *The integrity of the science advisory process:* Lewis M. Branscomb, "Science, politics, and U.S. democracy," *Issues in Science and Technology,* fall 2004.

266 *in a 2004 paper:* Maxwell T. Boykoff and Jules M. Boykoff, "Balance as bias: global warming and the U.S. prestige press," *Global Environmental Change,* vol. 14, 2004, pp. 125–136.

268 *There are rarely criminal penalties:* David Egilman, speech at the Center for Science and the Public Interest "Integrity in Science" project conference, July 12, 2004, Washington, D.C. Audio available at http://www.cspinet.org/integrity/cf_panelb.html (accessed December 25, 2004).

INDEX